Network Science for Military Coalition Operations:
Information Exchange and Interaction

Dinesh Verma
IBM Thomas J. Watson Research Center, USA

Information Science
REFERENCE

INFORMATION SCIENCE REFERENCE
Hershey • New York

Director of Editorial Content: Kristin Klinger
Director of Book Publications: Julia Mosemann
Acquisitions Editor: Lindsay Johnston
Development Editor: Joel Gamon
Publishing Assistant: Sean Woznicki
Typesetter: Myla Harty
Production Editor: Jamie Snavely
Cover Design: Lisa Tosheff
Printed at: Yurchak Printing Inc.

Published in the United States of America by
Information Science Reference (an imprint of IGI Global)
701 E. Chocolate Avenue
Hershey PA 17033
Tel: 717-533-8845
Fax: 717-533-8661
E-mail: cust@igi-global.com
Web site: http://www.igi-global.com/reference

Library of Congress Cataloging-in-Publication Data

Network science for military coalition operations : information exchange and interaction / Dinesh Verma, editor.
p. cm.

Includes bibliographical references and index.
Summary: "This book is structured into sections that look at some of the challenges related to coalition operations in different types of networks, such as communications and information networks and human and cognitive networks, and looks at other issues that impact the operations of coalitions, the management and use of policies across different organizations"--Provided by publisher.

ISBN 978-1-61520-855-5 (hardcover) -- ISBN 978-1-61520-856-2 (ebook) 1. Combined operations (Military science) 2. Communications, Military. 3. Computer networks. 4. Information networks. 5. Cognitive radio networks. 6. Social networks. 7. United States--Armed Forces--Information services. 8. Great Britain--Armed Forces--Information services. I. Verma, Dinesh, 1964-
U260.N48 2010
355.6'88--dc22

2009042517

British Cataloguing in Publication Data
A Cataloguing in Publication record for this book is available from the British Library.

Seraphin Calo, *IBM, USA*
Somnath Banerjea, *ABES Institute of Technology, India*
Suresh Chari, *IBM, USA*
Theodore Brown, *City University of New York, USA*
Vasileios Pappas, *IBM, USA*
Yuri Demenchko, *University of Amsterdam, Netherlands*

Table of Contents

Foreword xiv

Preface xvi

Acknowledgment xxi

Section 1
Coalition Communication and Information Networks

Chapter 1
Inter-Domain Routing in Mobile Ad Hoc Networks: Challenges and New Directions 1
Chi-Kin Chau, University of Cambridge, UK
Jon Crowcroft, University of Cambridge, UK
Kang-Won Lee, T.J. Watson Research Center, USA
Starsky H.Y. Wong, IBM T.J. Watson Research Center, USA

Chapter 2
Quality of Sensor-Originated Information in Coalition Information Networks 15
Chatschik Bisdikian, IBM Research, USA
Lance M. Kaplan, U.S. Army Research Laboratory, USA
Mani B. Srivastava, University of California at Los Angeles, USA
David J. Thornley, Imperial College London, UK
Dinesh Verma, IBM Research, USA
Robert I. Young, Defence Science and Technology Laboratory, UK

Chapter 3
Graph Mining Techniques for Networking Applications: A Review 42
Charu C. Aggarwal, IBM T. J. Watson Research Center, USA
Eric Yu-En Lu, Cambridge University, UK

Chapter 4
Region Coverage and Protection with Sensors: A Survey 63
Ted Brown, City University of New York, USA
Peter Brass, City University of New York, USA
Matthew P. Johnson, City University of New York, USA
Simon Shamoun, City University of New York, USA

Section 2
Security and Policy Technologies for Coalitions

Chapter 5
Security across Disparate Management Domains in Coalition MANETs 79
Mudhakar Srivatsa, IBM T.J. Watson Research Center, USA
Dakshi Agrawal, IBM T.J. Watson Research Center, USA
Andrew D. McDonald, Roke Manor Research Ltd, UK

Chapter 6
Policy Technologies for Federation and Interoperation of Coalition Networks 103
Mandis S. Beigi, IBM Research, USA
Seraphin B. Calo, IBM Research, USA
David A. Wood, IBM Research, USA
Petros Zerfos, IBM Research, USA

Chapter 7
Policies to Enable Secure Dynamic Community Establishment 121
Alberto Schaeffer-Filho, Imperial College London, UK
Emil Lupu, Imperial College London, UK
Morris Sloman, Imperial College London, UK

Chapter 8
Policy Technologies for Security Management in Coalition Networks 146
Seraphin B. Calo, IBM Research, USA
Clare-Marie Karat, IBM Research, USA
John Karat, IBM Research, USA
Jorge Lobo, IBM Research, USA
Robert Craven, Imperial College, UK
Emil Lupu, Imperial College, UK
Jiefei Ma, Imperial College, UK
Alessandra Russo, Imperial College, UK
Morris Sloman, Imperial College, UK
Arosha Bandara, The Open University, UK

Section 3
Human and Cognitive Issues in Coalitions

Chapter 9
A Model for Culturally Adaptive Policy Management in Ad Hoc Collaborative Contexts................ 174
John Karat, IBM TJ Watson Research, USA
Winston Sieck, Applied Research Associates, USA
Timothy J. Norman, University of Aberdeen, UK
Clare-Marie Karat, IBM TJ Watson Research, USA
Carolyn Brodie, IBM TJ Watson Research, USA
Louise Rasmussen, Applied Research Associates, USA
Katia Sycara, Carnegie Mellon University, USA

Chapter 10
The Network-Extended Mind 191
Paul R. Smart, University of Southampton, UK
Paula C. Engelbrecht, University of Southampton, UK
Dave Braines, Emerging Technology Services, UK
Michael Strub, Dstl Porton Down, UK
Cheryl Giammanco, Human Research & Engineering Directorate, USA

Chapter 11
Cultural Network Analysis: A Cognitive Approach to Cultural Modeling 237
Winston R. Sieck, Applied Research Associates, USA
Louise J. Rasmussen, Applied Research Associates, USA
Paul Smart, University of Southampton, UK

Chapter 12
Conversation Analysis of Coalition Communication in Network Centric Operations 256
Ping Xue, Boeing Research & Technology, USA
Stephen Poteet, Boeing Research & Technology, USA
Anne Kao, Boeing Research & Technology, USA

Compilation of References 275

About the Contributors 301

Index... 311

Detailed Table of Contents

Foreword xiv

Preface xvi

Acknowledgment xxi

Section 1
Coalition Communication and Information Networks

The network infrastructure available to the coalition partners has a significant impact on the effectiveness of joint operations. Coalition operations require interoperation and information exchange across networks that have evolved independently, using different technologies and different set of practices. The chapters in this section provide an overview of the challenges involved in coalition communication and information networks, and provide approaches to address those challenges.

Chapter 1
Inter-Domain Routing in Mobile Ad Hoc Networks: Challenges and New Directions 1
Chi-Kin Chau, University of Cambridge, UK
Jon Crowcroft, University of Cambridge, UK
Kang-Won Lee, T.J. Watson Research Center, USA
Starsky H.Y. Wong, IBM T.J. Watson Research Center, USA

During tactical operations in the field, mobile ad-hoc networks (MANETs) provide the primary connectivity mechanism for communications among the personnel involved. However, when coalitions are operating together, two or more of such MANETs need to communicate with each other. This chapter looks at the issues of interoperation among heterogeneous MANETs, and identifies new directions for development of seamless inter-domain routing among two MANETS.

Network Centric operation is about managing information that may originate from coalition partners, sensors and people, each of which may have a different level of trustworthiness and accuracy. Understanding the quality of information that is transmitted from different sources is hard in these contexts. This paper presents a summary of issues related with defining quality of information and an approach to characterize it which can be used across a wide range of applications of sensor generated information.

Information, people and computing devices form a myriad of interconnected networks in coalition contexts. In these cases, discovery of the graphs that represent the different interconnections among the various streams of information available to a coalition commander can yield significant insights into the operational state of any coalition operation. This chapter provides a comprehensive survey of graph mining techniques that can be used to understand these interactions.

An important component of coalition information networks is provided by their Intelligence, Surveillance and Reconnaissance (ISR) assets, or sensor networks that are deployed to protect a theater of operations. One important problem related to sensor networking is determining whether an area is adequately monitored and protected by means of sensors. This chapter provides a survey of the different approaches for covering a region using sensors, also discussing variations of the problem such as protecting a barrier using sensors.

Section 2
Security and Policy Technologies for Coalitions

Security is an important aspect that cuts across all types of networks in a coalition setting. One innovative way to manage security is by using policy technology. A policy is a high level guideline or directive that is provided to influence the operation of an organization or an IT system. Policy management is of critical importance in coalition operations because the policies in two organizations may differ significantly, and in some cases conflict with each other. Proper management of policies is essential for smooth coalition operations. Technologies that allow management, manipulation and use of policies can be extremely valuable to provide new capabilities for coalition operations. In this section, we look at security management issues in coalitions, with a particular emphasis on the use of policy technology for solving coalition networks security concerns.

Establishing secure communications in dynamic mobile ad-hoc networks involving coalition partners has significant challenges associated with boot-strapping the security and trust relationships while minimizing bandwidth, power and compute cycle consumption. Inter-domain routing and managing identities across partners create other security challenges. Some approaches to address these challenges are described in this chapter.

In order for policy technologies to be used in a coalition context, a life-cycle model for policies and architecture for policy management are needed. This chapter expands upon the life-cycle model in chapter 5 by an architecture customized for coalition operations. It also provides a review of some key policy enabled functions that can be activated in the context of coalition operations.

Establishing a community of interest in an agile manner is a key requirement of coalition operations that can react quickly to changing conditions during tactical operations. This chapter proposes an approach to build a secure network supporting such a dynamic community using policy based techniques. The basic framework used is that of a policy-enabled self-managing cell which can dynamically adapt to meet the needs of the community.

Chapter 8
Policy Technologies for Security Management in Coalition Networks.. 146
Seraphin B. Calo, IBM Research, USA
Clare-Marie Karat, IBM Research, USA
John Karat, IBM Research, USA
Jorge Lobo, IBM Research, USA
Robert Craven, Imperial College, UK
Emil Lupu, Imperial College, UK
Jiefei Ma, Imperial College, UK
Alessandra Russo, Imperial College, UK
Morris Sloman, Imperial College, UK
Arosha Bandara, The Open University, UK

This chapter provides an overview of a policy management framework and a policy life-cycle that can be used in the context of coalition networks. After providing the lifecycle and the policy management framework, the chapter illustrates how the framework can be used for coalition policy management, and enabling inter-operation of Intelligence, Surveillance and Reconnaissance (ISR) networks for coalitions.

Section 3
Human and Cognitive Issues in Coalitions

Coalition operations are influenced heavily by the interactions of people involved in the activity, perhaps more so that the interactions among individual networks or information assets. In this context, it is important to understand the different manners in which soldiers interact in coalitions. The human network is the collection of people, relationships and interactions that take place during a coalition operation. In this section, we look at some of the issues involved in understanding and modeling the human network, understanding cultural and communication differences, and leveraging the human network to extend the world view we carry in our mind.

Chapter 9
A Model for Culturally Adaptive Policy Management in Ad Hoc Collaborative Contexts................ 174
John Karat, IBM TJ Watson Research, USA
Winston Sieck, Applied Research Associates, USA
Timothy J. Norman, University of Aberdeen, UK
Clare-Marie Karat, IBM TJ Watson Research, USA
Carolyn Brodie, IBM TJ Watson Research, USA
Louise Rasmussen, Applied Research Associates, USA
Katia Sycara, Carnegie Mellon University, USA

An important aspect of coalition operations is the difference in the cultures and interaction behavior that exists among different members of the coalition. In this chapter, the authors describe different areas of research looking at resolving cultural and policy differences among coalitions, and provide a comprehensive list of research questions that lie within the intersection of these domains.

Chapter 10
The Network-Extended Mind 191
Paul R. Smart, University of Southampton, UK
Paula C. Engelbrecht, University of Southampton, UK
Dave Braines, Emerging Technology Services, UK
Michael Strub, Dstl Porton Down, UK
Cheryl Giammanco, Human Research & Engineering Directorate, USA

A new thesis in cognition community is that some human mental states and processes stem from complex webs of causal influence involving extra-neural resources, most notably the resources of our social and technological environments. This chapter explores the hypothesis that contemporary and near-future network systems are poised to extend and perhaps transform our human cognitive potential. It discusses the relevance of information and network sciences to the formation, maintenance and functioning of extended cognitive systems in network-enabled environments, and their applicability in a coalition military context.

Chapter 11
Cultural Network Analysis: A Cognitive Approach to Cultural Modeling 237
Winston R. Sieck, Applied Research Associates, USA
Louise J. Rasmussen, Applied Research Associates, USA
Paul Smart, University of Southampton, UK

This chapter describes a methodology for modeling culture as networks of ideas distributed among members of a population. The method, Cultural Network Analysis can be used to develop cultural models for groups and populations, typically depicted as a network representation of the culturally shared concepts, causal beliefs, and values that influence key decisions. It also discusses how Cultural Network Analysis can be used to develop tools to support multinational collaborative planning and decision making, and other applications of the methodology.

Chapter 12
Conversation Analysis of Coalition Communication in Network Centric Operations 256
Ping Xue, Boeing Research & Technology, USA
Stephen Poteet, Boeing Research & Technology, USA
Anne Kao, Boeing Research & Technology, USA

Team verbal communication is a central component of coalition operations. This paper presents an analysis of the nature of communication, or rather miscommunication that can happen in coalition operations. The analysis shows that pragmatic aspects of language usage are important to understand communication happening across organizational boundaries. Based on the results, the authors propose a new computation framework that can be used to automate the analysis of coalition communications.

Foreword

Welcome to this IGI Global book on Network Science for coalition operations. Network Science is a new term to describe the network in terms of its impact on the human interactions of organizations. The term implies a comprehensive approach to developing network theory from the viewpoint of the information network, the social/cognitive network in addition to the more traditional physical communications network. This comprehensive approach is to develop network science as an integrated whole, combining three previously disparate intellectual pursuits in the information social and physical networks. As currently used by the Army, the network refers to the information moving among physical communications nodes and the humans that use them. The emphasis in this definition is on the vision of a network of networks comprised of the connections among humans organized and interacting through technology and the transfer of information

The Network Science Program at the U.S. Army Research Laboratory (ARL) encompasses research in the information network, the social and cognitive interaction of humans using the network, and the underlying physical wireless network connecting human nodes together. The ARL network science program incorporates both new and existing ARL research activities, blending them into coherent program. Network Science is especially important in Coalition Operations where cultural differences and language differences can easily lead to misunderstandings. The Army created the Network and Information Science International Technology Alliance to perform basic research into all aspects of the network connecting decision makers for coalition operations. The Alliance performs research in four technical areas; network theory, security across a systems of systems, sensor information processing and delivery, and distributed coalition planning and decision making. The unique aspect of the Alliance is these technical areas are studied in combination to research how they interact rather than studying them individually. This is a significant difference from past programs.

Most of the authors who have contributed to this book are actively involved in researching the challenges of coalition operations in the context of the Alliance, the largest ever collaborative fundamental research program between the U.S. and UK militaries focused on network science. These authors have made several fundamental contributions to network science in the context of coalition operations, and they have provided a good summary of the advances in the different chapters of the field.

Network Science is an important emerging scientific discipline related to network centric operations, and its various aspects related to coalition operations are comprehensively described by active researchers in the field. I am excited that this book is coming to the scientific community through IGI Global. I know you will enjoy it.

John Gowens
Director, Computational and Information Sciences Directorate
U.S. Army Research Laboratories

John Gowens *was appointed as the Director, Computational and Information Sciences Directorate (CISD) at the U.S. Army Research Laboratory (ARL), the Army's corporate Research and Development laboratory in March 2005. As the Director, he is responsible for the entirety of the CISD programs which include conducting research for the Army in Computational Science, Computer Science, Communications, and Weather. The current budget for the Directorate is approximately $155M. About 310 government employees and another 410 on-site contractors execute this budget. Dr. Gowens was previously the Chief, Computer and Communications Sciences Division that performs basic and applied research in information systems and networking infrastructure for brigade and below operations. The research program is focused entirely on the challenge of supporting the individual soldier in small unit operations. Research activities include the Intelligent Optics Laboratory that is the premier free space optical communications test bed in the world, machine translation systems to provide soldiers with translation capabilities in the field (deployed to Balkans, Afghanistan, and Iraq), small robots (Packbots) for clearing caves (deployed to Afghanistan and Iraq), and high performance sensor radios and tracking algorithms for sensors (deployed to Afghanistan and Iraq). Under Dr. Gowens' leadership, the team was awarded the Hammer award for the Federated Laboratory, an innovative government program to harvest commercial communications technology for Army applications. Prior to joining the government in 1989, Dr. Gowens spent 20 years as a consultant and university faculty member. At PricewaterhouseCoopers, Dr. Gowens was Principal Investigator on an Expert Systems project with U.S. Army Forces Command, Ft. McPherson, GA and Headquarters Department of the Army Artificial Intelligence Center to design and prototype an expert system for mobilizing Army National Guard and Reserve forces in times of emergency. The system supported faster and more efficient use of resources when mobilizing reserve forces. As a faculty member at the Georgia Institute of Technology, Dr. Gowens led a project to design and construct a 7-axis robotic arm and workstation to support the daily business operations of a quadriplegic working at the Georgia Department of Human Resources. The work station enabled the quadriplegic to conduct business as a handicap access administrator. The workstation was featured on a CNN health and fitness show in 1988 as well as local news broadcasts. After graduation from the U.S. Naval Academy, Dr. Gowens was commissioned a 2nd LT in the U.S. Army Corps of Engineers. He served in various duty assignments in Germany and Thailand. Dr. Gowens resigned his commission in February 1968 after returning from Thailand and entered graduate school at Arizona State University where he completed an MBA and a Ph.D.*

Preface

Modern military operations are based around the principle of network-centricity, in which information collected by different sources, both human as well as mechanical, are transmitted over the network, analyzed and processed, and the resulting intelligence provided to the troops involved in tactical operations. The growing importance of network-centric operations has led to interest in the emerging field of network science, which studies the properties and behavior of the different types of networks that arise in military operations.

The understanding of networks is important to effectively use the distributed information and processing capability available to different troops involved in any aspect of network centric operations. However, the scope of network science is not limited to that of the military alone. The concepts of network science can be applied successfully to the networks that are encountered in commercial enterprises, telecommunications networks, as well as the general public which is interconnected to an ever-increasing extent due to the global pervasiveness of the Internet. Thus, network science and its advances are likely to benefit not just the domain of modern warfare, but society in general.

A large number of military operations in the present age require the collaboration of multiple countries establishing a coalition. If we examine the number of military operations world-wide occurring since the end of the cold war to the present time, we will find that a preponderance of them involve a coalition of armies established in an asymmetric operation against insurgents or a rouge nation. Due to the current geopolitical situation and the increasing importance of using diplomacy and building international consensus before launching any military operation, coalition are likely to increase in importance and significance.

As described further in this preface, coalition networks have some characteristics which are unique to the fact that they are composed after-the-fact from two or more networks that have evolved independently. In this book, we look at some of the advances made in the understanding of coalition networks. The chapters deal with issues in different types of networks that are found in the context of coalition operations, and follow the structure of communication networks, information networks, and human networks that are used in the basic field of network sciences.

An understanding of the science behind coalition operations can benefit not just military operations, but any context in the modern world where two independent organizations need to collaborate together for a shared goal. In this age of globalization, commercial enterprises frequently outsource operations that are not considered to be within their core competency. Mergers and acquisitions require commercial enterprises to revisit their network and information processing infrastructure architecture and design. Furthermore, in an era of ever-changing technology, new partnerships and alliances frequently emerge with the alliance or partnership drawing upon the skills available among the various members.

Most of the chapters that are included in this book describe research that was done within the context of the International Technology Alliance in Network and Information Sciences, a collaborative research program sponsored by the U.S. Army Research Laboratories and the UK Ministry of Defence established to promote the science behind coalition networks. In an interesting application of some of the principles discussed in this book, the alliance itself is a conglomeration of different research labs, universities and government agencies. The authors of the various chapters have come across some of the challenges described in interconnecting and managing different computing elements, information sources and cultural differences while conducting their research in the context of the alliance.

DIFFERENT TYPES OF NETWORKS

Networks of many types can be found in different organizations, including the military, commercial enterprises, telecommunications and in the daily lives of the public. Thanks to the ubiquitous reach of the Internet, being connected in different types of network is a natural part of modern life. Network science, in its general definition, attempts to understand, characterize and analyze these various types of networks.

In order to categorize and understand the properties of the many different types of networks, the field of network science typically classifies networks into three types - communication networks, information networks and human network.

Communication networks provide the infrastructure which interconnects different computers, mobile devices and other hardware and software elements available to different personnel involved in an operation. The primary function of communication networks is to provide the ability to carry bits of information between the different elements of the networks, e.g. an email sent on a PDA is delivered to the email program running on a geographically distant laptop. The phone network, the Internet, the satellite systems used for military surveillance, the different flavors of email and instant messaging are all examples of communication networks that are found in the modern technological society.

Information networks provide the infrastructure that interconnects different elements of data that is available in a distributed communication network. While communication networks provide the ability to transfer the data among different distributed components, information networks provide the ability to analyze, process, and convert the data to yield useful information to the users of the network. Information networks include software systems that perform data analysis, data mining, knowledge extraction, problem diagnosis and the infrastructure that aims to make the conversion of data into knowledge more efficient. Such infrastructure can include technologies such as content distribution, caching, databases and data warehousing.

The human network consists of the relations that exist between the users of the entire network and characterize their interactions with other users, information sources and computing elements that are present in the overall system. Human networks may be aided by some applications, e.g. social networking sites, but the human network is focused more generally on the set of users who are interacting with the information and/or communication networks. Human networking looks at the properties of social interaction among human users, and uses information about these interactions to improve the effectiveness of humans, either individually or as part of a collaborative team.

All these different types of networks have some similarities, common characteristics and properties. These common properties that hold across all types of networks, e.g. the small world property of networks, are part of the scope of network science. However, each network has its own set of idiosyncrasies and properties that are valid only for that type of network. The study of properties that are specific to different types of networks also lies within the scope of network science.

COALITION NETWORKS

In a coalition environment, the communication, information, and human networks need to satisfy some additional constraints which may be less pronounced or less prominent within the context of single-organization networks. There are also several scientific challenges that apply equally well to coalition and single-organization networks.

Looking at the unique challenges that exist in the domain of communication networks, coalition networks need joint command and control of networks that may have evolved independently, use different technologies, and many not have adequate level of interoperability support between themselves. The different networks in the coalition may want to share some information about themselves to permit a joint view of operations and a joint management, but they may not be able to share full information about themselves to each other due to restrictions on the sharing of some sensitive information. As the different networks evolve over time, the best manner for them to interconnect may also change dynamically. Technologies need to be developed to address these challenges, and new theories are needed to understand the implications of interoperability among the different networks.

In the context of information networks, information needs to be disseminated seamlessly across two or more different coalition partners, while complying with any policies or restrictions that may be imposed on the networks. This creates new challenges in understanding how the quality of information coming from different coalition partners can be characterized and understood, how the different assets belonging to the many partners can be brought together in a seamless whole, and how the information can be processed and analyzed efficiently across different organizations. The security attributes associated with the different elements of information and the differences in security practices of different coalition partners make designing and securing coalition information networks a particularly challenging exercise.

In the context of human networks, differences in culture, communication patterns and behavior styles of different members coming from two different organizations can have a significant impact on coalition operations. An understanding of these differences and how they can be overcome to obtain effective operations is crucial for any endeavor that spans more than one organization.

A big challenge associated with coalition operations that spans all three types of networks is the existence of different policies for different groups that make up the coalition. The differences in policies affect the operations of all aspects of communications, information, and human networks in coalition operations.

Some of the recent advances made in improving the knowledge of network science for coalition operations are compiled in this book. The next section explains the structure of the book and the content of the different chapters.

STRUCTURE OF THE BOOK

Following the division of network science into the understanding of three different types of networks dealing with communication, information, and humans, the book is structured into sections that look at some of the challenges related to coalition operations in these different types of networks. The first section of the book looks at the issues of communications and information network, while the last (third) section looks at the issues in human and cognitive networks. The second section looks at another issue which impacts the operations of coalitions, security and policy management across different organizations.

In the communications and information networks section, this book includes four chapters. The first chapter examines the challenges in inter-domain routing. The next chapter discusses approaches that can characterize quality of information in sensor networks. The third chapter provides a comprehensive survey of techniques used for mining of graph patterns in information networks. The final chapter in this section provides a survey of various approaches that address an important of problem in information networks, ensuring that a theater of operations can be covered by means of sensors.

The second section of the book deals with the security issues in coalition contexts, with a specific focus on the exploitation of policy technologies to solve the unique security problems arising in coalition contexts. This section consists of four papers, with the first chapter looking at security issues that arise in coalition networks. The second chapter proposes a method for federating multiple coalition networks into a single one while resolving their policy conflicts. The third chapter discusses how dynamic communities of interests can be established rapidly using policy based techniques, The fourth and the final chapter in this section discusses aspects of policy based security management in coalition environments.

The third and the final section of the book deals with issues related to the human and cognitive networks. The first chapter in this section provides a bridge between policy technology and human issues. It examines how cultural differences among members can affect security policies and the open research problems in the intersection of culture and policy. The second chapter in this section discusses the concept of the network-extended mind, how the cognitive model one develops in mind depends on the information available in one's network. The third chapter provides an overview of cultural network analysis, a technique that can be used to effectively analyze differences in behavior of different coalition members. The fourth and final chapter presents some results in analyzing the communication patterns of coalition members, and discusses the insights one can gain by examining the conversations of coalition partners.

Taken together, the twelve chapters in this book provide a current snap-shot of network science research looking at interoperability issues in coalition networks.

WHO IS THE BOOK FOR?

This book is intended for researchers in the academia, industry, and governments who want to understand the issues in coalition operations, and obtain an overview of the recent advances in the field of network science that are pertinent to network-centric operations for coalitions. This book will introduce some new advances in network science. Researchers in the field of communication networks, information management systems and human/cognitive issues will find the chapters in this book to be of particular relevance.

If you are a military or industrial architect or planner looking to reap the benefits of network-centric operations in your operations, either military or commercial, you will find the advances described in this book to be of relevance. The book provides a broad survey of many issues that arise in different types of networks across organizations and an awareness of these issues will benefit you in developing a better plan for the cross-organization network.

If you are a graduate student or researcher in the networking area, this book will provide a good snapshot of contemporary research in the field of network science.

WHO IS THE BOOK NOT FOR?

This book focuses on aspects of network science that are related to coalition operations or issues that span more than one organization communicating together. As a result, the book does not provide a detailed treatment of network science issues which are related to understanding networks in general. Thus, this book is not appropriate for those who are looking for a broad overview of network science, but more suitable for those who want to look at specific areas within network science.

This book is a compendium of research papers and surveys. As such, it is not a comprehensive introduction to the subject of network science for coalition operations for someone unfamiliar with the field. It is instead targeted for researchers who already have some understanding of the area and are looking for focused detailed research papers on specific aspects of network science.

Acknowledgment

The following acknowledgement applies to all of the chapters of this book except Chapter 3.

This research was sponsored by the U.S. Army Research Laboratory and the U.K. Ministry of Defence and was accomplished under Agreement Number W911NF-06-3-0001. The views and conclusions contained in this document are those of the author(s) and should not be interpreted as representing the official policies, either expressed or implied, of the U.S. Army Research Laboratory, the U.S. Government, the U.K. Ministry of Defence or the U.K. Government. The U.S. and U.K. Governments are authorized to reproduce and distribute reprints for Government purposes notwithstanding any copyright notation hereon.

Additionally: The editor would further like to acknowledge the Agreement Number W911NF-06-3-0001 for the overall content of the book, and for facilitating the exchange of information among different researchers that were enabled due to the program. Neither the idea of the book, nor its contents would have been feasible without the existence of the U.S. UK International Technology Alliance in Network and Information Sciences, which was established under the above agreement number.

Section 1

Coalition Communication and Information Networks

Chapter 1
Inter-Domain Routing in Mobile Ad Hoc Networks:
Challenges and New Directions

Chi-Kin Chau
University of Cambridge, UK

Jon Crowcroft
University of Cambridge, UK

Kang-Won Lee
T.J. Watson Research Center, USA

Starsky H.Y. Wong
IBM T.J. Watson Research Center, USA

ABSTRACT

Because of the increasing number of diverse routing protocols proposed to deal with the network dynamics in mobile ad hoc networks (MANETs), the heterogeneity of MANETs has increased dramatically. While many of these extant proposals only concern enhancing routing in a single domain, little attention has been given to the interoperations among heterogeneous MANETs. Moreover, the existing inter-domain routing protocols (i.e., BGP) that have been designed for the Internet cannot cope with the new challenges derived from the MANETs, such as (1) the rapid dynamic changes of network topology due to mobility, and (2) the larger diversity in intra-domain ad hoc routing protocols. In this chapter, the authors address some of the major challenges and identify new directions to the development of seamless inter-domain routing for enabling end-to-end communications over heterogeneous MANET domains.

INTRODUCTION

The value of a network lies in the ability of interconnecting its systems. Often a network is used to interconnect multiple heterogeneous sub-networks, which may be controlled by different administrative domains, adapted to different missions, or optimized for different environments. The internetwork consisting of heterogeneous sub-networks of different administrative domains is fundamentally different from the interconnection in a homogeneous network, where the later is usually controlled to optimize a

DOI: 10.4018/978-1-61520-855-5.ch001

certain global objective, while the former does not always optimize any global objective. In the Internet, the *Border Gateway Protocol* (BGP) (Rekhter & T. Li, 1995) provides a mechanism for interconnecting heterogeneous administrative domains of different intra-domain routing protocols, mechanisms, and policies. The mechanism to interconnect heterogeneous administrative domains is known as *inter-domain routing*, and the domains are called autonomous systems (AS) in the Internet.

Mobile ad hoc networks (MANETs) also require effective interconnections of multiple domains in dynamic uncertain environments, such as coalition military operations, emergency operations for disaster recovery, and communication in vehicular ad hoc networks (VANETs). In these situations, multiple organizations in different administrative domains often need to communicate and cooperate with each other. For example, in a disaster recovery scenario, the local police force may need to coordinate with fire fighters, military forces, and medical crews by sharing information and communicating with each other regardless of particular networking technologies that are used by different groups. In another example, a group of cars on the road may instantly form a communication network for sharing traffic information, preventing accidents, and sharing data. In practice, it is unlikely that these different MANETs support the same network technologies, routing protocols and routing policies. These applications call for the development of inter-domain routing over heterogeneous MANET domains.

Although inter-domain routing is a well-developed technology in the Internet, the inter-domain routing problem in MANETs is fundamentally different from that of the Internet. In MANETs, the network connectivity changes dynamically due to user mobility, and these changes can happen in the order of minutes or even seconds. On the other hand, in BGP, these network dynamics usually happen in the order of days (Oliveira & Zhang & Zhang, 2007), and the static gateway assignment in BGP cannot adapt to the frequent connectivity changes in the MANET environments. In addition, there are no clear boundaries among network domains, and multiple domains can overlap in the same geographic region. Moreover, different routing protocols for MANETs, such as reactive routing protocols (Perkins & Royer & Das, 2003; Johnson & Hu & Maltz, 2007), proactive routing protocols (Clausen & Jacquet, 2003; Perkins & Bhagwat, 1994), geo-routing protocols (Basagni & Chlamtac & Syrotiuk & Woodward, 1998; Karp & Kung, 2000), etc., may be employed by different domains. A well-designed inter-domain routing framework should be able to handle the interaction among different intra-domain MANET routing protocols.

In this chapter, we study some of the major research challenges of providing inter-domain routing over MANETs, and identify new directions and ideas for a solution. In particular, we discuss the related issues of inter-domain routing over MANETs as follows:

i) *Function of BGP*: We clarify the function of BGP as an interoperation mechanism, rather than a unified routing protocol over heterogeneous routing protocols. This motivates our design of an inter-domain routing mechanism over MANETs.
ii) *Issues of BGP over MANETs*: We discuss some of issues when we directly apply BGP on MANETs, and the source of the problems.
iii) *Issues of Existing Adhoc Routing Mechanisms*: There are a number of ad hoc routing mechanisms akin to integrating heterogeneous routing protocols. We compare these with the problem of inter-domain routing, and identify the differences.
iv) *Issues of Mobile Ad hoc Inter-domain Routing Protocols*: We present some general challenging issues in various aspects of protocol design and related implementations.

v) *A Solution Framework for Mobile Ad hoc Inter-domain Routing*: As a viable solution to mobile ad hoc inter-domain routing, we present a framework for the mobile ad hoc inter-domain routing protocol design and a set of important protocol operations.
vi) *Issues of Mobile Ad hoc Inter-domain Routing Policies*: Finally, we discuss other challenging issues in specifying the inter-domain routing policies, and applying them in dynamic network environment.

We remark that this chapter draws on the discussions from our prior work (Chau & Crowcoft & Lee & Wong, 2008).

FUNCTION OF BGP

The RFC 1771 (Rekhter & T. Li, 1995) introduced Border Gateway Protocol (BGP) for inter-domain routing over the Internet. We clarify that the major function of BGP used in today's networks is as a signaling protocol for interoperations among heterogeneous domains.

Decision vs. Signaling

By a routing protocol, we mean a decision mechanism to decide the proper forwarding paths at routing systems after sufficient exchange of connectivity information. By a signaling protocol, we mean a specification of formats and interpretations of messages for the exchange of connectivity information, without enforcing a decision at the systems. Hence, a routing protocol should output a proper decision that satisfies the objectives of its designer, while a signaling protocol only ensures the communications among systems are carried out properly, which may or may not be relevant to a decision.

BGP specifies various message fields for exchanging connectivity information among routing systems. BGP also includes a list of recommendations of deciding forwarding paths based on the connectivity information. For instance, it is recommended that a selected path should have minimal hop counts, and the priority of multi-exit discriminators (MEDs) is higher than interior gateway protocol (IGP) costs. However, Internet service providers (ISPs) are observed to implement various tweaks that work around these recommendations. For instance, they often pretend redundant AS numbers, or refuse to accept MEDs.

Therefore, BGP is more suitably regarded as a signaling protocol for exchanging information among domains, because the global decision mechanism of forwarding path is not defined, and ISPs can override the decision mechanisms in their own networks.

Unification vs. Interoperation

We then clarify the notion of interoperation. Due to the nature of heterogeneity of the Internet, it is likely that different ISPs are employing different decision mechanisms for selecting forwarding paths. Given a set of decision mechanisms (e.g. routing protocols), we might unify them or interoperate them. If we unify these decision mechanisms, we replace them with a single decision mechanism that can work as either one of them. If we interoperate these decision mechanisms, we make them work together without the knowledge of how they exactly work. Hence, the detail of decision mechanisms is transparent to interoperation, while unification necessarily involves control and re-engineering over decision mechanisms. That is, interoperation respects heterogeneity, whereas unification imposes homogeneity. It is unlikely that all different ISPs would agree on a global decision mechanism, or would be willing to reveal their own decision mechanisms. Hence, it is unrealistic to design an inter-domain routing protocol for providing a unified routing service. However, BGP can still provide interoperation among different ISPs, regardless the details of routing policies and protocols of different ISPs.

Figure 1. The MANET of domain A is partitioned due to mobility

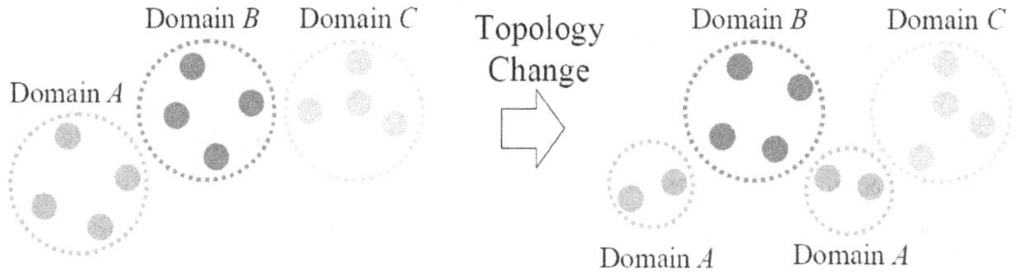

Connectivity vs. Commitment

Although one may state that BGP exchanges connectivity information (the physical interconnection of the network), in broader sense we argue that BGP actually exchanges *commitment* information (the logical relation among systems of the network). To explain the difference between connectivity and commitment, we note that one routing system may commit to transmit packets for only some routing systems, but not for others. This explains that even when the AS-level network topology is connected, it is not true that every AS can send packets to any other AS. Also, commitment can be regarded as a logical abstraction of connectivity, suitable for more sophisticated networks even without physical infrastructure.

ISSUES OF BGP OVER MANETS

In this section, we explain why BGP or a BGP-like protocol is inapplicable to the MANETs environment.

First, let us consider Figure 1, which consists of three MANET domains. One might apply a BGP-like protocol to this scenario as in Figure 2. However, there are several issues that render such protocol inapplicable. First, the path vector protocol in BGP implicitly assumes the availability of the following functions:

i) *Internal Gateway Detection*: The internal gateways within the same domain can detect the presence of each other so that they can communicate about the information of external routes.
ii) *Internal Network Knowledge*: The gateways know the reachable destinations and the internal routes to the destinations within the domain.

These functions are normally supported by the proactive intra-domain routing protocols through continual maintenance of network state information. However, we cannot assume the availability of this information in the MANETs that use a reactive routing protocol in their domains. Also a direct application of a path vector protocol over MANETs

Figure 2. A similar setting in terms of topology change in BGP

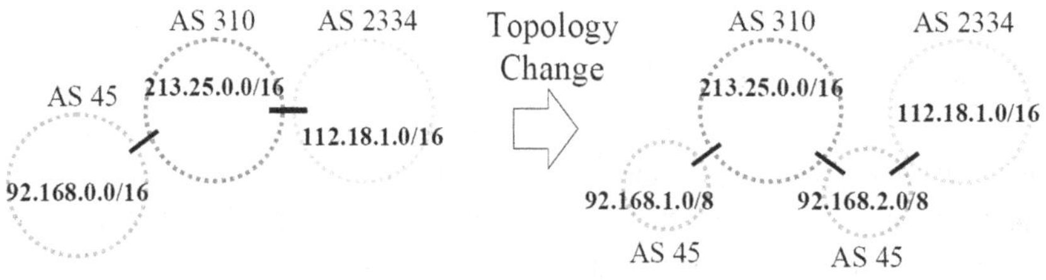

to support these functions may be undesirable to MANETs, because of the dynamic node mobility and scarce wireless communication bandwidth.

Second, in BGP every destination is identified by an IP address, which follows a certain network hierarchy. To announce the destinations in a domain, gateways will aggregate the IP addresses in the domain by suitable IP prefixes (e.g., 92.168.0.0/16). However, in MANETs, mobility and ad hoc deployment can create arbitrary network partition, unlike the perfect split of IP addresses as in Figure 2. Hence, IP prefixes do not suitably aggregate the IP addresses in partitioned MANETs and thus we cannot use the prefix-based routing of BGP.

Third, BGP relies on a path vector protocol that filters the paths consisting of repeated AS numbers to prevent looping. For example, in Figure 2, after topology change, the inter-domain level path from a source in AS 45 (92.168.1.0/24) to AS 2334 (112.18.0.0/16) is AS 45→AS 310→AS 45→AS 2334. This path will be filtered by the BGP path vector protocol, and hence it will prevent the nodes in AS 45 (92.168.1.0/24) from reaching AS 2334 (112.18.0.0/16).

In general, the design considerations for inter-domain routing in MANETs are fundamentally different from that of BGP. The main challenge of BGP is to cope with the extreme scale of the Internet; however, the scale of the network is not the main concern in MANETs since they will be relatively small due to physical/wireless, technical, and geographical constraints. Rather the main challenge here is to handle the constant changes in the network connectivity at both individual node level and at network domain level due to node mobility.

ISSUES OF EXISTING AD HOC ROUTING MECHANISMS

In the wireless ad hoc networks context, there are other proposals to enable communication akin to integrating heterogeneous routing protocols. Hybrid routing protocols such as SHARP (Ramasubramanian & Haas & Sirer, 2003) combine the proactive and the reactive strategies to improve the network performance. The basic idea of SHARP is to create proactive routing zones around the nodes with lots of data traffic, and use reactive routing in other areas. Although the hybrid routing protocols enable communication among proactive and reactive routing protocols, they require that all nodes are controlled by the same administrative policies thus precluding autonomous operations in multi-domain scenarios. This is a severe and somewhat unrealistic restriction. Inter-domain routing should require seamless deployment and interoperation of different routing protocols among heterogeneous, independent domains.

In MANET environments, there also exist other routing protocols such as the Cluster-based routing and hierarchical routing in MANETs (Chen & Liestman & Liu, 2004). The goal of these routing protocols is to provide a scalable routing solution in a single domain. The basic idea is to create small clusters from a single domain, select heads for each cluster, and then form a routing backbone among those cluster heads. Although it may seem that cluster-based routing strategies are similar to inter-domain routing in the sense that they elect cluster heads (similar to gateways in inter-domain routing) and form a high level routing structure among the cluster heads, they are fundamentally different. Cluster-based routing and hierarchical routing are designed to be operated in *homogeneous* environments, where all nodes share the same set of policies and protocols. On the other hand, the goal of inter-domain routing is to enable interoperations on *heterogeneous* domains with autonomous control, in which each domain can use different routing protocols and policies.

There are multiple proposals (Crowcroft & Hand & Mortier, et al 2003; Schmid & Eggert & Brunner, et al 2004; Ma & Chuah, 2005) that discussed how to enable communication among heterogeneous wireless networks. All of these

Table 1. Comparison of different routing mechanisms.

	BGP	*Hybrid-based Routing*	*Cluster-based Routing*
Supporting Mobility	No	Yes	Yes
Supporting Heterogeneity	Yes	Yes	No
Supporting Autonomy	Yes	No	No

schemes focus on very high level architectures and only provide a *sketch* of the required components (e.g., translation of different naming spaces, and different protocols). Plutarch (Crowcroft & Hand & Mortier, et al 2003) is an architecture that translates address spaces and transports protocols among domains to support interoperation of heterogeneous networks. TurfNet (Schmid & Eggert & Brunner & Quittek, 2004) proposes inter-domain networking without requiring global network addressing or a common network protocol. The authors in (Ma & Chuah, 2005) compared different design choices of inter-domain routing based on a strong assumption that all nodes can act as gateway nodes. While these related works have considered various issues regarding interoperation of multiple networks, none of them provided a complete solution. For example, they did not consider practical issues such as policy-based interaction among domains. Table 1 compares the main features of different routing mechanisms.

ISSUES OF MOBILE AD HOC INTER-DOMAIN ROUTING PROTOCOLS

After understanding the limitations of BGP and existing ad hoc routing protocols, we identify that there are a number of challenging issues that we need to address in the MANET environment when we devise a mechanism for mobile ad hoc inter-domain routing. These issues are:

i) *Node Mobility*: In the Internet, the border of a domain is well-defined so that at the edge routers/gateways can be easily setup. However, due to node mobility, the border of a domain is unclear in MANETs. In addition, the domains can split into multiple partitions and overlap with other domains. This introduces a problem of deciding which nodes should become gateways. In general, a well-designed inter-domain routing protocol for MANETs must be able to handle such network topology changes.

ii) *Heterogeneous Domains*: In MANETs, there are multiple different types of routing protocols with very different characteristics, and it is not possible for them to communicate with each other without extra support. For example, in a coalition environment, some domain may use a reactive intra-domain routing protocol such as Dynamic Source Routing (Johnson & Hu & Maltz, 2007) and Ad hoc On Demand Distance Vector (Perkins & Royer & Das, 2003), and some other domain may use a proactive intra-domain routing protocol such as Destination-Sequenced Distance Vector routing (Perkins & Bhagwat, 1994) and Optimized Link State Routing Protocol (Clausen & Jacquet, 2003). In a reactive domain, the path to a destination will not be available before there is some traffic going to that destination; while a proactive routing protocol will try to compute all paths regardless of demand. Thus we need to design a framework that can seamlessly connect different types of domains (without requiring changes to original routing protocols).

iii) *Support for Domain Autonomy*: As in the Internet, one of the main design considerations is that each domain should retain its autonomy and it should be able to enforce its own domain-specific policy. For example, domain *A* may choose to use domain *B* as its transit domain rather using domain *C* (for various reasons such as cost, reliability, security). Existing ad hoc routing protocols that have some structural resemblance such as Hybrid Routing or Cluster Routing do not support this concept of autonomy. Thus it is necessary to design an inter-domain routing protocol that can support domain autonomy and domain-specific routing policies.

A SOLUTION FRAMEWORK FOR MOBILE AD HOC INTER-DOMAIN ROUTING

In this section, we present *IDRM* (Inter-domain Routing protocol for MANETs) as a basic solution framework, which can be used to support inter-domain routing functionalities to address the challenging issues presented in the previous chapter. We note that developing a fully operational inter-domain routing solution for MANETs will require a huge amount of concerted efforts by different organizations from academia, industry, government, and standard bodies. Our goal is provide a basic framework to that is flexible enough to develop a practical solution of mobile ad hoc inter-domain routing.

In IDRM, each domain retains administrative control within its own domain while participating in collaboration. We denote a domain as a logical entity (e.g., belonging to the same organization). One the other hand, we call a MANET a connected directed graph with nodes belonging to the *same* domain. Thus a domain may consist of multiple MANETs. For example, there may be multiple MANETs of a single domain due to network partition as illustrated in Figure 2.

To enable inter-domain communications, our proposed framework requires special nodes as *gateways*. The role of gateways is more than just handling inter-domain routing; they need to bridge any technical seam that may exist between MANETs at physical, MAC, and network layers. For example, they may need to speak multiple radio technologies, understand different MAC layer interaction, or translate between different protocols. However, our main focus is limited to the inter-domain routing functions of the gateways. For simplicity, we assume that a non-gateway node does directly communicate with the nodes in other domains. Thus multiple MANET domains may operate in the same region. We illustrate this idea in Figure 3.

There are several issues that we need to handle: (1) partition and merge of MANETs, (2) membership announcement, (3) support for policy-based routing, and (4) inter-domain routing in data plane. The first two points are due to node mobility and dynamic topology, and the latter two are general issues with inter-domain routing with autonomy of each domain.

Handling Domain-level Topology Changes

As discussed in the previous section, one of the key challenges for inter-domain routing in MANET is the dynamically changing network topology. In particular, a single domain may be partitioned into multiple MANETs due to node mobility, and in order to handle such changes, gateways in the domain should detect the events. Consider the case in Figure 1, when a domain *A* has split into two networks. In this case, the gateways in domain *A* first need to discover the partition. In a domain where the intra-domain routing protocol is proactive, this event will be automatically detected via route updates. However, for a domain with a reactive intra-domain routing protocol, this event may not be detected for a long time. To handle this problem, in IDRM, we require the gateways maintain soft state by periodically sending beacons to each other. The period of beacon can be adaptively set based on the mobility of the nodes and the rate of topology change.

Figure 3. An example of how inter-domain routing connects different domain in MANET

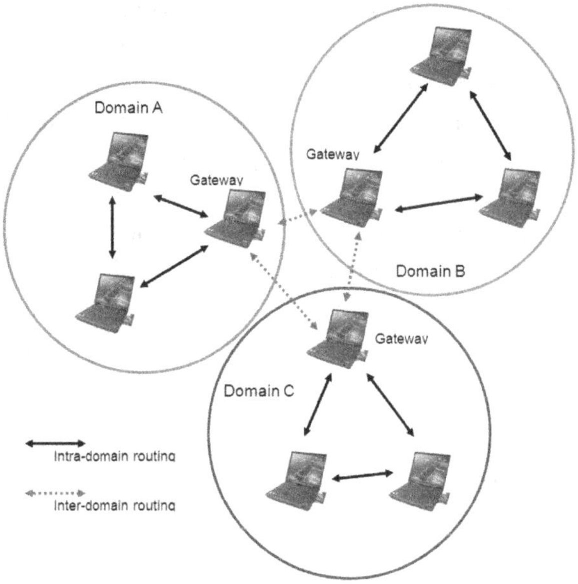

After detecting a partition, the gateways in the same partition should generate a new MANET ID so that the new partition can be uniquely identified. By dynamically assigning a new ID, we can prevent the path vector routing algorithm from mistakenly considering the route via partitioned networks as a loop. It is possible to extend this basic protocol to include a leader election process and let the leader of a domain coordinate intra-domain operations (e.g., hierarchical beaconing among gateways, or MANET ID generation). However, we do not consider such schemes in the first design for simplicity.

This computation should be performed independently at each gateway in the way that (1) all the gateways in the same partition to generate the same ID, and (2) the collision of IDs of different networks to be as low as possible. One way to achieve these goals is to use a pseudo random number generator to create a new ID using the IDs of all the gateways in the network as input. The gateways in the same partition use a simple hash function (e.g., MD5) to generate a random number, and then prefix it by the domain ID to get a new MANET ID. We encode the domain ID in the new MANET to support a dynamic policy translation (as discussed in Sec. 6). Conversely, when multiple partitioned MANETs come close and get re-connected, this condition should be detected by the gateways and a new ID for the merged MANET should be generated. This follows the same process as the case of network partitioning.

Membership Management and Announcement

Periodically the gateways should advertise the IDs of the nodes that they can reach; for this operation, the gateways need to collect the IDs of all the nodes in the MANET for advertisement of the membership to other domains. As we pointed out earlier, in MANETs we cannot rely on IP prefix for routing between domains due to arbitrary partitions and merges. There are two possible approaches to deal with the situation. First, the gateways can coordinate and reassign the node IDs so that each MANET can have a unique prefix every time a topology change occurs. However, this will incur significant management overhead (e.g., to generate unique prefix, generate unique node IDs, to update name-to-ID mapping) and thus will only be useful when the new topology will remain unchanged for a relatively long time.

Second, a more practical approach to handle topology changes is to let the gateways in partitioned networks advertise the membership information, and this membership digest is used for inter-domain routing. For a reasonable size MANET with less than 1000 nodes, we find that a plain membership digest containing a set of node IDs (e.g., IP addresses) without any compression is better than a more scalable solution such as Bloom filter (Broder & Mitzenmacher, 2002). Obviously, the second approach (based on membership digest) can cope with network dynamics in a better way and is more graceful when partitioned MANETs merge (by just merging the memberships). Hence, we employ the second approach in IDRM.

Keeping track of the non-gateway membership in a domain poses a similar challenge to network partition detection; in a reactive routing domain, a gateway may have a stale view of its membership, and can only discover the membership change when it has data to transfer. Although we can periodically perform a membership query, this can be potentially expensive. Thus instead, we let a reactive domain only initiate a membership query when there is an indication that its membership may have changed, e.g., a node in the membership digest cannot be reached, and a timeout period has passed.

Policy Support

Inter-domain routing policy is enforced in a similar way as in BGP. By exchanging route updates (announcements and withdrawal) in a path vector protocol, inter-domain routing policies will be translated as the decisions of filtering and selecting routes at gateways. One important issue to address in MANETs is that these routing policies are defined by network operators as static rules. Now in a MANET environment, a single domain may partition into multiple networks (e.g., a domain A breaks down into A_1 and A_2). Thus it is necessary to have a mechanism to automatically translate the original policy when such topology change happens. In Section 7, we will discuss some issues on how to translate the static policies when a domain partitions under the next-hop-based specification and the cost-based specification.

Data Plane Operations

When a node sends data packets to an external destination (in another domain or in another partitioned network), it forwards the packets to one of the reachable intra-domain gateways. In a reactive domain, the sending node will first initiate a route discovery, and a gateway node that has a route to the destination will respond. In a proactive domain, the sending node will have a list of intra-domain gateways, and select one of them based on its own preferences. Note that the list of reachable intra-domain gateways will be learned from the regular route updates.

In either case, the gateway will first see if it is directly connected to the domain that contains the destination. If this is the case, the gateway will send the packet to the destination; otherwise, it will forward the packets to a gateway connected to the destination domain based on the inter-domain routing information.

For incoming packets, the gateway performs protocol translation and invokes the intra-domain routing protocol. In a reactive domain, the gateway will initiate a route discovery process if it does not already have the route in the cache. In a proactive domain, the gateway can determine if the destination is reachable from the local routing table.

If for some reasons the destination cannot be reached (e.g., the node may have been disconnected from any domain) IDRM does not provide feedback for unreachable destination. Following the design principles of the Internet, the problem should be handled at a higher layer (i.e., the network only provides best effort delivery). Although we only discuss proactive and reactive routing protocols in this paper, it is straightforward to extend this framework can support other types of intra-domain routing protocols (e.g., geo-routing and hybrid routing).

Finally, we remark that we have prototyped all these protocol operations a well-adopted simulation framework, ns-2, and extensively evaluated its performance in different MANET scenarios (Lee & Wong & Chau, et al 2009). Furthermore, we also obtained overhead analysis to justify the feasibility of the proposed mechanism.

ISSUES OF MOBILE AD HOC INTER-DOMAIN ROUTING POLICIES

The specification of inter-domain routing policy by network administrators is a key component in the policy-based interactions. In this section, we first review different approaches that are used in practice for specifying routing policies, and discuss the issues of how we can accommodate these specifications in dynamic networks.

Common Policy Specifications for Inter-Domain Routing

In the networking community, the common approaches to specify inter-domain routing policies are as follows.

i) *Next-Hop Specification*: A simple approach to specify inter-domain routing policies is based on the next-hop domains in the routes to a specific destination, and ignored the rest of downstream domains. The next hop specification is widely used in practice to denote the mutual commercial relations among Internet service providers, which are coarsely classified as customer-provider or peering relations. A common practice by the Internet service providers is that the routes from customers are preferred to peers and providers, and routes from peers are preferred to providers.

ii) *Enumerative Specification*: A more sophisticated approach that enumerates all possible routes of the domains in the network, and ranks those routes by a complete order of preference. For example, domain A may specify the following ordered list ($A \to D$, $A \to B \to D$, $A \to C \to D$, $A \to C \to B \to D$), for a destination domain D. Routes with a higher rank will be selected if available, and backup routes will be given a lower rank, which will only be selected when no other routes are available. This approach is feasible when only simple routes are considered.

iii) *Cost-Based Specification*: A more practical and flexible approach is to assign a numerical cost, which is subjective to a local domain, to all other domains. The subjective cost captures the local evaluation of the reward or penalty of forwarding packets through the respective domain, such as hop counts, available bandwidth, and reliability of the path. In this case, the routes with the minimum total subjective cost of all the downstream domains will be selected.

iv) *Forbidden-Set Specification*: Another practical approach is to specify local policies is to identify a set of forbidden domains to traverse. The forbidden set captures the security concerns as to traversing insecure domains. Only the routes consisting of domains that are not from the forbidden set will be selected among available ones. Usually, an additional tie-breaking mechanism (e.g. random tie-breaking or lexicographical tie-breaking) is required for selecting a route among multiple candidates.

Issues of Dynamic Policy Refinement over MANETs

Inter-domain routing policies will be relatively long-lived compared to the dynamisms of MANETs such as topology changes. Typically network administrators will need to specify the inter-domain routing policies a priori to the operations of MANETs (although on demand policy modification will be supported). As discussed earlier, a salient characteristic of MANETs is the dynamic nature of network topology due to node mobility. Thus we need to examine how the aforementioned approaches of policy specifications can be supported in the dynamic MANET environments.

The path vector routing protocol relies on the uniqueness of domain IDs to specify policy-based inter-domain routes and identify the existence of cyclic paths. To cope with the challenge of dynamics in MANETs, IDRM gateways generate new MANET IDs for a partitioned network when a topology change is detected. Now the question is how we can translate the static policy that administrator defined for the original network configuration when such network change occurs. In other words, we need to address the *policy refinement* problem where a static routing policy must be dynamically translated into a more specific policy for the new network configuration. As presented earlier, when IDRM generates a new domain ID, it retains the original domain ID, and this characteristic is useful when writing a meta-rule for the policy refinement. We now explain the necessary policy refinement procedures using a simple illustrative example.

Consider Figure 4 with MANETs consisting of three domains *A*, *B*, *C*. Because of ad hoc mobility, there is a change of the underlying network topology, such that nodes in domain *A* are partitioned into two sets, each of which falls outside the wireless communication range of each other. However, by traversing the nodes in domains *B*, *C* nodes in domain A_1 and domain A_2 can still communicate with each other.

After the partition, the use of inter-domain connectivity (e.g. through domains *B*, *C*) to enable intra-domain communications (e.g. between two partitions of domain *A*) may be necessary. Inter-domain routing policies for MANETs, therefore, also need to specify the actions for this situation.

Firstly, let us consider the next-hop policy specification. Assume that domain *B* had a next-hop policy for domain *C* that reads $C < A <$ others, where $C < A$ means *C* is more preferable to *A*. Now after the partition of *A*, it can refine the original policy into $C < A_1 \approx A_2 <$ others, where $\approx$ denotes an equivalent relation. Thus now it reads C is more preferable to either A_1 or A_2 and there is no preference between A_1 and A_2. This is a straightforward translation and one can specify this rule using some standard policy refinement logics such as the one proposed in (Bandara & Lupu & Moffett, et al 2004). In other words, for the next-hop policy, we can specify a meta-rule that simply replaces the original domain with new sub-domains. We present this meta-rule in a pseudo code format in (Chau & Crowcroft & Lee & Wong, 2008). Note that it is not straightforward

Figure 4. An example of network partitioned in a MANETs of three domains

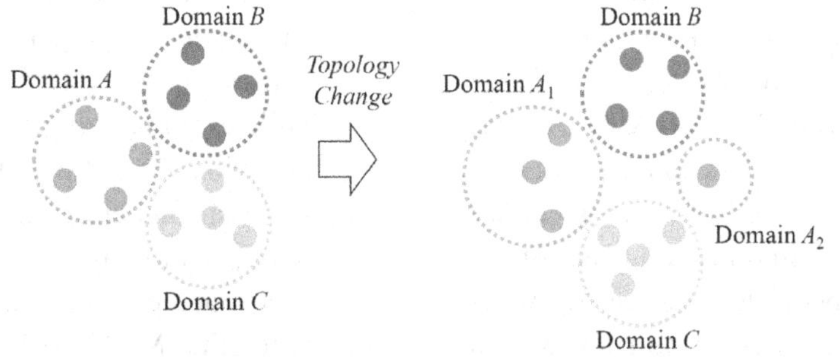

to specify the preference between A_1 and A_2 in the next-hop policy because their preferences cannot be determined a priori.

FUTURE RESEARCH DIRECTIONS

While we have proposed a basic solution framework to enable coalition operations in MANETs, there are still many open research issues in this area. In this section, we will like to discuss four open questions.

i) *How to utilize heterogeneous routing information*? As we discussed before, there are multiple types of intra-domain routing protocols in MANETs. The characteristics of these intra-domain routing protocols are very different. While some of the protocols optimize for the number of hops the data will transmit, other protocols may optimize for different goals such as adaptation to node mobility. For some performance metrics, such as hop counts and propagation delay, they share some similarities and there may be ways to incorporate these performance metrics together. For other metrics, such as hop counts and bandwidth, there may be no direct relationship. One research challenge here is how can these intra-domain routing protocols utilize the domain-specific optimization metrics from other domains? We believe a well-designed inter-domain routing protocol should able to utilize these heterogeneous routing information as much as possible, and make reasonable translation if the information cannot be used directly.

ii) *How to optimize gateway assignment?* As we mentioned in section 4, domains in MANETs do not have clear borders. In this case, how can the network administrator assign nodes (or employ some automated mechanism to direct some nodes) to become gateways such that the assignment of the gateways can improve network performance such as connectivity, delay and throughput? Also, due to node mobility, the topology of the MANETs is changing from time to time. How can the MANETs self-organize themselves to adapt these network dynamics? We believe a good inter-domain routing protocol should intelligent enough to assign gateways dynamically in order to adapt network changes and improve the network performance.

iii) *How to ensure convergence?* We note that in the Internet, IGPs tend to converge reasonably fast (sub-millisecond *Open Shortest Path First* convergence has been discussed). However, in the MANET environment, there is no systematic study on the convergence aspect of the intra-domain routing protocols. In this case, we cannot assume that the underlying intra-domain routing can provide update-to-date intra-domain routing information. Also, the MANETs will assign their gateways according to their optimization objectives, which imply we cannot assume each MANETs are running homogenous algorithm to assign gateways. We believe a sophisticated and more robust design of inter-domain routing in MANET should be able to handle the delayed changes reported by the intra-domain routing protocols.

iv) *How to support policies?* We note that in the Internet, policies for the inter-domain routing are supported by setting a next hop of a destination, and the next hop is selected according to the domain's preferences. However, due to node mobility in MANETs, it is possible for a domain, e.g. domain A, surrounded by other domains whose are conflicting with domain A's policies (the policies do not allow domain A to use any neighbouring partitions to forward data). One research challenge here is how can the isolated partitions relax their policy constraints such that they can maintain communication with other MANETs?

CONCLUSION

With the increasing usage and the technology advancement, we foresee that MANET will play a significant role in the future wireless network. Similar to the case in the Internet, one important aspect to provide seamless communication is to connect heterogeneous components (i.e., network partitions). In this chapter, we reviewed the role of the inter-domain routing protocol (i.e., BGP) of the Internet, and identified the problems that we encounter when we try to directly apply BGP in the MANET environment. By understanding the limitations of BGP and other ad hoc routing protocols, we identified the challenges to design an inter-domain routing protocol in heterogeneous MANETs. Namely, we identified the main challenges are to support node mobility, heterogeneous domains, and autonomy of domains.

We then presented a basic solution framework for inter-domain routing in MANET called IDRM. The proposed framework provides the following essential operations: (i) coping with domain-level topology changes, (ii) managing the members in the domain and announcing the information, (iii) supporting policy-based inter-domain routing, and (iv) supporting actual data exchange between domains.

REFERENCES

Bandara, A. K., Lupu, E. C., Moffett, J., & Russo, A. (2004). A Goal-based Approach to Policy Refinement. In *Proceedings of the IEEE Policy Workshop.*

Basagni, S., Chlamtac, I., Syrotiuk, V. R., & Woodward, B. A. (1998). A Distance Routing Effect Algorithm for Mobility (DREAM). In *Proceedings of the ACM/IEEE Mobicom.*

Broder, A., & Mitzenmacher, M. (2002). Network applications of bloom filters: A survey. *Internet Mathematics, 1*(4), 485–509.

Chau, C. K., Crowcroft, J., Lee, K. W., & Wong, S. H. Y. (2008). IDRM: Inter-Domain Routing Protocol for Mobile Ad Hoc Networks. University of Cambridge Computer Laboratory Technical Report, UCAM-CL-TR-708, Cambridge, UK.

Chau, C. K., Crowcroft, J., Lee, K. W., & Wong, S. H. Y. (2008). How to Enable Policy-based Interactions in Dynamic Wireless Networks. In *Proceedings of the IEEE Policy Workshop.*

Chen, Y., Liestman, A., & Liu, J. (2004). Clustering Algorithms for Ad Hoc Wireless Networks. In *Proceedings of the Ad Hoc and Sensor Networks.*

Clausen, T. & Jacquet, P. (2003). *Optimized Link State Routing Protocol (OLSR)*, RFC 3626.

Crowcroft, J., Hand, S., Mortier, R., Roscoe, T., & Warfield, A. (2003). *Plutarch: an argument for network pluralism.* ACM Computer Communication Review.

Johnson, D., Hu, Y., & Maltz, D. (2007). *The Dynamic Source Routing Protocol (DSR) for Mobile Ad Hoc Networks for IPv4. RFC 4728.*

Karp, B. N., & Kung, H. T. (2000). GPSR: Greedy Perimeter Stateless Routing for Wireless Networks. In *Proceedings of the sixth annual ACM/IEEE International Conference on Mobile computing and networking (Mobicom).*

Lee, S. H., Wong, S. H. Y., Chau, C. K., Varadarajan, S., Lee, K. W., Crowcroft, J., & Gerla, M. (2009). *Self-organizing Inter-Domain Routing for Heterogeneous MANETs.* UCLA Technical Report, Los Angeles, CA. Retrieved May 2009, from http://www.cs.ucla.edu/~shlee/papers/idrm.pdf

Ma, W., & Chuah, M. C. (2005). Comparisons of Inter-domain Routing Schemes for Heterogeneous Ad Hoc Networks. In *Proceedings of the IEEE WOWMOM.*

Oliveira, R. V., Zhang, B., & Zhang, L. (2007). Observing the evolution of internet as topology. New York, NY, USA. In *Proceedings of the ACM SIGCOMM.*

Perkins, C. Royer, E., & Das, S. (2003). *Ad hoc On-demand Distance Vector (AODV) Routing, RFC 3561.*

Perkins, C. E., & Bhagwat, P. (1994). Highly Dynamic Destination-Sequenced Distance Vector (DSDV) for Mobile Computers. In *Proceedings of the ACM SIGCOMM Conference on Communications Architectures, Protocols and Applications.*

Ramasubramanian, V., Haas, Z. J., & Sirer, E. G. (2003). SHARP: A Hybrid Adaptive Routing Protocol for Mobile Ad Hoc Networks. In *Proceedings of the ACM MOBIHOC.*

Rekhter, Y. & Li, T. (1995). *A Border Gateway Protocol 4 (BGP-4). RFC 1771.*

Schmid, S., Eggert, L., Brunner, M., & Quittek, J. (2004). TurfNet: An Architecture for Dynamically Composable Networks. In *Proceedings of the 1st IFIP International Workshop on Autonomic Communication (WAC).*

Chapter 2
Quality of Sensor-Originated Information in Coalition Information Networks

Chatschik Bisdikian
IBM Research, USA

Lance M. Kaplan
U.S. Army Research Laboratory, USA

Mani B. Srivastava
University of California at Los Angeles, USA

David J. Thornley
Imperial College London, UK

Dinesh Verma
IBM Research, USA

Robert I. Young
Defence Science and Technology Laboratory, UK

ABSTRACT

Coalition operations greatly benefit from the exchange of information collected from a plethora of wirelessly communicating sensors deployed in the theater of operation. However, to make judicious use of this information and then act effectively based on this information, knowledge of its quality, and a common means of expressing and communicating this quality are necessary. This chapter considers quality of information (QoI) for sensor networks starting with building a definition of QoI from first principles and by exploiting industry efforts to define quality in other domains, notably quality of service. The definition touches upon two complementary yet distinct aspects of quality: (a) the inherent quality attributes that characterize information; and (b) the assessment of these attributes within the context of applications utilizing the information, referred as the value of information (VoI). It then discusses information as a service and the various QoI viewpoints that associate sensor-originated information providers with sensor-enabled applications in dynamic coalition environments. Then, information pro-

DOI: 10.4018/978-1-61520-855-5.ch002

cessors and operators are introduced, which are functional modules in end-to-end systems that process information based on the QoI attributes. Finally, a data model for QoI metadata, describing the QoI attributes, is presented. The data model provides a common means to describe and communicate QoI attributes among information processors. It enables the indexing and searching of the most pertinent information, of desired quality, and information sources and the on-demand binding of applications to sources that is necessary to support operations in multiparty coalitions.

INTRODUCTION

Achieving and then exploiting critical information advantage over adversaries is a key objective of military *network-centric operations* (NCO) (Alberts, Garstka, Hayes, & Signori, 2001). Such military operations may be executed by either single-nation units or, of particular interest to this book and this chapter, by multi-nation coalition partners. Critical information advantage allows coalition partners to gain a deep, common understanding of situations, i.e., *shared situation awareness*, at an accelerated pace. It allows the timely support of decision making regarding actions towards the situations of concern that are both effective with respect to their outcome and efficient with respect to the resources spent for and assets affected by these actions.

In pursuit of information advantage and, more broadly, information superiority, an intertwined web of various types of assets is utilized for the creation, manipulation, and dissemination of pertinent information and ensued knowledge. These assets–including physical, i.e., hardware; virtual, i.e., software; and human assets– are organized in a collection of layered networks including (from bottom-to-top):

- *physical networks*: comprising data sources, e.g., sensors and communication networks;
- *information networks*: comprising information technology (IT) systems and the processes and protocols they support for managing information, including processing, storing, indexing, retrieving, disseminating, and so on; and
- *user* (or *social*) *networks*: comprising the information consumers (computer applications, humans, and human-computer interfaces (HCI)) on behalf of which information is collected and knowledge about situations is shared.

The scope of an information network may extend to include not only the hardware and software system infrastructure but also the collection of all the information pieces that reside or are accessible through these systems as well as any linkages and references that exist to or can be made amongst these pieces.

Sensing systems of various types and configurations, including in situ or remote sensors, stationary or mobile multi-sensor platforms, sensor networks, and combinations thereof, are key enablers of the NCO objective for information superiority. With their ever increasing capabilities, with respect to sensing, communications, processing, operational lifetime, sensing systems are deployed to provide real (or, near real) time information of situations in ways that were never possible before or exposed humans to heightened danger otherwise. Note, however, that sensor-originated information is not necessarily a substitute for all kinds of human-collected intelligence –sensors are, for the foreseeable future, incapable of interpreting, on their own, the social and cultural contexts surrounding the situations they observe. Instead, depending on the circumstances, the two types of gathered in-

Figure 1. Decision making and action taking with and without quality consideration

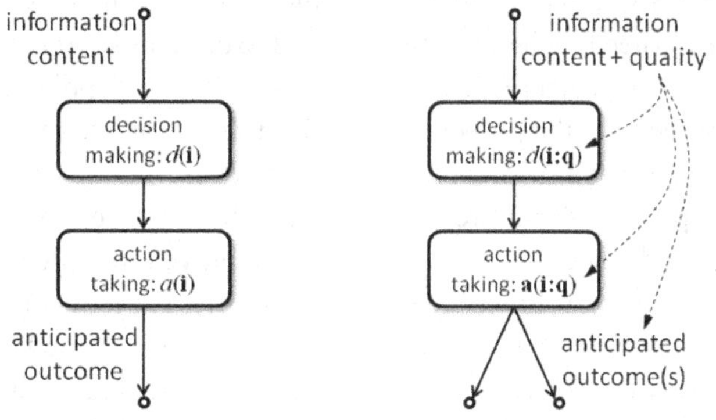

formation will coalesce to supplement, support, and confirm each other.

The collection and dissemination of information among coalition partners is necessary for achieving shared situation awareness. However, the mere availability of information content is not necessarily sufficient to support effective decision making and action taking that attain anticipated levels of success. It is of little value if the understanding of a situation that is shared by the parties tasked for the situation is built upon the wrong information. It is even worse if the parties are not aware of the possibility of misunderstanding of the situation, or its extend, and hence be unprepared of the consequences of their actions, or inactions. The results may range from suboptimal (wasteful) usage of resources to unanticipated catastrophic outcomes both militarily and humanely at various levels. Thus, not only the information *content* but also its *quality* is important. Just as the information content itself, the quality of the information, which relates in some sense to how *pertinent* the information may be to a task at hand, needs also to be shared and taken into consideration in decision making and in devising of alternative action plans. This is highlighted in Figure 1, where the notation **i:q** stands for information **i** of quality **q**, and the bold lettering represents vector, i.e., multidimensional, components; the expression **a(i:q)** represents the multiple action plans considered as result of deciding using information **i:q**.

In the past several years, a considerable research effort has been made and results published on the intelligent deployment and operation of sensing systems and, in particular, *wireless sensor networks* (WSN). This effort has been concentrating on the operational challenges of these systems, such as energy consumption, limited processing and communication device capabilities, unattended operation, packet routing and transport, and so on, (Akyildiz, Su, Sankarasubramaniam, & Cayirci, 2002). These are all very important and challenging problems looking *inward* at the WSNs and their operations. Significantly less effort has been made on the relationships –the *outward* relationships– that these systems have with the applications whose information needs these systems are to support. Quality of information is one such outward relationship and, despite its high importance, it has experienced a very limited exposure within the context of WSNs and even more so within the context of WSNs for coalition operations.

But, what is *quality of information* (QoI)? How does it manifest itself? What role can it play? How can it be described and communicated? These are some of the questions that this chapter answers. The topic of QoI is both broad and deep, something the authors themselves experienced firsthand, and cannot be covered in its entirety in a single chapter. Our aim is to expose the topic and provide a framework that could serve as a focal point around which a robust and concerted research effort on QoI for WSNs can be pursued. The authors feel that there is a need for increasing research efforts in the area, and whether readers of this chapter agree or disagree with the proposed framework is secondary to putting a stake on the ground and establishing a point of departure out of which such research can advance.

The chapter is organized as follows: We start with a brief presentation about sensing and information in general terms that serves as the backdrop for the developments in the rest of this chapter. Then, we tackle the QoI definition within the context of sensor networks, along with an example of how QoI affects the design of a perimeter protection system. Following, we present challenges of QoI for dynamic environments including multifaceted viewpoints of QoI. We then discuss information processors and operators that act upon information content and QoI metadata and present a data model for QoI metadata used to communicate QoI information among sources, processors and applications. We conclude with a summary of the chapter.

BACKGROUND

Sensing, within the context of this chapter, is defined as the act of detecting automatically especially in response to some physical stimulus.[1] Hence, sensing systems are deployed to detect physical phenomena. As a consequence, sensing systems are deployed to augment our capability to observe what happens in areas of interest we are incapable of entering or it is too costly or dangerous to do so, such as observing the seas for enemy submarines and the skies for enemy airplanes and missiles, or entering enemy territory to observe his intentions. Sensing systems enable a broad collection of intelligent, sense-and-respond type of applications such as biometric security systems, intrusion detection, location-based services, environmental control, sensor-to-shooter systems, remote healthcare monitoring, computerized supply-chain systems, and so on. This short list underscores the vastness and diversity of applications that sensing enables from the minute to the grand.

End-users collect and use information to gain an understanding of the world. Sensing systems are deployed to collect this information which, hopefully, is pertinent to needs and interests of the end-users. Armed with this understanding, end-users, including humans or computer applications, make decisions and take actions in response to the state of affairs in the world. For example, an automated door system will automatically open the door if it senses[2] the presence of a human approaching it; an automated sprinkler system will cut-off watering the back-yard lawn if it senses that it is raining; an intelligent notification system will alert the necessary units if it senses enemy actions consistent with missile deployment or pre-launch procedures, or if it senses an incoming missile, or if it senses the gathering of people or assets of concern at a particular location, and so on.

Sensing systems are deployed to collect pertinent information and enhance the ability of end-users to make decisions and take actions that add value at desirable levels to the outcome of the actions taken. For the class of *sensor-enabled applications*, i.e., applications that operate on sensor-originated data and derived information, these actions are influenced by what the sensing systems sense happening in the world, or, more accurately, what the interpreters of the sensor-originated information "believe" is happening in the world. The distinction between *data* and *information* is a source of quite a philosophical debate and quite often the two are used interchangeably. For the purpose of this chapter, we will also consider the two interchangeably; computerized systems operate on (input) data after all. However, we will keep in mind some mild differences that imply a level of hierarchy between the terms:

- *Data* represents a collection of pieces of facts, such as, for example, "10:30pm," "60°C," "bathtub," or "Joe."
- *Information* represents the placement and interpretation (processing or fusion) of individual pieces of facts (data) within a common context, "the time now is 10:30pm," "the water temperature is 60°C," "Joe is in the bathtub."

The above distinction has been noted by other as well, see for example Alberts *et al.* (2001) and Driscoll, Tortorella, and Pohl (2005). However, we note here that the scope of context underlying the definition of information is not (and, we believe, cannot be unambiguously) specified. As a result, one man's information can be another man's data, which produces a recursive intermingling of the terms. For example, the aforementioned pieces of information represent pieces of facts in their own right that can be further merged together under a broader common context to produce the (composite) piece of information: "The time is 10:30pm and Joe is now in the bathtub where the water temperature is 60°C." Without further delving in additional terminology, we note here that Alberts *et al.* (2001) also consider the even higher (cognitive) primitives of *knowledge*, *awareness*, and *understanding*, that ultimately guide decision making and action taking. In the particular example, these would relate to the conclusion that (provided that the information is correct) Joe is in big trouble, because of the dangerously scolding 60°C water in the bathtub, requiring actions such as checking on Joe, shutting down the water, calling the paramedics, and so on.

In the broad scheme of things, information represents available pieces of facts upon which knowledge (awareness and understanding) about situations of interest can be built. A more specific interpretation of information within the context of sensor networks will be given later on in the chapter.

QUALITY OF INFORMATION (QOI)

So what is QoI? Defining this easily understood term is a challenging task. Robert Pirsing's quote summarizes this point very well "...Even though quality cannot be defined, you know what it is...," (Pirsing as cited in Gomes 1996, p. 22). The quote notwithstanding, a formal definition of QoI is necessary to serve as the foundation that provides coherency for and bridges the efforts (research and otherwise) that pursue and advance the QoI cause. To this end, we present QoI in two different levels. At first we present a definition of QoI at a broad level that is derived by studying other "quality" efforts, including industry standards efforts on Quality of Service (QoS). We then apply this definition more specifically to QoI within the context of sensor-originated information. Both of these are important as QoI could apply at different scales and scopes, even within the context of the same application design and deployment, as outlined in the section on QoI in dynamic environments.

The Quality of Information

The Oxford dictionary defines quality, among others, as (*a*) the degree of excellence of something when measured against other similar things, and also as (*b*) a distinct attribute or characteristic.[3] In building our QoI definition, we use both these definitions to imply that quality relates to attributes or characteristics based on which similar "things" can be compared against. The communications industry (both computer and telecom) has put a lot of effort in dealing with quality about the services they provide (QoS). ITU standards recommendation for Quality of Digital Telecommunication Services Service (ITU-T Rec. E.800) defines QoS as the:

The collective effect of service performances, which determine the degree of satisfaction of a user of the service.

These service performances include both service management performances such as effectiveness and responsiveness of customer and technical support, as well as more traditional connection quality performances such as efficient connection establishment and information transfer. Interestingly, despite the vast literature on the subject, QoS in the TCP/IP world has not been formally defined. However, a glossary entry in RFC 2386 (Crawley, Nair, Rajagopalan, & Sandick, 1998) specifies QoS as:

A set of service requirements to be met by the network while transporting a flow.

What the above two definitions of QoS provide –the ITU-T definition more directly than the TCP/IP one– is to acknowledge the role of the user, or the recipient of the service, on behalf of which the quality of the service provided is to be assessed. The TCP/IP definition does not refer to a user explicitly, but it can be construed as the entity that imposes the "service requirements" referenced in the definition. In the case of QoI, the role of the user is played by the application context under which the information is to be used. Furthermore, both definitions point to quality as a multi-dimensional entity represented by a collection of quality characteristics, parameters, or attributes.

In defining QoI for sensor networks, we combine the essence of the above industry standards definitions with the aforementioned relation of quality to attributes and characteristics. Research in information quality in the enterprise world (including many information intensive applications and industries like marketing, manufacturing, retail, customer relationship management, information technology system management, Web-based information search and access, and so on) has identified a very large number of attributes for comparing information (well over a hundred) (Knight, 2007; Wang & Strong, 1996). The attributes cover both inherent (objective) quality properties related to the information content, e.g., measurement precision, as well as subjective ones that are interpreted within the context of a use, e.g., usability. Extended from these, QoI relates to:

1. The inherent characteristics of information, as when we state that:
    - The sensor measurements have a ±3% margin of error.
    - The localization outcome (based on the sensor measurements) is accurate to within ±5 meters.
    - This image is of high resolution (has lots and lots of pixels).
    - The video feed has a five-second delay.
2. The information's role within the context of its use, as when we are questioning:
    - Is the margin of error really±3%? (an accuracy question)
    - Who said that a sensor measurement has a ±3% margin of error? (a trust-related question)
    - Do the sensor measurements cover the extend of my information needs? (a completeness-related question)
    - Is this high resolution image of any use? (a relevancy-related, or operational context, question)
    - Is the video feed too late for my needs? (a timeliness-related question)

While acknowledging the fact that QoI becomes meaningful and is eventually assessed within the context of a use, we also recognize that if we were to allow QoI to "monolithically" subscribe to this premise, owing to the immeasurable variety of information uses, it would have been very hard to make any progress on the subject. In defining QoS, the use case of the service in question, i.e., providing telecommunication services to subscribers or providing differentiated network flow services, is well defined, or at least sufficiently scoped. In contrast, in defining QoI, there exists no (and there might never be a) single or a deterministic set of use cases that can serve as a benchmark for QoI assessment. The meaning of information is broad and its uses abound. The same piece of information may find many different uses, and, hence, its quality be assessed and reassessed over and over again as needed and valued differently by different applications. We do not discount the GIGO (garbage in, garbage out) principle. However, given the multiple uses of information, having *judged* a particular piece of information as low quality and "unfit" within the context of a particular use (e.g., this high resolution image is irrelevant to known needs) shall not necessarily prejudice us and others in considering the same piece of information again in a different occasion.

Therefore, instead of considering QoI as a single (holistic) domain, we have opted for a split, layered interpretation of QoI as highlighted in Figure 2. According to this interpretation, the entire QoI domain comprises:

- A *QoI (sub-)domain* (in an abusive reuse of the term QoI) representing the inherent quality characteristics of information, i.e., the collection of factual statements that can be said about the information along various quality dimensions such as *accuracy*, *latency*, and *provenance*; and
- A *VoI* (for *value of information*) *(sub-)domain* representing the assessment of QoI, i.e., the value interpretations of QoI within various application contexts.

This layered division of the QoI domain allows us to separate the ability of performing the process of judging information regarding its fit for our needs from the potential outcome ("yes," "no," "maybe") of the judging process. We may repeat the judgment process for the same piece of information as often as necessary without prejudice. Note that the concept of fitness of use for information is widely accepted in studies of information quality (Wang & Strong, 1996).

The division of the QoI domain lends itself to a corresponding split definition for QoI as well. We will overlay this definition on top of the reference abstract information flow model shown in Figure 3. According to the reference information flow model, information reports are provided from an information *source* to a *receiver* within an *information stream* carried over a *path* linking the source and the receiver. We do not determine the duration of a stream upfront; a stream could be instantaneous, comprising of a

Figure 2. Split vs. holistic view of QoI

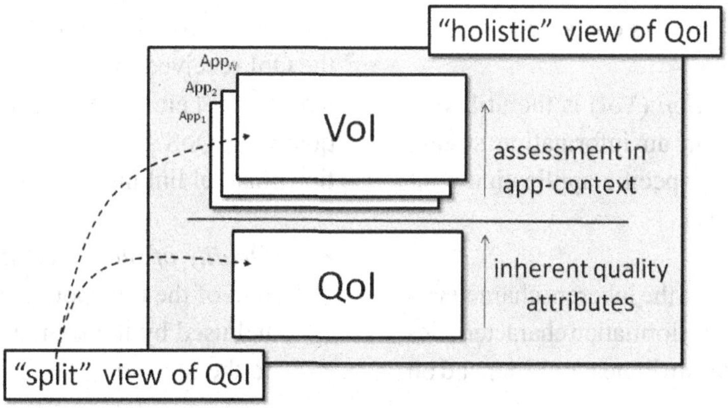

Figure 3. The abstract information flow model

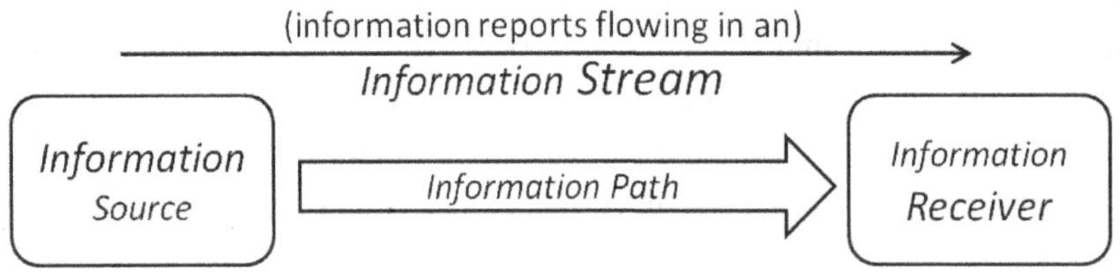

single information report, intermittent, seasonal, persistent, and so on.

The source and the receiver in the flow model are functional entities residing in nodes accessible over collections of interconnected physical networks. They may represent *end-points* either on the source side, e.g., sensors of various types, including humans, or the receiver side, e.g., human analysts or software agents operating on behalf or for the benefit of human analysts. They may also represent intermediary information processing points residing in the midst of the end-to-end path between the aforementioned source and receiver end-points. In the latter case, the entire path can be represented by a daisy-chained, or, more generally, graph, construct of source and receiver functions linking end-sources to end-receivers.

Given the abstract information model in Figure 3, we define QoI and VoI as follows:

- *Quality of information* (QoI) is the body of evidence (described by information quality *attributes*) available that can be used to make judgments about the *fitness-of-use* of the information reports contained in an information stream.
- *Value of information* (VoI) is the utility of the information in an information stream when used in the specific application context of the receiver.

QoI attributes capture the inherent characteristics of information. The information characteristics and corresponding QoI attributes could depend on the type of information, e.g., imagery, acoustic, seismic, etc. They exist independently of the specific application context in which the receiver will use the information and are used to assess the VoI in an application-specific manner. For example, suppose a sensor system is able to provide information reports about the (x, y) coordinates of the location of a perimeter breach with error $\varepsilon = (\pm\varepsilon_x, \pm\varepsilon_y)$. This is a fact (or, a piece of evidence) that inherently characterizes (or, qualifies) these reports. This fact will be used to judge how much additional value does the information content (the (x, y) coordinates) provides to a situation assessment. Clearly, for the error value ε, the location estimates provide more "valuable" information the larger the area the perimeter encloses. Hence, assuming that the error estimates along the two dimensions are not grossly different, the value added by the location reports increases with a decreasing ratio of the error area (the product of the errors in the x and y dimensions, $\varepsilon_x \cdot \varepsilon_y$) to the area of the enclosed region.

Impairments of a communication channel cause transmission delays, errors (intentional or unintentional), disruptions, and so on. Hence, the QoI received by a receiver could be different than the QoI emitted by a source as a result of the quality of QoS service provided by the information channel linking the two, where

- *Quality of service* (QoS) is a characterization of the transport properties of the channel used by the source to send information to the receiver.

Figure 4. The QoI/QoS/VoI linkage

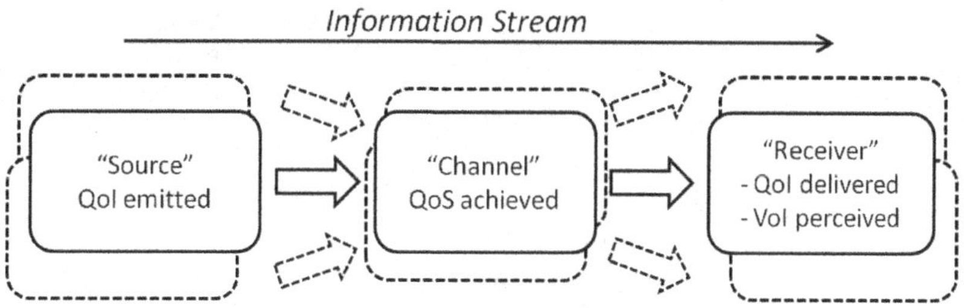

This is shown in Figure 4, which is a generalization of the abstract flow model in Figure 3 emphasizing the fact that information streams from multiple sources to multiple destinations may share channels and hence affect each other's QoI.

Next we provide a refinement of information and QoI that targets sensor-enabled applications and systems, and bridges QoI to traditional computational techniques such as estimation and hypothesis testing.

Information and QoI for Sensor-Enabled Applications

As previously mentioned, information is collected to gain an understanding of the world of interest. How the world of interest is supposed to be conceived and information about it collected lies at the core of using sensors and sensor-originated information. The world of interest is understood, i.e., can be reasoned against, on the basis of world models comprising:

- collections of *objects* possessing *state*, comprising one or more (possibly time varying) *elements*; and
- *relationships* between these objects.

For example, Figure 5 shows the world comprising objects, such as vehicles and troops, related by their affinity to a spatiotemporal region of interest. These objects have states comprising elements, which could be *numerical* in nature, such as location and velocity, or *categorical*, such as type (e.g., tank, personnel carrier) and allegiance (e.g., friend, foe). Both numerical and categorical state elements could also be time-varying (friends can become foes with time or vice versa!), thus the state itself can be time-varying and, hence, exhibit a temporal behavior. The prediction of this behavior and the assessment of its implications are essential objectives of intelligence gathering and information fusion, e.g., as described by the JDL model (Hall & McMullen, 2004).

In this setting, data sources, e.g., sensors, are used to collect and provide data that populate the values of the state elements and thus provide an understanding of the state and behavior of the objects of interest. These data are transported through communication networks; processed through processing centers, where they may be correlated and fused with data from other sources, and be enriched with context; and eventually delivered as information to end-users. Thus, given a model of the world as previously discussed:

- *Information* is an estimate of the elements of the object states that is derived from data collected from the sensory sources;
- *QoI* is a measure of the goodness of this estimate along various quality dimensions; and
- *VoI* is the importance of the information of the given quality for a particular task.

Figure 5. The end-to-end information path

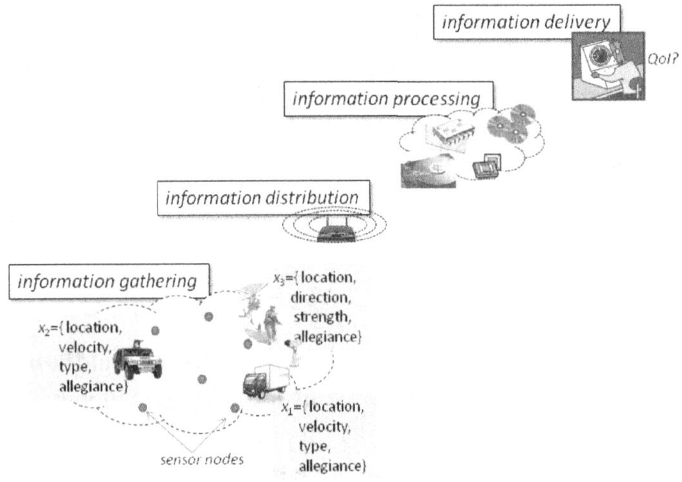

The importance in the VoI characterization above should be interpreted as the benefit (i.e., value) that a piece of information brings into one's ability to ultimately take actions (with predictable enough outcomes).

QoI attributes of interest include information accuracy, latency, and provenance, where:

- *Accuracy* represents the spread, biases, and granularity in of the inferred elements of the world object states;
- *Latency* (or, *age*) is the time interval between the arrival of the information report (i.e., the arrival of the state estimates) and the time that the state presumably had the reported value; and
- *Provenance* is the history of measurements and processes contributed to the creation of the reported state estimate.

Accuracy and latency are the natural QoI attributes and are present in all listings of QoI-related attributes, see for example (Blasch & Plano, 2005; Knight, 2007; Wang & Strong, 1996) and references therein. They are used to assess how satisfactory are the accuracy level and timeliness of the information provided. In addition to these two natural QoI attributes, we also consider provenance. Provenance allows one to ultimately judge the supporting meta-information (supplied directly or indirectly) about the information, be it the reputation of the original source, the level of trust that can be placed on subsequent information fusion operations, and so on. One may argue that provenance may stand on its own as the only QoI attribute of importance, for if the entire history of information processing is known then the QoI parameters can, in principle, be evaluated as needed. However, complete knowledge of such history may not always be available. Provenance is typically implied, especially, when the information gathering, distribution, processing, and delivery phases in Figure 5 belong to a single administrative domain, e.g., belong to a single nation. In coalition operations, however, these (or even parts of these) phases may belong to or executed by different administrative domains. Hence, in this case, provenance will need to be externalized and communicated as part of the body of evidence that can be used to judge fitness-of-use of a particular information report.

The aforementioned state-oriented interpretation of the QoI domain lends to an interesting map metaphor for QoI that supports the view that "...

the map underlies man's ability to perceive..." (R. S. Wurman as cited in Markoff, 2009). Consider a map drawing of the aforementioned region of the world of interest defined by the spatiotemporal boundaries of the information sought. Collections of objects related to this region can be superimposed, or painted, on the map drawing as layers of detail. These layers are created using the information collected from sensor, and possibly other sources, which, for example, place an armored vehicle in location (x_v, y_v) or a structure, e.g., a building, in location (x_b, y_b) relative to the coordinate system of the map. Subsequently, these objects are also tracked via their time-evolving state, e.g., the armored vehicle moves in NW direction at a speed of 60km/h. Within this context, the QoI of the sensor-collected information relates to the ability to draw on, place on, and annotate the map of the world of interest as accurately and quickly as necessary. As we use the collected information to enrich the state of affairs of the world depicted on the map, questions of uncertainty inevitably rise whose answers reflect positively or negatively on the map's quality and eventually value for the purpose it was created for. Such uncertainty questions include:

- *Estimation uncertainty*: How accurate is the reported location of objects and velocity of vehicles?
- *Classification uncertainty*: How high is the probability of misjudging the vehicle as friend or foe, tracked or wheeled, military or civilian?
- *Detection uncertainty*: How high is the probability that there is a vehicle in the reported location or region in the first place, or that there is just one vehicle?
- *Freshness uncertainty*: How current is this depiction of state of affairs in the world.

In describing the importance of information, we have used the terms value and utility. This comes from the fact that benefits gained by using a particular piece of information may be translated to some form of monetary gain or loss. This is also in-line with the term "information value" as used in the area of decision making under uncertainty conditions. In the latter case, the cost of acquiring a particular piece of information to reduce the uncertainty in a decision parameter is compared against the monetary benefit of executing, say, a business decision more effectively in light of this information. For example, in his seminal paper in the area, Howard (1966) illustrates the fundamentals of the *information value theory* using a contract bidding scenario where "our company" and a competitor bids for a contract. The objective is to decide on our bidding offer that will maximize our expected benefit from the contract. The parameters of uncertainty are the competitor's bidding offer and the cost of producing and providing the goods and services that are included in our offer in response to the contract requirements. Associated with these uncertainties is the aforementioned cost of acquiring relevant information that will reduce the uncertainties in these parameters and, hopefully, increase the expected benefit. The trade-off between the cost of acquiring the information and the increase in expected benefit is indicative of the value of the sought information. The higher the return on (the information gathering) investment, the more valuable the information becomes.

Next we present a simple example of a design for a sensor-enabled system that takes quality and value of information into consideration.

QoI-Aware Perimeter Protection

Suppose we are tasked to design a perimeter protection (or alert) system separating friendly from unfriendly areas, see Figure 6. The system is supposed to trigger a "yellow alert" whenever a potential trespasser enters the "yellow zone," which extends d [length units] from the perimeter, referred to as the zone depth. A sensor-enabled localization mechanism is used that localizes objects to within, say, $\pm\varepsilon$ [length units] from the yellow zone. As a result, when an alert is triggered, the potential trespasser may already have entered the yellow zone, or may still be far from it. Suppose that there is cost associated with triggering an alert belatedly, which increases the closer the trespasser is to the perimeter due to, for example, the increased possibility that the trespasser could damage assets in the friendly area. Likewise, suppose there is cost associated with triggering an alert early, which increases the further the trespasser is from the perimeter due to, for example, the cost of dispatching troops to monitor and, if needed, engage the trespasser. Given the accuracy of the localization system, we would like to calibrate the alert system, i.e., adjust the depth d, so that the expected cost incurred with each alert is minimized.

Figure 7 shows the costs incurred under the simplifying assumptions that they are linear with distance; also, the maximum damage cost c_{dm} is assumed to be larger than the minimum dispatch cost c_{ds} ($c_{dm} > c_{ds}$) For notational brevity and convenience, the origin of the axes has been moved to the point of intersection of the cost lines. With perfect localization ($\varepsilon = 0$), the depth of the yellow zone should be set where the cost lines intersect, i.e., at the reference depth $d_0 = l$. As ε increases, so does the depth d of the yellow zone as well; hereafter, we will write $d(\varepsilon)$ instead of d. The depth $d(\varepsilon)$ is chosen to balance the expected damage and dispatch costs. Assuming that the localization error is uniformly distributed in the closed interval [$-\varepsilon$, $+\varepsilon$], we obtain, after some algebra

$$d(\varepsilon) = d_0 + g(\varepsilon) = d_0 + \frac{c_{dm} - c_{ds}}{c_{dm} + c_{ds}} \varepsilon = d_0 + \frac{1-H}{1+H} \varepsilon,$$

where, $H = c_{ds} / c_{dm} \leq 1$.

Figure 8 shows the amount of depth adjustment $f_a(\varepsilon) = d(\varepsilon) - d_0$ as a function of the costs ratio H. Note that as $H \to 0$, i.e., as the damage cost becomes significantly higher than the dispatch cost, $f_a(\varepsilon) \to \varepsilon$. This is the case where the system protects assets of significant value and, thus, cannot afford the possibility of any error in localization and, as expected, triggers alerts as soon as a potential trespasser is believed to be within the maximum error ε from the nominal yellow zone depth, see Figure 7. On the other hand, when $H \to 1$, i.e., the damage and dispatch costs become

Figure 6. A perimeter protection system

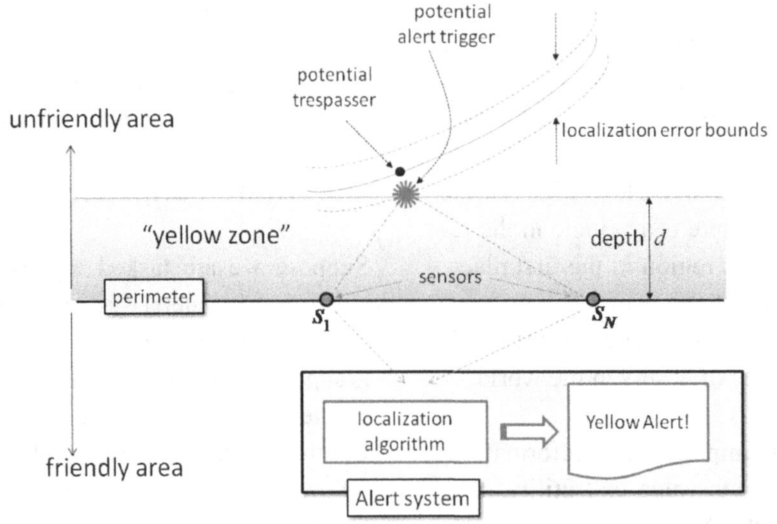

comparable, $f_a(\varepsilon) \to 0$, which implies that the alert system will require no further depth adjustment and will trigger an alert when the potential trespasser is believed to be at the boundary of the nominal yellow zone.

The simplifying stated or implied assumptions notwithstanding, the example just presented showcases how the QoI of the sensing system is taken into consideration in designing the entire alert system to minimize a cost objective. Furthermore, the QoI, captured here by the error ε, is directly used to evaluate the value that the information reports bring to the system and its operation through the adjustment function $f_a(\varepsilon)$–as

Figure 7. A simple cost model for VoI assessment in the perimeter protection system

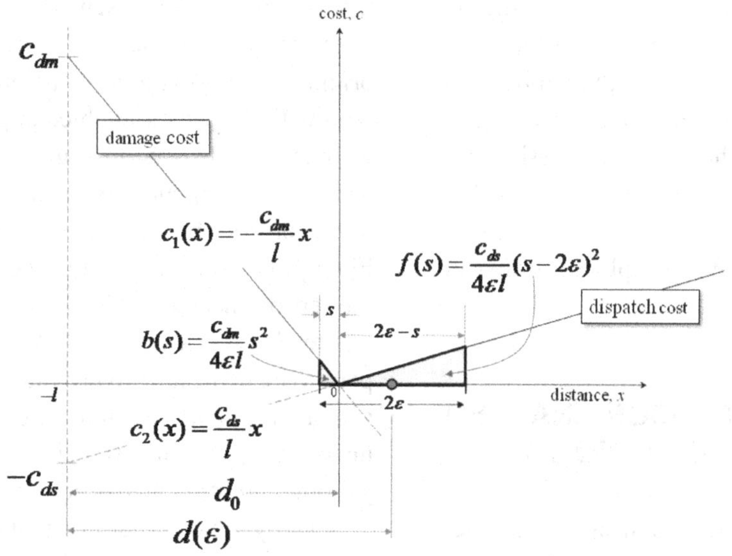

Figure 8. Depth adjustment for the alert system $f_a(\varepsilon)$ as a function of the ratio H of damage and dispatch costs at various sensor error levels ε

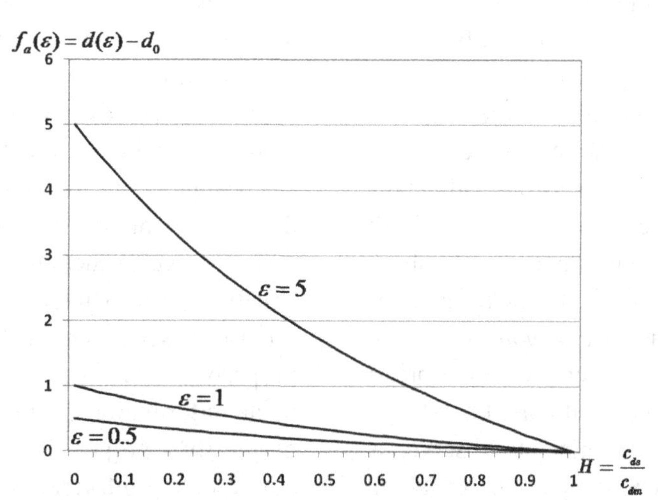

the cost of the assets protected changes, the yellow zone depth $d(\varepsilon)$ can be adjusted accordingly. The range of error ε afforded depends on the quality of the sensor system itself, and the localization algorithm. They too can be associated with a cost, the cost of procuring sensors of different qualities and computational capabilities.

In general, all the aforementioned costs and associated QoI capabilities and their impact on the value of the alert system and sensor-enabled applications can be used as search criteria to guide dynamic searching for and binding to the most appropriate sensing systems for the application tasks at hand. This search and bind process can be repeated as often as the application design and operational needs dictate. Therefore, QoI can play roles at different stages of the lifecycle of sensor-enabled applications. This is explored further in the following section.

QOI IN DYNAMIC SENSOR-ENABLED APPLICATION ENVIRONMENTS

Typically, sensor-enabled applications are deployed together with the sensing resources that support their information needs; we refer to this as the monolithic or *closed* deployment model. Building, structure (e.g., bridges), soil, habitat, traffic monitoring, property security, remote patient care, and street surveillance are but a few examples of such closed applications. Even when these are configurable and extendable (e.g., we are able to update an application module here, or add a new camera there), these are by all means closed systems that were planned, designed, built and deployed for a particular purpose. Applications and sensors are for all practical purposes statically bound sharing a *common context of operation* (CCO) where applications have a priori (or, at least auditable and verifiable) knowledge as to *what* the sensors are trained to observe, as well as to *when* and *where* observations are taken. The what, when, and where *primitives* are part of the 5WH ({why, when, where, what, who, how}) principle for organizing sensor-originated information in dynamic environments (Bisdikian, Branch, Leung, & Young, 2009).

However, in dynamic coalition environments and depending on the situation, sensing, computational, communication, etc., capabilities may be provided by different coalition partners. For example, consider a case where a small multinational coalition unit is urgently sent to a territory that is "serviced" by a sensor field deployed and operated by some other coalition member. Sensor-originated information must find its way to and bind with the sensor-enabled application (e.g., a visor system displaying aggregated surveillance data to its user) supplied to the unit on its way to the territory by yet another coalition member. Finally, the necessary communication support for the sensor-enabled application is provided by an unattended aerial vehicle (UAV). The UAV happens to be in the area to provide communication, and other support, to a number of additional tasks underway at the same time in the area, involving yet other additional coalition members.

Clearly, a closed deployment model is not appropriate for the aforementioned situation. There can be no pre-existing common context of operation. Sensors and applications cannot know each other prior to the beginning of the mission, and they may not all know each other even after the mission is underway. What such dynamic situations require is an *open* model of sensor-enabled application deployment according to which the constituent components of the end-to-end system are designed and deployed separately but they dynamically bind to each other as needed. Service discovery techniques are commonly used in these situations (Zhu, Mutka, & Ni, 2005). However, in order for a sensor-enabled application to operate properly, it does not only need to find and bind to an information provider, but also establish on-the-fly a proper CCO (the what, when, and where of the information content and with the desired QoI).

Figure 9 highlights the aforementioned open deployment model. The figure shows sensor-originated information content being enriched with pertinent metadata to assist indexing and searching for the purpose of achieving CCO (more on QoI metadata later on). Information providers collect and process sensor-originated information and provide it to applications that need it. The three-step process (I need..., I provide..., I select...) shown at the top end of the figure is part of a bigger set-up that enables the QoI-based open deployment model; this bigger set-up, referred to as the four viewpoints of QoI, will be discussed shortly. The open deployment model enforces a clear separation of responsibilities between the roles of sensor-enabled application developers and sensor system designers, facilitating effective skill and development reuse and ensuing economies of scale.

Of course, a closed deployment model could be viewed as a special, degenerate case of an open one, for example, by eliminating the explicit presence of information providers, and, hence, the necessity of executing the aforementioned three-step process. An implication of Figure 9 for the open deployment model, and, hence by extension, for the closed deployment model as well, is that it is reasonable to think of information and its quality as a commodity, such as a product or service offering, provided to end-users for consumption. End-users may seek an exclusive, shared, permanent, or transient association with this information service offering. For example, for a sensor-enabled application built and operated according to the closed deployment model, the application has an exclusive and permanent association with the information that "its own" sensors produce. On the other hand, a user that searches and retrieves information from the Web (the ultimate open deployment model) has a transient and shared association with the information, as multiple end-users may access and retrieve the same information at the same time (and for different purposes).

We note here that Driscoll *et al.* (2005) advocate considering information as the outcome of a manufacturing process, i.e., a manufactured information product. This is a very appealing notion that when applied to information can leverage centuries-old experiences with manufacturing

Figure 9. The open deployment model for dynamic sensor-enabled systems

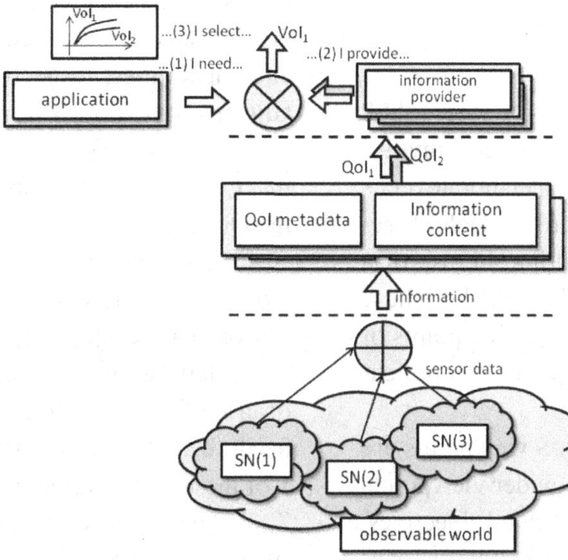

processes. Although not considering information as a manufactured product, the study in (Blasch, & Plano, 2005) on information quality and situation awareness for JDL-based fusion information systems is quite complementary to this view. However, owing to the high dynamicity and variability of its creation procedures (not all of which is under our control or we have full knowledge of) and uses, we prefer to view information as a service offering that could be dynamically created to user (and use) specification with anticipated performance levels (its QoI) and assessed at runtime within the context of its applications (its VoI).

Treating information as a service that originates from sensor information providers, or even value-add information brokers (that build upon the information offerings of other providers or brokers), subscribed to and consumed by end-users, information and its quality exhibits a lifecycle spanning the phases of design, creation, deployment, execution, and eventually withdrawal. QoI plays a role at each of these lifecycle phases. In considering quality of service (QoS), ITU considered a multifaceted lifecycle viewpoint for QoS too. We adopt a similar viewpoint for QoI as well. We discuss this next, setting the stage first by briefly highlighting the four viewpoints of QoS.

The Four Viewpoints of QoS

While ITU recommendation ITU-T Rec. E.800 provided a workable definition for QoS involving providers and users, it did not elaborate on its variations with respect to advertised QoS objectives and QoS experienced during the use of a service. ITU-T Rec. G.1000 elaborated on these variations by introducing the four viewpoints of QoS, see Figure 10, relating quality to application needs.

As Figure 10 shows, the QoS viewpoints are split between customer and provider viewpoints taking into consideration both planned objectives, advertised prior to a (potential) customer subscribing to a service, and operational attainments after the customer subscribes to it. The four viewpoints briefly are:

- *Customer's QoS requirements*: pertains to a customer's requirements regarding the levels of QoS it expects and seeks receiving.
- *QoS offered by provider*: pertains to the level of QoS expected to be provided by the service provider.
- *QoS achieved by provider*: pertains to the level of QoS achieved by the provider during the offering of the service to a customer.
- *QoS perceived by customer*: pertains to the level of satisfaction, e.g., as accessed via customer surveys, a customer experiences using the service.

The four viewpoints circle leading toward a service provision equilibrium between what a customer expects to receive and what a provider actually offers, both of which could change and adjust as needed and as possible with time.

The Four Viewpoints of QoI

The QoS viewpoints highlight, if not a perpetual, at least a transient give-and-take between customers and providers as each announce their desires and capabilities, respectively, and then making adjustments to these at runtime. We adopt similar viewpoints for QoI bringing together the roles that QoI and VoI play during different phases of the QoI lifecycle.

Figure 11 highlights the four viewpoints of QoI, where the role of a customer is played by the sensor-enabled application. The QoI viewpoints circle starting with a statement (or, announcement) of an application's information and related quality needs, e.g., "…I need information about the state **x** of the world in region A over the time horizon T…". The state vector **x** comprises state elements such as, see for example Figure 5, location, veloc-

Figure 10. The four viewpoints of QoS

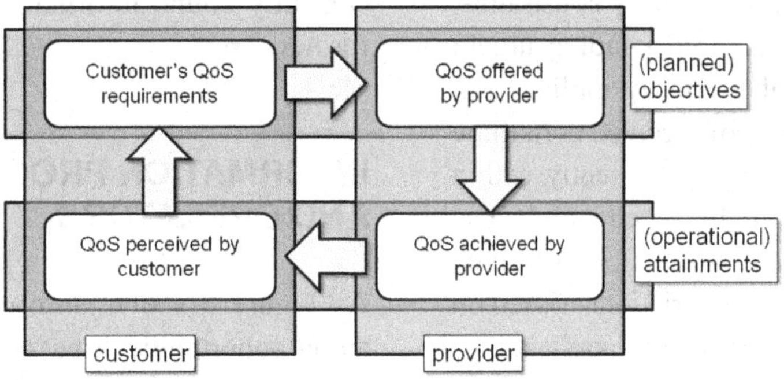

Figure 11. The four viewpoint cycle of QoI

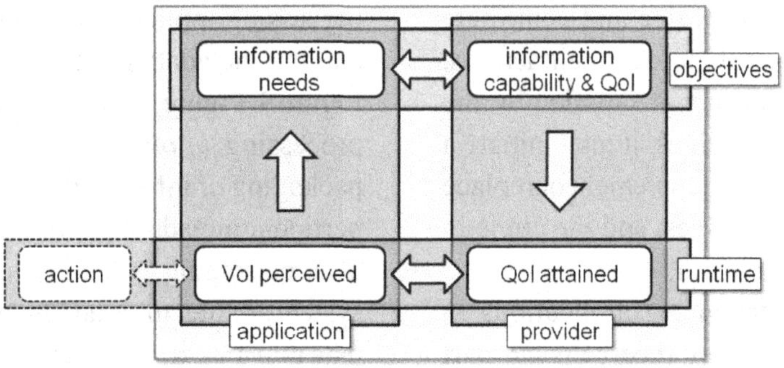

ity, type, allegiance, and so on, of objects sensed in the aforementioned region and time horizon. The state elements sought must also satisfy desired QoI levels, e.g., be accurate and current to within stated tolerances specified by the application. Corresponding to an applications' statement of information needs, sensor information providers state their information provision capabilities advertising the information content they are able to provide along with the QoI levels they can provide for this content. Using these provider statements, applications may execute searches for appropriate providers that best satisfy their needs. It is possible that during searching, applications may negotiate with providers based on number of criteria including information content, QoI levels, cost, energy needs, and so on. Note, the aforementioned sequence of operations alternating between an application statement (…I need…), a provider statement (…I provide…), and an application action (…I select…) is only one possibility and implementations may opt for alternatives, for example providers advertise their capability statements, and applications simply select providers based on these unsolicited advertisements.

Having found providers that satisfy the information needs at an acceptable level, applications bind to them to receive their sensor-originated information. At this point, we move from the domain of the stated QoI objectives to the domain of runtime QoI experiences. Specifically, based on the operational realities experienced by the providers at runtime, the providers attain certain levels of QoI for the content delivered to the applications they support. The QoI attained could be dynamically affected by a number of environmental and operational parameters such as weather and terrain conditions, volatility of sensor resources, e.g., due to availability, energy reserves, or faulty operation, communication channel constraints, e.g., due to sharing of bandwidth or fading, urgent service request from higher priority applications, and so on. Therefore the QoI received at runtime by applications may not fall persistently within the expected levels originally stated and, hence, the value of the received information perceived by applications may exhibit variations. Based on the VoI perceived by the receiving applications, applications may still decide to take an action either utilizing or ignoring the received information. They may also opt to negotiate at runtime with the information provider for "better" information that could add higher value and contribute more effectively to the outcome of their actions. Finally, if an application is not satisfied by the information services it receives, it may initiate a new search for providers to supplement or replace the services it already receives, and the circle in Figure 11 starts all over again.

Note that actions taken by applications in light of the information they receive is not part of the QoI viewpoints. However, VoI itself is assessed with respect to the benefits the information (with its QoI) brings to these actions. The effectiveness of the actions taken, or the projected effectiveness of these actions, could be reasons for applications to initiate QoI negotiations with currently bound information providers or seek for new ones.

INFORMATION PROCESSORS AND OPERATORS

As Figure 5 shows, information streams have ample opportunities to be "touched" and processed (e.g., altered, combined, refined, etc.) along the path from the sensor sources to the end-users; the JDL information fusion model layers several such processes (Hall, & McMullen, 2004). This processing applies both to the information content and the metadata used to enrich this content, *cf.* Figure 9. Figure 12 shows information processors processing *information objects* comprising the packaging of information content (*iCont*(*s*)) with pertinent metadata (*mtdt*(*s*)), e.g., QoI metadata. This packaging is logical in that, depending on system architecture and implementation, the metadata may be carried together with the information content in the information streams or may reside at repositories to be retrieved from processors as needed. The latter is signified by the dashed line surrounding the metadata elements of information objects in the figure.

Figure 12. Information processors

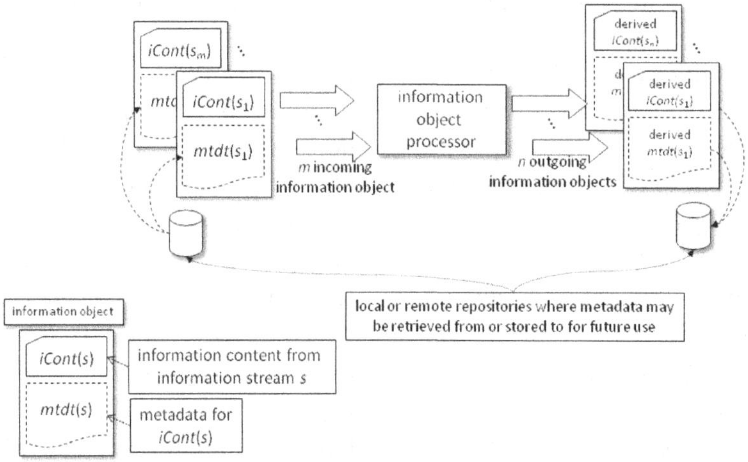

The metadata in an information object may apply to the entire underlying information stream *s*, such as the owner of the source of the information stream, or to individual stream items, such as the timestamp of each sensor measurement carried by the stream. Furthermore, individual stream items may be derived from "parent" streams, e.g., a newly derived (child) stream that carries the averages of successive groups of *N* sensor measurements extracted from another (parent) stream. In this case, the metadata of the derived stream may also be applicable to collections of stream items from the parent streams. Such is the case with the data quality metadata in (Klein, Do, Hackenbroich, Karnstedt, & Lehner, 2007), where, for example, measurements that are recorded regularly are periodically aggregated, e.g., averaged. A quality attribute is assigned for each aggregated result that signifies the number of missing measurement recordings experienced in the parent stream during each aggregation period.

There is a variety of operations that an information processor may perform on information objects, e.g., aggregation, modification, etc. Hence, there is a variety of *information operators* that may reside within an information processor. These operators will operate and modify in some form the nature of either the information content or the metadata or both. More than one operator may reside in a processor, but, otherwise, we make no assumption regarding the physical association of operators to processors, which is left as a design choice for system implementers. Operators act on individual information stream items, e.g., individual measurements, or collections of them. Implemented by software agents, operators can be thought as acting on inputted objects encapsulating an appropriate subset of items from the underlying information streams. This subtle difference aside, in the sequel, we use the terms information streams and information objects interchangeably.

Figure 13 shows the following three classes of information operators for information streams (metadata shown in parentheses implies their optional presence in the stream):

- *Annotation operator* (or, *annotator*): a single-input-single-output operator that attaches metadata to an information object.

Figure 13. Information operators

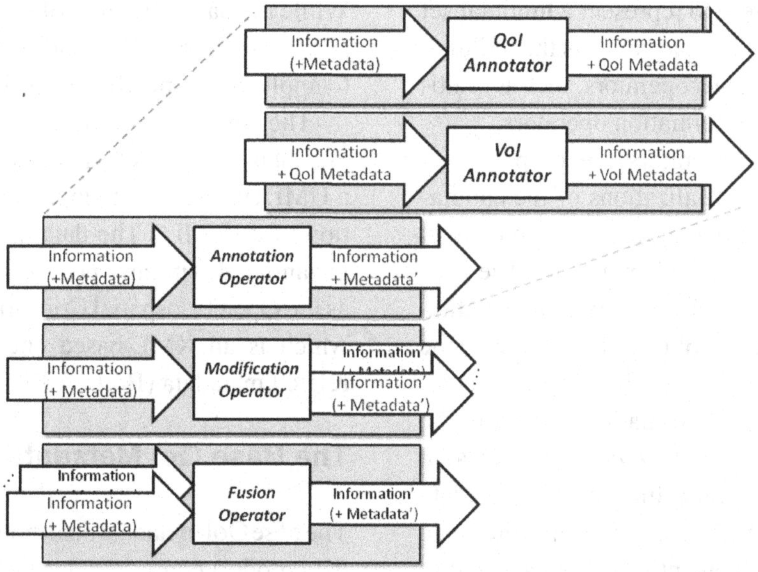

Annotators primarily enrich incoming information objects by augmenting their metadata entries.

- *Modification operator* (or *modifier*): a single-input-multiple-output operator that modifies the information content or metadata in an object. Traditional stream operations like obtaining moving averages, deriving maxima and minima, stream splitting, etc., are typically performed by modifiers.
- *Fusion operator* (or *combiner*): a multi-input-single-output operator that uses multiple information objects to create a new one, such as when a stream of humidity measurements and a stream of temperature measurements combine to create a stream of heat index measurements.

Clearly, an annotator can be viewed as a special case of a modifier (a "null" information modifier). Likewise, a modifier can be viewed as a special case of a combiner (a "unary" combiner). Nonetheless, because of their intuitive distinctions, and to avoid confusing verbiage, such as an "annotator combiner," we consider the above three operator classes as representative of three broad classes of operations performed on information streams. We further consider these to represent a minimal set of operator classes that can serve as the building blocks for deriving other operators, such as multi-input-multi-output information operators.

Two classes of operators are of particular interest. They are specializations of the annotator and deal with enriching information content with QoI and VoI related metadata. The *QoI annotator* operates on an incoming information stream attaching to it QoI metadata that describe (or can be used to describe) the inherent quality characteristics of the information content in the stream. Likewise, the *VoI annotator* operates on an information stream attaching to it VoI metadata that describe (or can be used to describe) the value of the information content in the stream within the application context of the stream's receiver. While QoI annotators may exist throughout the end-to-end information flow, VoI annotators will typically exist only at the end points of the flow, interpreting the value of the received information stream within the context of its end-use.

Next we present an information model to facilitate the interoperable exchange of QoI and VoI metadata and hence enable information processing along the path from sources to users in dynamic, multi-domain, coalition environments.

QUALITY METADATA MODEL

To enable the exchange of QoI related-information between information seekers, providers, and processors means are needed for expressing and communicating QoI in a commonly acceptable manner. For this, we have introduced a data model that encompasses the collection of QoI attributes and their values that can be tagged to each particular piece of information. The data model comprises an application-agnostic structure that contains: (*a*) means to index information agnostically, e.g., according to time and space; (*b*) a collection of placeholders for the quality attributes; and (*c*) a collection of quality statements for these attributes. While the basic structure of the data model is application agnostic, the model itself can be extended to application-specific instances as necessary.

The application-agnostic structure of the QoI metadata data model will be described next using a UML (unified modeling language) representation of it (UML). The data model itself may be instantiated with any pertinent technology such as the Open Geospatial Consortium's SensorML, which is an XML-based encoding for sensor-related metadata (Botts & Robin, 2007).

The Base QoI Metadata Data Model

The base QoI metadata data model (or, simply, QoI data model) comprises a set of common classes that provide the structure of the model, see Figure 14. Context dependent instances of the classes in the base model provide the QoI metadata for a particular piece of information content. The core QoI metadata class is represented by the BaseQoI class. This class relates to: (*a*) a piece of information content, represented by the Information class; (*b*) a metadata context, represented by the Context class; and (*c*) a collection of QoI attributes, represented by the abstract QoIBaseAttr class. Finally, the BaseQoI class also relates with a QoIQualifier class which represents the metadata information about instances of the data model itself, such as its author, time of authoring, publisher, and so on.

The Context class is an abstract representation of context attribute classes. It is a superclass of the physical (spatiotemporal) context of the information content under consideration and relates to the logical (application) context of the information. The physical context comprises the spatial (QoISpatialCntxAttr) and temporal (QoITemporalCntxAttr) classes that contain the spatial bounds (geographical horizon) and temporal bounds (time horizon) for which the QoI metadata are applicable for the information. The logical context, represented by the AppDomainCntx class, relates the information content and its QoI metadata with the specific application domain for which the QoI data model is valid, any industry standards followed, etc. For example, for imagery, the AppDomainCntx class may reference, among others, a related URL of a NIIRS specification (NIIRS) used to encode quality information in an instance of the model. Likewise, for an integrated air picture, it may reference a SIAP methodology (SIAP, 2003).

The actual QoI metadata are represented by a collection of attribute classes subclassed from the QoIBaseAttr class. Figure 15 shows four QoI attributes: *accuracy* (QoIAccuracyAttr), *latency* (QoILatencyAttr), *integrity* or *provenance* (QoIIntegrityAttr), and *format* (QoIFormatAttr). The QoIFormatAttr class can be viewed as a representation of the quality of data, which measures quality related to the formatting of the information as data. Each of these classes comprises a collection of pertinent QoI attribute class parameters, such as error rate, bias, various types of latency, and other shown in the figure. The collection of attribute classes and associated parameters in the figure is not meant to be exhaustive. These classes can both be augmented or subclassed depending on the needs of the application domain specified by the AppDomainCntx class. Again depending on the application, instances of these attribute

Figure 14. The base QoI data model

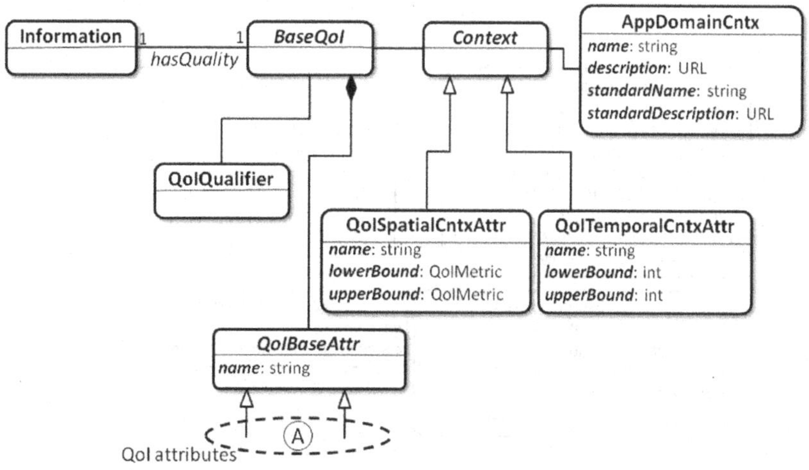

classes may use a modified set, e.g., a superset or a subset, of the parameters, see Wang and Strong (1996) or Knight (2007) and for a collection of potential quality parameters.

The metadata parameter values in the attribute classes can be represented by primitive types (e.g., a double), or by instances of the abstract QoIMetric class, or a URL class, see Figure 16. A QoIMetric parameter can be of quantitative type (QoIQuantitativeMetric) described by a value and an error range. It may also be of enumerated type (QoIEnumeratedMetric) described by a collection of categories or tags. For example the *sourceIntegrity* parameter in the QoIIntegrityAttr class may assume either quantitative (numerical) values between 1 and 10, or enumerated values from, say, the set {untrusted, erratic, trusted, authoritative}. The URL class is used to point to a URL location, typically used to provide addition information, e.g., about an application, or to point to industry standards used to describe quality parameters in an instance of the data model. The latter permits the encapsulation of existing industry efforts on quality in the data model.

The Base VoI Metadata Data Model

Similarly to the base QoI data model, a base VoI data model can be defined as shown in Figure 17. The model has a similar class structure to the QoI data model and even inherits some of the latter's constituent classes. Since VoI is assessed from the QoI parameters found in a corresponding QoI data model instance, the BaseVoI class relates to a corresponding BaseQoI class. In the figure, we have grayed out the context classes to underscore that these can be inherited from the corresponding classes in the QoI data model. To reduce cluttering in the figure, we represent the spatiotemporal context classes by the grayed-out PhysicalCntx class.

The VoIBaseAttr class represents an abstract placeholder for the VoI attribute classes. Three example classes are shown in Figure 17 that relate to the *trust*, *usefulness* and *convenience* of using the information, the classes and parameters in them are self-evident, but more detailed information can be found in Bisdikian *et al.* (2009). As in the case with the QoI attributes, the collection of VoI attribute classes and associated parameters in the figure is not meant to be exhaustive. Depending on

Figure 15. The QoI attribute classes

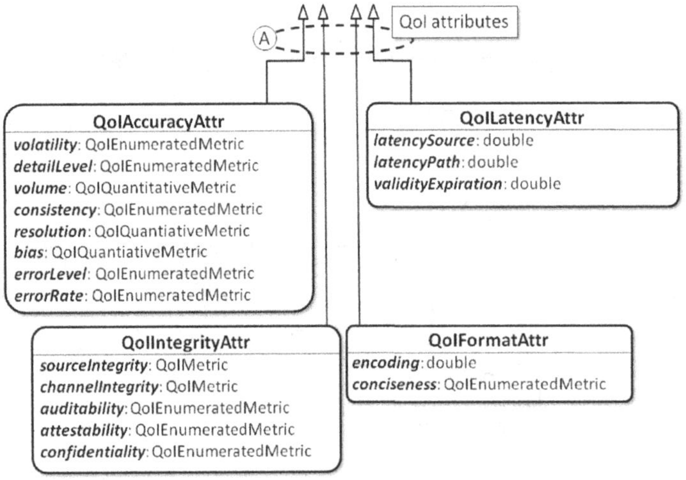

the application, collections of quality parameters such as those in Wang and Strong (1996) or Knight (2007) can be used as necessary.

In closing, we note that the AppDomainCntx class in the VoI data model may be instantiated differently than the corresponding one in the QoI data model. For example, the AppDomainCntx class in the QoI data model may indicate that the information content is imagery and its QoI attributes are described according to the NIIRS specification. On the other hand, the AppDomainCntx class in the VoI data model may indicate that the end-user application, on behalf of which this piece of information content is used, is an intrusion detection application. Hence, the value assessment (or, judgment of fitness-of-use) as to whether the NIIRS value of the imagery is acceptable or not is made relative to this end-user application.

CONCLUSION

The effective sharing of sensor-generated quality information is one of the cornerstones of network-centric coalition operations. However, despite its high-importance and the explosion in sensor-network-related research in recent years, the area of quality-driven, on-demand, sensor-generated

Figure 16. The QoI metric data model

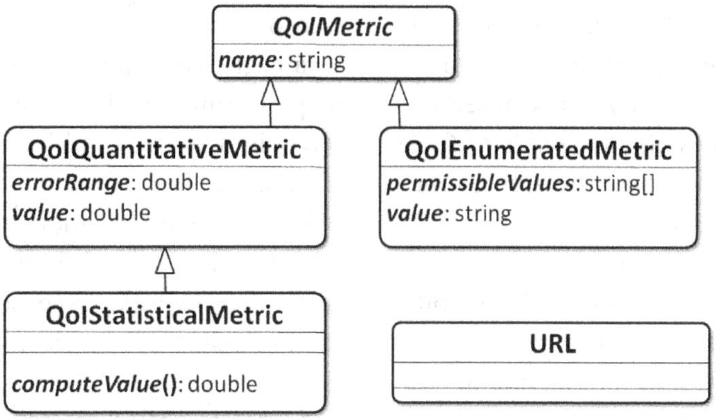

Figure 17. The base VoI data model

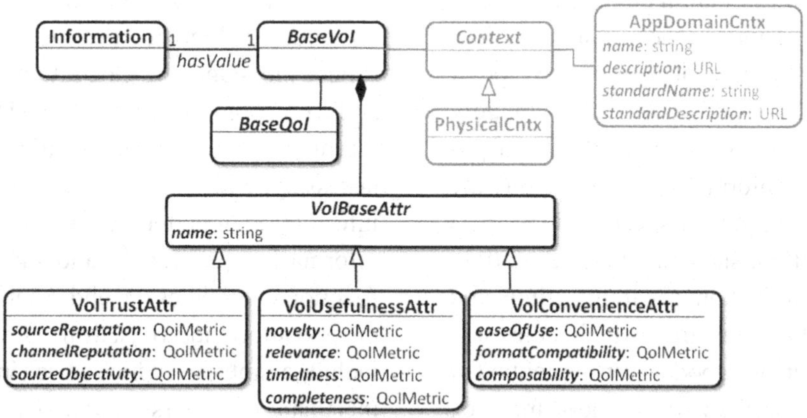

information has received relatively limited attention from the sensor network research community so far. This is in stark contrast with extensive research in the areas of information quality for enterprise-grade systems, e.g., relating to information residing in databases or data warehouses, that are deployed in both the commercial and military arenas. With this in mind, this chapter had a dual objective: (*a*) to juxtapose research in quality of information (QoI) to ongoing research in wireless sensor networks; and (*b*) to highlight the many and diverse aspects of the topic. The ultimate purpose of these objectives were to underscore that the QoI research domain is a broad, not a single-issue topic, e.g., relating merely to the numerical accuracy of a sensor-originated estimate of sorts, but rather a very important, multi-issue, multi-dimensional research topic ranging from fundamental characterizations of QoI to system building and operation architectural principles. The chapter covers only one aspect of the authors' research pursuits in the area of QoI for sensor networks; additional research in the area is summarized later on.

Specifically, in this chapter, we developed a definition of QoI that allows us to combine both the "traditional" quantitative understanding of QoI, such as the goodness of an estimate or the resolution of an image, with the difficulties of interpreting these within the context of myriads of information end-uses. We have presented a split definition of QoI that stacks application-specific interpretations of information on top of a collection of inherent characteristics of information. These characteristics may depend on the type of information under consideration, such as imagery or acoustic, but they exist independently of potential uses of the information. This permits the separation of the "technical specifications" of a piece of information, such as image resolution, object-size discrimination capability, time and place of image taking, etc., from whether these are judged to be sufficiently good within the context of a specific application. We associate the former, i.e., the technical specifications of information, with the quality of information (QoI), while the latter, i.e., the outcome of information judgment, with the value of information (VoI). We subsequently interpreted this broad definition of QoI within the context of sensor-enabled applications that try to reconstruct a portion of the world that is of interest such as a specific region of concern in a theater of military operations. This interpretation considers models of the world comprising interrelated objects whose state needs to be constructed sufficiently accurately using the sensor-originated information. We also provided a simple, insightful example of how QoI influences a sensor system design and the cost of actions taken using information of given QoI.

We then discussed the need of establishing a common context of operation (CCO) between sensor sources and sensor-enabled applications to enable their dynamic coalesce in supporting open deployment models for sensor-enabled applications in dynamic, multi-domain, coalition environments. The CCO captures the what, when, and where of the information sought by applications and provided by sensors information providers. Extending the notion of sensor-originated information as a dynamic service, we also presented the four viewpoints of QoI that describe the lifecycle of information provision in dynamic environments. The four viewpoints reflect the stages where applications advertise their information and QoI needs and providers advertise their information and QoI provision capabilities. Upon selecting providers to bind to, the QoI viewpoints also reflect the stages of QoI attained by providers and corresponding VoI perceived by applications during runtime, and the ensuing negotiation, if necessary, with information providers to receive information of better QoI. We then presented the information processors and operators that are functional modules in end-to-end systems that process information based on QoI attributes. These include annotators, which enrich information with metadata; modifiers, which alter information or its metadata; and combiners, which fuse multiple pieces of information or their metadata to a new piece of information.

To facilitate the exchange of QoI related-information between applications, providers, and information processors means are needed for expressing and communicating QoI in a commonly understandable manner. For this, we have presented a data model representing the collection of QoI and VoI attributes and their values. The data model comprises an application-agnostic structure that contains means to index information agnostically, e.g., according to time and space, a collection of placeholders for the attributes, and a collection of quality and value statements for these attributes. While the basic structure of the data model is application agnostic, the model itself can be extended to application-specific instances as necessary.

The research work in this chapter complements other QoI-related research that authors of the chapter have been pursuing in the past few years. These include mechanisms for improving sensor data fidelity by identifying and compensating for faulty sensors (Zahedi *et al.*, 2008); system-oriented issues for QoI computation in sensor-network-based detection systems (Zahedi, Srivastava, & Bisdikian, 2008); QoI computations for shooter localization systems (Kaplan, Damarla, & Pham, 2008); end-to-end mission to sensor modeling techniques for QoI (Gillies, Thornley, & Bisdikian, 2009) and their use in developing QoI to VoI utility mappings (Thornley, Young, & Richardson, 2009); QoI-aware sensor network management (He, Zafer, & Bisdikian, 2008; Liu, Leung, Bisdikian, & Branch, 2009); and the impact of communication impediments, such as transmission delay and lost packets, in transferring QoI-related information over networks (Verma *et al.*, 2009). Research in all these areas is still open and ongoing. Despite their broad coverage, they touch only a portion of how QoI may relate and influence "traditional" research in sensor networks, or spearhead brand new research paths.

In closing, we are encouraged to see a recent uptake in QoI activities for sensor networks (QoISN, 2008; IQ2S, 2009; IQ2S, 2010). We hope to see this trend accelerating to further increase research efforts in QoI to lead to a comprehensive, systematic study of the multifaceted aspects of this important component for effective coalition operations.

ACKNOWLEDGMENT

Research was sponsored by US Army Research laboratory and the UK Ministry of Defence and was accomplished under Agreement Number W911NF-06-3-0001. The views and conclusions contained in this document are those of the authors and should not be interpreted as representing the official policies, either expressed or implied, of the US Army Research Laboratory, the U.S. Government, the UK Ministry of Defense, or the UK Government. The US and UK Governments are authorized to reproduce and distribute reprints for Government purposes notwithstanding any copyright notation hereon.

REFERENCES

Akyildiz, I. F., Su, W., Sankarasubramaniam, Y., & Cayirci, E. (2002). A Survey on Sensor Networks. *IEEE Communications Magazine*, *40*(8), 102–116. doi:10.1109/MCOM.2002.1024422

Alberts, D. A., Garstka, J. J., Hayes, R. E., & Signori, D. A. (2001). *Understanding Information Age Warfare*. Washington, DC: CCRP.

Bisdikian, C., Branch, J., Leung, K. K., & Young, R. I. (2009). A Letter Soup for the Quality of Information in Sensor Networks. *First Int'l Workshop on Information Quality and Quality of Service* (IQ2S'09), Galveston, TX, USA, March 9.

Bisdikian, C., Kaplan, L. M., Srivastava, M. B., Thornley, D. J., Verma, D., & Young, R. I. (2009). Building Principles for a Quality of Information Specification for Sensor Information. In *Twelfth Int'l Conf. on Information Fusion* (FUSION 2009), Seattle, WA, USA, July 6-9.

Blasch, E., & Plano, S. (2005). DFIG level 5 (user refinement) issues supporting situational assessment reasoning. In Eights Int'l Conf. on Information Fusion (FUSION), Philadelphia, PA, July 25-28.

Botts, M., & Robin, A. (2007). *Open Geospatial Consortium Inc: OpenGIS® Sensor Model Language (SensorML)*. Implementation Specification.

Cisco Systems. (2000). *Internetworking Technologies Handbook* (3rd ed.). Indianapolis, IN: Cisco Press.

Crawley, E., Nair, R., Rajagopalan, B., & Sandick, H. (1998). *A Framework for QoS-based Routing in the Internet* (RFC 2386). Internet Engineering Task Force (IETF).

Driscoll, P. J., Tortorella, M., & Pohl, E. (2005). *Information Product Quality in Network Centric Operations* (Tech. Rep. No. DSE-TR-0516). West Point, NY: United States Military Academy, Operations Research Center of Excellence.

Gillies, D., Thornley, D., & Bisdikian, C. (2009). Probabilistic Approaches to Estimating the Quality of Information in Military Sensor Networks. *British Computer Society: The Computer Journal*. doi:10.1093/comjnl/bxp047.

Gomes, H. (1996). *Quality Quotes*. Chicago: Irwin Professional Publishing.

Gozdecki, J., Jajszczyk, A., & Stankiewicz, R. (2003). Quality of Service Terminology in IP Networks. *IEEE Communications Magazine*, *41*(3), 153–159. doi:10.1109/MCOM.2003.1186560

Hall, D. L., & McMullen, S. A. H. (2004). *Mathematical Techniques in Multisensor Data Fusion* (2nd ed.). Boston: Artech House.

He, T., Zafer, M., & Bisdikian, C. (2008). Detecting Transient Signals with Incomplete Observations. In MILCOM 2008, San Diego, CA, Nov. 17-19.

Howard, R. D. (1966). Information Value Theory. *IEEE Trans. on Systems Science and Cybernetics*, *2*(1), 22–26. doi:10.1109/TSSC.1966.300074

IQ2S. (2009). In *First Int'l Workshop on Information Quality and Quality of Service for Pervasive Computing*. Part of the *Seventh Annual IEEE Int'l Conf. on Pervasive Computing and Communications* (PerCom 2009), Galveston, TX, USA, March 9.

IQ2S (2010). In *Second Int'l Workshop on Information Quality and Quality of Service for Pervasive Computing*. Part of the *Eighth Annual IEEE Int'l Conf. on Pervasive Computing and Communications* (PerCom 2010), Manheim, Germany, March 29.

Kaplan, L. M., Damarla, T., & Pham, T. (2008). QoI for Passive Acoustic Gunfire Localization. *First Int'l Workshop on Quality of Information for Sensor Networks* (QoISN 2008). Atlanta, GA, Sept. 29.

Klein, A., Do, H. H., Hackenbroich, G., Karnstedt, M., & Lehner, W. (2007). Representing Data Quality for Streaming and Static Data. In *23rd Int'l Conf. on Data Engineering Workshops* (ICDE 2007), Istanbul, Turkey, April 15-20.

Knight, S.-A. (2007). *User perceptions of information quality in World Wide Web information retrieval behavior*. PhD Thesis, Edith Cowan University, Perth, Australia.

Liu, C. H., Leung, K. K., Bisdikian, C., & Branch, J. W. (2009). A New Approach to Architecture of Sensor Networks for Mission-Oriented Applications. *SPIE Defense & Security Symposium,* (DSS 2009), Orlando, FL, April 13-17.

Markoff, J. (2009). The Cellphone, Navigating Our Lives. *New York Times*. Retrieved May 29, 2009, from http://www.nytimes.com/2009/02/17/science/17map.html

Meddeb, A. (2010). Internet QoS: Pieces of the Puzzle. *IEEE Communications Magazine*, *48*(1), 86–94. doi:10.1109/MCOM.2010.5394035

National Imagery Interpretability Rating Scale (NIIRS). (n.d.). Imagery Resolution Assessments and Reporting Standards (IRARS) Committee. Retrieved from http://www.fas.org/irp/imint/niirs.htm

QoISN. (2008). *First Int'l Workshop on Quality of Information for Sensor Networks*. Part of the *Fifth IEEE Int'l Conf. on Mobile Ad-hoc and Sensor Systems* (MASS 2008), Atlanta, GA, Sept. 29.

Single Integrated Air Picture *(SIAP) Attributes, ver. 2.0.* (2003). SIAP System Engineering Task Force (SE TE), Technical Report 2003-029, Aug. 2003.

Thornley, D. J., Young, R. I., & Richardson, J. (2009, April 13-17). Toward Mission-specific Service Utility Estimation Using Analytic Stochastic Process Models. In *SPIE Defense & Security Symposium*, (DSS 2009), Orlando, FL.

Unified Modeling Language (UML). (n.d.). Open Management Group (OMG). Retrieved from http://www.uml.org/

Verma, D., Ko, B. J., Zerfos, P., Lee, K.-w., He, T., Duggan, M., Stewart, K., Swami, A., & Sofra, N. (2009). Understanding the Quality of Monitoring for Network Management. *British Computer Society: The Computer Journal*. doi: 10.1093/comjnl/bxp058

Wang, R. Y., & Strong, D. M. (1996). Beyond accuracy: What data quality means to data consumers. *Journal of Management Information Systems*, *12*(4), 5–33.

Zahedi, S., Srivastava, M. B., & Bisdikian, C. (2008, Nov. 17-19). A Computational Framework for Quality of Information Analysis for Detection-oriented Sensor Networks. San Diego, CA: MILCOM.

Zahedi, S., Szczodrak, M., Ji, P., Mylaraswamy, D., Srivastava, M. B., & Young, R. (2008, Nov. 17-19). Tiered Architecture for On-Line Detection, Isolation and Repair of Faults in Wireless Sensor Networks. *MILCOM 2008*, San Diego, CA.

Zhu, F., Mutka, M. W., & Ni, L. M. (2005). Service discovery in pervasive computing environments. *IEEE Pervasive Computing / IEEE Computer Society [and] IEEE Communications Society*, *4*(4), 81–90. doi:10.1109/MPRV.2005.87

ENDNOTES

1. See, for example, in *Merriam-Webster On-Line*, http://www.merriam-webster.com.
2. More accurately, "...if its sensor detects..."
3. See in *AskOxford.com*, http://www.askoxford.com.

Chapter 3
Graph Mining Techniques for Networking Applications: A Review

Charu C. Aggarwal
IBM T. J. Watson Research Center, USA

Eric Yu-En Lu
Cambridge University, UK

ABSTRACT

Computer networks are inherently graphical in structure, since they contain multiple geographically distributed nodes, which are connected by communication links. In many cases, such as large military networks, this also leads to a network-centric view of the data. This naturally leads to an information network representation of the data. In information networks, the nodes represent the different elements of information, and the links represent the logical relationships between them. An even higher layer of logical organization is a social network, in which the nodes correspond to the different entities in the network, and the links correspond to the interactions. The common element among the different kinds of networks which are encountered in practical network-centric representation of the data is that they can all be represented in the form of a structural graph. This structural behavior can be used in order to glean different kinds of practical insights. This chapter will provide an overview of these different graph mining techniques and their applications to the various kinds of networks.

INTRODUCTION

This chapter will study a number of graph mining applications in the context of network-centric applications. Such graph mining applications are relevant to different levels of conceptualization of network-centric applications. We will study three different levels of conceptualization, which are referred to as *information*, *social*, and *communication* networks respectively. The lowest level of conceptualization for network-centric applications are communication networks which carry the flow of information between different nodes. It has recently been noticed that the structural behavior (Chakrabarti et. al. 2008) (Chakrabarti et. al. 2004) of communication networks is very useful in characterizing the behavior of network-centric applications which are built on top of these networks. Higher

DOI: 10.4018/978-1-61520-855-5.ch003

levels of logical abstraction are built on *top of* large communication networks which carry the actual flow of information. Some examples of such abstractions are *information networks* and *social networks*.

An information network is a logical network of data, information, and knowledge objects, which may be acquired and extracted from disparate sources such as documents, images, sensors, as well as advanced data analytical processes. Information networks are thus defined as conceptual or physical organizations of information in which nodes represent information from a variety of sources, and the links represent the *conceptual* and *logical* relationships between these different pieces of information. Information networks essentially provide a network-centric view of data organization. The links in an information network are distinct from the physical links in a communication network, though the underlying physical links may also affect the conceptual organization represented by the information network links. Furthermore, the underlying links may be very different for different kinds of information networks. For example, in the case of the web, this may correspond to pointers between web pages (or hyperlinks), and in the case of military networks, these links may refer to how the information elements in different entities *interact with* one another. In general, *information networks* refer to a *network-centric organization of the data* and their relationships with one another.

Social networks are closely related to information networks, can can be considered a higher layer representation. In social networks, nodes correspond to entities, which may interact with one another and the links between them may constitute a variety of interactions between the different participants. Examples of social networks include Facebook, LinkedIn, and Myspace. We note that many intra-organization and multi-organization applications share a number of characteristics with social networks. For example, the email communication between different organizations may have similar structure to a social network. In such cases, it may be useful to determine the important *communities of interaction* in the social network, and other structural behavior which provides feedback about the flow of information in the social network.

Since communication, information and social networks have a natural graph structure, it is interesting to explore whether some of the recently proposed techniques for graph mining can be used in order to extract and manage the underlying data in the information networks more effectively. In this chapter, we will examine this relationship. Furthermore, we will also examine how the underlying physical structure of the communication network may affect the network-centric organization. For example, large organizations or military coalitions may have large networks which they rely on for the flow of information between them. Similarly, the entities in a military battlefield, such as tanks, battalions, and other physical assets may form a *logical* information network, in which the nodes correspond to the *information elements* in the different objects, and the links correspond to the relationships between them. In many cases, the structural relationships can be used to isolate important regions of the network which may contain information which is relevant to a specific task. This can be used to determine the nature of the *information flows* in the network, and the importance of the different nodes in such information flows.

This chapter will study the relevance of graph mining methods to a variety of networking applications. Specifically, we will study the relevance of recent graph and structural data mining techniques to these methods. This paper is organized as follows. In the next subsection, we will discuss the concept of information networking and its applications to multi-organization and coalition applications. In section 2, we will discuss different graph mining techniques, and their relevance to different information networking applications. Specifically, we will discuss recent mining techniques such as data reduction, dense subgraph extraction, and graph query processing techniques. In section 3, we will discuss a summary of the chapter along with future research directions.

Networking Applications

As discussed earlier in this chapter, the common thread linking the different kinds of the networks is their *massive structural behavior*, and its importance in analyzing the interactions between the entities in the corresponding network. We would like to mention that the applications of the techniques discussed in this paper extend beyond computer networks, because of the general applicability of graphical data representations.

Graph mining applications can be useful in analyzing different *layers* of the network. This could represent the underlying communication network which carries the information flow, the information network which is a logical representation of the underlying information, and the social network which represent the interactions between the different entities.

- The structure of the communication network plays a key role in defining the flows of information in a network. The structural behavior can be used to determine the underlying bottlenecks at the different nodes. In such cases, graph mining algorithms can be used to analyze the lower layer of the information network which regulates the flow of information. This can also be used for more robust network design in order to ensure that the flow of information between different nodes is not impeded.
- When the information network defines the logical organization of information, graph structural analysis can be used in order to derive the complex relationships between the different data elements. Graph mining helps in isolating important structural relationships between the different entities in the information network, and how the relationships among these entities can be leveraged for practical applications.
- In social networks, graph mining techniques can be used in order to identify the relationships between the interacting entities of the network. This includes methods such as mining communities in the network, analyzing information flows in the network, and mining *central representative points* in the network. This is also closely related to the problem of finding significant linkages and dense communities in web graphs (Gibson et. al. 2005).
- Social networks are closely related to blogging communities and many forms of online marketing, which use the structure of the network in order to propagate the underlying information. A closely related problem to these techniques is that of *viral marketing*, which uses the underlying network structure in order to determine the nodes at which the release of a piece of information maximizes the total influence.

The above applications are quite diverse, and both the organization and interpretation of underlying elements in information networks may be very different. However, we will see that there are some common elements between the different applications, which map to the same underlying applications. One common feature between various kinds of such applications is the nature of the graphs which are often encountered for mining purposes. These features are as follows:

- The underlying graphs are quite large and may contain millions of nodes. The number of edges are typically even larger, and may not be easily held in main memory.
- The degree distributions of the different nodes can vary considerably. Typically such degree distributions are governed by the power-law (Faloustos et. al. 1999)

which suggests that the graphs may show huge differences in the underlying behavior of the node degrees. These huge differences may result in a number of challenges for structural mining algorithms such as clustering and community detection.

Central Problems in Networking Applications

Most of the networking problems relate to dynamic behavior on the underlying structure of the graph. However, many of these problems are tied together by common threads. For example, in a social networking application, the edges in the stream may arrive sequentially. The structural behavior of these edges may be used in order to determine the communities in the underlying network. The same problem often arises in the context of telecommunication networks in which we wish to determine the common participants who talk with one another. Some common examples of the problems which are often encountered in networking applications are as follows:

- **Community Analysis:** In this case, we attempt to determine closely grouped clusters in the underlying data.
- **Centrality Analysis:** In massive networks, we attempt to determine central nodes in the data. This is helpful in determining critical regions in the underlying network.
- **Viral Spread in Networks:** Viral spread is a critical problem in many networking applications. For example, computer viruses rely on the structure of the network for the purposes of propagation.
- **Schema Matching:** Large networked organizations often carry structured data in many different formats. It is useful to utilize schema matching techniques in order to determine the correspondence of the attributes in the data.
- **Network Compression:** For large networks such as the web, it is helpful to utilize compression techniques for reducing the size of the network representation. The compressed representation may be used for a variety of querying and mining applications.

In the next section, we will discuss the different kinds of graph mining algorithms and their applications to different network scenarios.

GRAPH MINING PROBLEMS FOR NETWORKING APPLICATIONS

In this section, we will describe the different graph mining techniques, and their applications to network applications. Some examples of graph mining techniques which are frequently studied in the literature include frequent pattern mining, indexing, clustering, classification, shortest paths, and connectivity applications. We will describe each of these problems in detail and their applications to various networking applications.

Graph Classification

The problem of graph classification is frequently encountered in the graph domain. The classification is of two types:

- **Vertex Classification:** In this case, we attempt to classify the node labels of the different vertices from training data. In other words, we have a set of graphs in which nodes are labeled, and another set of graphs in which the nodes are not labeled. We use the labeled graphs in the training data in order to label the vertices of the unlabeled graphs.
- **Full Graph Classification:** In this case, we associate labels with the entire graphs rather than the individual vertices. As in

the previous case, the labels in the test data are predicted from the labels in the training data.

Both of these different kinds of classification have been studied extensively in the literature. The problem of *full graph classification* is frequently encountered in the context of XML data. A common approach to solving this problem is to attempt to learn structures which are specific to particular kinds of classes. For example, in the methods discussed in (Zaki and Aggarwal 2003) rules are designed in order to map the underlying graph structures to different classes. For a given test graph, we determine the most discriminative graph structures, which are indicative of classification behavior. The rules for these graph structures are used in order to perform the classification. While the work for full graph classification (Zaki and Aggarwal 2003) has focused on the case for XML data, it is also useful in a number of network applications. An example is that of a web tracking application, in which we attempt to classify the customers at a web site depending upon their browsing behavior. The browsing pattern of a customer can be represented as a graph of path traversal patterns, which is superimposed on the site graph. The structural patterns of browsing of these customers can be used in order to make predictions about their buying behavior.

The problem of vertex classification has been studied extensively (Cook and Holder 2007). Typical methods for vertex classification (Cook and Holder 2007) attempt to learn the structure around different vertices, and use them for classification purposes. The vertex based classification methods are quite useful for information networking applications. This is because vertex-classification methods can be used in order to learn the effect of different communication patterns between different entities. For example, consider an information network in which queries may originate at different nodes, and the information is distributed throughout the graph. For a given query, the relevant data may be drawn from a variety of nodes, which may not be known in advance. Rather, this information may need to be *learned* from the underlying data. In order to learn this information, we label the nodes with the corresponding query types, along with the nodes which specified the data which is most relevant for the query. For a given query, we would like to determine the nodes containing the most relevant data for it. We use the historical information on the vertex labels in order to perform this particular classification, and determine the most relevant nodes.

Another important application of classification arises in the context of node labeling in social networks. In many social networks, we would like to use the information available at the different nodes in order to identify important characteristics such as the *authorities* on a particular subject. The graph may already be partially labeled on some of the nodes. Clearly, the adjacency information can be very useful in determining important relationships between the different nodes, and assigning the corresponding labels. In such cases, applications can be designed for learning the vertex labels. These are then used for classification purposes.

Graph classification techniques can also be used to isolate important regions of failure diagnosis in communication networks, especially when such failures are preceded by unusual levels of activity in specific regions of the communication network. In such cases, the failure labels on the nodes can be learned using training data which relates the structural activity behavior in different regions of the network to the corresponding vertex labels. Such techniques can be used in order to supplement attribute-based root-cause analysis techniques in communication networks. More details on a variety of graph classification algorithms may be found in (Cook and Holder 2007).

Graph Clustering

A closely related problem to graph classification is that of clustering. In this section, we will discuss a variety of algorithms for clustering graph data. This includes both classical graph clustering algorithms as well as algorithms for determining dense regions in graphs. Clustering algorithms have significant applications in a variety of graph scenarios such as congestion detection, facility location, and XML data integration (Lee 2002). Within the context of graph algorithms, the clustering can be of two types:

- **Node Clustering Algorithms:** In this case, we have one large graph, and we attempt to cluster the underlying nodes with the use of a distance (or similarity) value on the edges. In this case, the edges of the graph are labeled with numerical distance values. These numerical distance values are used in order to create clusters of nodes. A particular case is once in which the presence of an edge refers to a similarity value of 1, whereas the absence of an edge refers to a similarity value of 0. We note that the problem of minimizing the inter-cluster similarity for a fixed number of clusters essentially reduces to the problem of *graph partitioning* or the *minimum multi-way cut problem*. This is also referred to the problem of mining dense graphs and pseudo-cliques. Recently, the problem has also been studied in the database literature as that of *quasi-clique determination*. In this problem, we determine groups of nodes which are "almost cliques". In other words, an edge exists between any pair of nodes in the set with high probability. We will study the different classes of node clustering algorithms in a different section.
- **Graph Clustering Algorithms:** In this case, we have a (possibly large) number of graphs which need to be clustered based on their underlying structural behavior. This problem is challenging because of the need to match the structures of the underlying graphs, and use these structures for clustering purposes. Such algorithms are discussed both in the context of classical graph data sets as well as semi-structured data. Therefore, we will discuss both of these variations.

Since node clustering algorithms are most relevant to information networks, we discuss them below. A number of algorithms for graph node clustering are discussed in (Glake et. al. 2003). In (Glake et. al. 2003), the graph clustering problem is related to the minimum cut and graph partitioning problems. In this case, it is assumed that the underlying graphs have weights on the edges. It is desired to partition the graph in such a way so as to minimize the weights of the edges across the partitions. The simplest case is the 2-way minimum cut problem, in which we wish to partition the graph into two clusters, so as to minimize the weight of the edges across the partitions. This version of the problem is efficiently solvable, and can be resolved by repeated applications of the *maximum flow problem* (Ahuja et. al. 1992). This is because the maximum flow between source s and sink t determines the minimum s-t cut. By using different source and sink combinations, it is also possible to find the global minimum cut. A second way of determining a minimum cut is by using a contraction-based edge-sampling approach. This is a probabilistic technique in which we successively sample edges in order to collapse nodes into larger sets of nodes. By successively sampling different sequences of edges and picking the optimum value (Tsay et. al. 1999) it is possible to determine a global minimum cut. Both of the above techniques are quite efficient and the time-complexity is polynomial in terms of the number of nodes and edges. An interesting discussion of this problem may be found in (Flake et. al. 2003).

The *multi-way graph partitioning problem* is significantly more difficult, and is NP-hard (Garey & Johnson 1979) In this case, we wish to partition a graph into k>2 components, so that the total weight of the edges whose ends lie in different partitions is minimized. A well known technique for graph partitioning is the Kernighan-Lin algorithm or KL-algorithm (Kernighan and Lin 1970). This classical algorithm is based on a hill-climbing (or more generally neighborhood-search technique) for determining the optimal graph partitioning. Initially, we start off with a random cut of the graph. In each iteration, we exchange a pair of vertices in two partitions, to see if the overall cut value is reduced. In the event that the cut value is reduced, then the interchange is performed. Otherwise, we pick another pair of vertices in order to perform the interchange. This process is repeated until we converge to a optimal solution. We note that this optimum may not be a global optimum, but may only be a local optimum of the underlying data. The main variation in different versions of the KL-algorithm is the policy which is used for performing the interchanges on the vertices. We note that the use of more sophisticated strategies allows a better improvement in the objective function for each interchange, but also requires more time for each interchange. This is a natural tradeoff which may work out differently depending upon the nature of the application at hand. We note that the problem of graph partitioning is studied widely in the literature. A detailed survey may be found in (Fjallstrom 1998).

The problem of graph clustering finds numerous applications in information networks because of the ability to determine closely related patterns from a group of underlying linkages. In an information network, the links between the entities may be represented as an underlying graph structure. In such cases, node clustering algorithms may be used in order to determine closely related groups of nodes. Such techniques can be used in order to determine close structural associations or *communities* between groups of entities. The problem of finding communities has also been studied in the context of the web and social networks (Gibson et. al. 2005). In this case, a min-hash approach is used in order to determine closely related communities from the underlying data. This min-hash approach essentially determines the communities, by using the common outlinks among the different entities.

The problem of graph clustering finds applicability in many forms of networked data. An important problem is that of determining the *s-t* connectivity in a communication network. The *s-t* connectivity essentially determines the minimum number of links required to disconnect a network. Since information networks are often built on top of communication networks which may be prone to link-based failure, it is often useful to determine the minimum connectivity in order to know the likelihood of failure. This also allows us to determine the important information bottlenecks which are caused as a result of the failures in the underlying communication network. We note that the minimum *s-t* cut problem is directly related to this problem of determining the information bottlenecks. While the minimum *s-t* cut problem is efficiently solvable for memory-resident graphs, the problem is far more challenging for the disk resident case. This is because most algorithms for finding the minimum cut may access the edges in random order. This can result in huge performance bottlenecks as result of random access of the underlying data. Therefore, a key challenge is to extend the algorithms to the disk-resident case. As far as we are aware, there are no known algorithms, which are specifically designed for determining minimum cuts in disk resident graphs. This is therefore an important problem for future research.

The determination of dense regions of the graph is closely related to the problem of graph summarization and dimensionality reduction. The process of dimensionality reduction on graphs can be used in order to represent them in a small space, so that they can be used effectively for indexing and retrieval. Furthermore, compressed graphs can be used in a variety of applications in which it is desirable to use the summary behavior in order to *estimate* the approximate structural properties of the network. These estimates can then be subsequently refined for more exact results at a later stage.

Many web applications and social networks can be typically represented as massive graphs. For example, the structure of the web is itself a graph (Gibson et. al. 2004), (Kumar et. al. 2000), (Raghvan and Garcia-Molina 2003), in which nodes represent web pages, and hyperlinks represent the edges of this graph. Similarly social networks are graphs in which nodes represent the members of the social network, and the friendship relationship between members represent the corresponding links. Node clustering algorithms are a natural fit for community detection in massive graphs. The communities have natural interpretations in the context of a variety of web applications:

- For the case of web applications such as web sites, communities typically refer to communities of closely linked pages. Such communities are typically linked because of common material in terms of topic, or similar interests in terms of readership.
- For the case of social networks, communities refer to groups of members who may know each other very well, and may therefore be closely linked with one another. This is useful in determining important associations in the underlying social network.
- Blogging communities often behave like social networks, and contain links between related blogs. The techniques discussed in this chapter are also useful for determining the closely related blogs with the use of community detection methods.

Many of the node clustering applications discussed in this chapter are used in the context of social networks (Gibson et. al. 2004), (Kumar et. al. 2000), (Raghvan and Garcia-Molina 2003). The min-hash approach (Broder et. al. 1997) (Gibson et. al. 2004) is commonly used when the underlying graph is massive in nature, such as that in the case of the web. This is because the min-hash approach is able to summarize the graph in a very small amount of space. This is very useful for practical applications in which it may be possible to represent the entire graph on disk. For example, the size of the web graph is so large, that it may not even be possible to store it on disk without the use of add-ons onto standard desktop hardware. Such situations lead to further constraints during the mining process, which are handled quite well by min-hash style approaches. This is because the min-hash summary is of extremely small size compared to the size of the graph itself. This compressed representation can even be maintained in main memory and used to determine the underlying communities in the network directly. It has been shown in (Broder et. al. 1997) (Gibson et. al. 2004), that such an approach is able to determine communities of very high quality.

Large telecommunication companies may have millions of customers who may make billions of phone calls to one another over a period of time. In this case, the individual phone numbers may be represented as node, and phone calls may be represented as edges. In such cases, it may be desirable to determine groups of customers who call each other frequently. This information can be very useful for target marketing purposes. Furthermore, we note that the graphs in a telecommunication network are represented in the form of *edge streams*, since the edges may be received continuously over time. This results in even greater challenges from the point of view of analysis, since the edges cannot be explicitly stored on disk, and it is difficult to infer global structure with the use of only one pass over the data set. The methods discussed in (Gibson et. al. 2004)are particularly useful in such scenarios.

An interesting application in the context of the Enron crisis was to determine important email interactions between groups of Enron employees. In this case, the individuals are represented as nodes, and the emails sent between them are represented as edges. Node clustering algorithms are very useful in order to isolate dense email interactions between different groups of customers. This approach can be used for a variety of intelligence applications such as that of determining suspicious communities in groups of interactions.

Relationship to Centrality Analysis

Node clustering algorithms are closely related to the concept of *centrality analysis* in networks. For example, the technique discussed in (Rattigan et. al. 2007) uses a k-medoids kind of approach which yields k central points of the network. This kind of approach is very useful in the different kinds of networks, though in different contexts. In the case of communication networks, these central points are typically the critical structural nodes in the network. It is useful to determine these critical structural nodes in order to determine the failure points of the underlying network. In the case of information network, these central nodes can be considered the *information hubs* of the underlying network. The information hubs are useful in determining the key sources of information in the network. In the case of social networks, centrality analysis can be used in order to determine the central points in information flows. Thus, it is clear that the same kind of structural analysis algorithm can lead to different kinds of insights in different networks.

Graph Matching

The graph matching problem is defined as follows. Let us consider a graph in which the vertices are labeled, and the edges may be either directed or undirected. For a given pair of graphs, we would like to determine a one-to-one mapping between the nodes in the two graphs, so that the labels and the edges are approximately matched. If desired, a similarity function may be constructed in order to quantify this level of similarity. In such cases, a one-to-one matching is not necessary, and the similarity function may quantify the overall quality of the matching. Such techniques lead to an *error-tolerant matching* between the graphs.

Graph matching is related to the problem of isomorphism, in which we try to perform an exact one-to-one matching between vertices. This version has been shown to be NP-hard (Garey & Johnson 1979). Since the error-tolerant version is a generalization of the one-to-one matching problem, it is NP-hard as well. The different kinds of error-tolerant methods for graph matching have lead to challenges for mining purposes. A common method for matching graphs is the *least squares method.* In this case, the optimal matching between the two graphs is determined by constructing an optimal permutation matrix for defining a permutation between the two sets. The error of the matching is defined in terms of the least squares function. A second class of techniques is that of spectral methods, in which optimal correspondences between the two graphs are derived with the use of subspace projections in the graph eigenspaces. More general heuristic algorithms are also possible, in which combinatorial algorithms are used in order to define labeling of graphs in terms of each other. Detailed discussions on graph matching algorithms are discussed in (Brugger et. al. 2008) and (Jiang and Bunke 2008).

Graph matching is particularly useful in matching the data from different sources in information networking applications (Melink et. al. 2002). In many information networking applications, the same data may often be derived from different distributed sources, which leads to redundancy. For example, multiple organizations may archive the same data in different forms, which can only be indirectly inferred by using the underlying graph structure. This redundancy needs to be eliminated in practical applications. Typically, the data from different sources are represented in the form of a *schema* which is specific to the particular data source. This schema is typically in the form of a graph structure, in which the nodes represent the entities and the edges representing the relationships between the different entities. The nodes are typically labeled with the attributes for the corresponding schema. We note that the same data from different sources may have slightly different schema. For example, one organization may store the names of people as (last-name, first-name, address) and a different organization may store the records as (name, mailing address, home address). In addition, the relationships between these different entities may be specified in the graph. In order to properly integrate the data from these different organizational sources, it is useful to utilize graph matching algorithms. Such algorithms typically use the structure of the data in the underlying graph in order to match the data from different sources. Schema matching is particularly challenging when different portions of the data are available from different data sources. In such cases, it may be necessary to perform partial matches on the relevant portions of the graph. This can be very challenging from a graph matching and isomorphism point of view.

Indexing Graphs

In this subsection, we will discuss algorithms for indexing graph data, and its application to information networking applications. The problem of indexing graphs has been studied in the recent literature (Yan et. al. 2004), (Yan et. al. 2005). The problem of indexing graphs is especially relevant for the disk resident case, in which the size of the underlying graph database is so large that it needs to be stored on disk. Indexing is often used in order to retrieve portions of the data on the disk, when the portions are too large to be held effectively on disk. A classic technique which is often used is that of *substructure similarity search*, in which we wish to determine substructures which are similar to a given target. Techniques for substructure similarity search are discussed in (Yan et. al. 2005).

The techniques in (Yan et. al. 2004) and (Yan et. al. 2005) use frequent pattern mining algorithms in order to uncover the corresponding structures. The idea is to construct *representative patterns* from the underlying graphs. These representative patterns form the fragments which are used in order to query for similarity. An index is constructed from the inclusion behavior of these fragments in different graphs. Specifically, an inverted representation is constructed which points to the different graph identifiers from the different structure fragments. For a given target graph, the common fragments are determined, and a similarity function is constructed based on this similarity. The graph with the greatest similarity is returned as the final result.

Such similarity search techniques are often quite relevant to information networking applications, since the logical design of the networked data may often be represented in the form of *ontologies* or semi-structured data. Substructure similarity search is a natural technique which can be leveraged to uncover similar patterns of relationships in the underlying data. The inclusion of structural information in similarity search applications can provide for a much richer set of semantics, than is possible with the use of attribute-based representations. Graph indexing can be particularly useful in such scenarios, especially when the ontologies are very large and may contain thousands of nodes.

Graph indexing techniques are also naturally geared towards a number of other applications such as connectivity and distance indexing. In (Aggarwal et. al. 2009), techniques have been proposed to determine connectivity indexes in massive disk resident graphs. These methods are useful for determining the critical points of failure in the underlying network. In addition, techniques are proposed in (Feigenbaum et. al. 2005) to determine distance indexes in disk resident graphs. Such indexes can be useful for making routing decisions in massive graphs.

Frequent Pattern Mining

In this section, we will discuss the relevance of frequent pattern mining algorithms for graph data sets. The problem of frequent pattern mining attempts to determine highly connected graphs which are frequent in nature. The first method for frequent pattern was proposed in (Inokuch et. al. 2000) . This technique is based on the Apriori algorithm and determines both connected and disconnected graphs. We note that the Apriori algorithm is easy to extend to the case of graph data by treating each edge as an item in the data and determining the frequent patterns on this set of items. Kuramochi (Kuramochi & Karypis 2001) further developed this idea with the adjacent representation of a graph, and an edge growing strategy. As in the case of the work of (Inokuchi et. al. 2000), this algorithm is also essentially based on the Apriori algorithm. Note that instead of the subset test in the case of frequent pattern mining algorithms, this algorithm needs to use a subgraph test. The subgraph test is much more difficult that the subset property, because the isomorphism property holds in the case of graphs. This test is NP-hard, and this increases the complexity of the overall process of determining the frequent graphs.

In order to reduce the costs associated with the subgraph isomorphism test, an algorithm called gSpan was proposed in (Yan and Han 2002). The algorithm can be applied in two ways, depending upon the nature of the underlying data:

- If the entire graph dataset is available in main memory, then gSpan can be applied directly.
- For the case of larger disk-based graphs, the prefix-based projection approach (Pei et. al. 2001) is used, and then gSpan is applied.

gSpan discovers all the frequent subgraphs without candidate generation and false positives pruning. It combines the growing and checking of frequent subgraphs into one procedure, thus accelerates the mining process.

The frequent pattern mining approach is very useful in a number of direct and indirect applications of information networks. Specifically, frequent pattern mining can be used in the following scenarios:

- Many of the information networking applications discussed earlier such as graph classification and indexing use the problem of frequent pattern mining as a kernel. Therefore, it is essential to develop efficient frequent pattern mining algorithms in order to enable efficient instantiations of these applications.
- Frequent pattern mining can be used in order to determining frequent structural patterns between groups of entities. This can be used in order to determine associations between groups of entities. This is useful in order to determine *global patterns of interaction* in large social networks. Such global patterns of interaction can often provide interesting insights about the relationships between the different entities in the information network. We note that this is closely related to the problem of finding (node) clusters in a graph. The main difference is that node clusters are typically represented as partitions over *individual graphs*, whereas frequent patterns can determine the the repeated patterns of activity over *multiple activity graphs which are superposed over a single base graph.* This approach can be used for applications beyond social networks. Some examples of such applications include the analysis of social interactions (such as electronic communications) between different entities.
- Frequent pattern mining techniques can be used to determine traffic bottlenecks in a large communication network. This is

because the frequent patterns of activity in the busy periods provide us with an idea of the typical patterns of activity which can create the bottlenecks.

Many of the frequent pattern mining techniques have also been extended to very large graphs, (He and Sing2007) (Ranu and King 2009). The extension to very large graphs is particularly useful for networking applications, because of the massive size of the networks which form a base for such graphical representations.

Graph Reachability Queries

The graph reachability problem is defined in the context of *directed graphs*. For a given node *s*, we would like to determine all nodes which are reachable from it. The reachability problem has been studied extensively in the context of very large graphs (Cheng et. al. 2006), (Cheng et. al. 2008), (Cohen 1997), (Wang et. al. 2006). A closely related version of the problem attempts to determine a directed path of length below a pre-defined threshold. Most of the underlying algorithms use Dijkstra-style labeling techniques in order to determine the vertices which are reachable from a given node.

The reachability problem is an important application from the perspective of determining linkages in social and information networks. For example, in an information network in which links represent transitive relationships, the reachability problem can help identify important information linkages in the underlying data. In the case of social networks, the edges may reflect directional flow of messages. In such cases, the reachability labels provide an idea of the chain of relationships in the underlying network.

A closely related problem is that of determining the shortest paths (Djikstra 1959) in large networks. While the shortest path problem has been efficiently solved in the memory resident case, the problem is extremely challenging in the case of disk-resident data. This is because disk-resident graphs typically make random accesses to the underlying data. Such random accesses will result in very low efficiency. A number of techniques (Papadis et. al. 2003), (Samet el. Al. 2008) have been recently developed for the disk-resident case. However, these techniques are designed for spatial databases, and further techniques are required for generalizing the method to the case of arbitrary networks.

Reachability problems are useful for all the three kinds of networks: information, communication and social. In the case of information networks, reachability distances can be used for *link analysis*, and in determine the relationships between different parts of the data. In the case of communication networks, shortest reachability distances are useful in determining the *optimal routing* of the data across the network. It is also useful as a subroutine for determining the optimal network design of the communication network. In the context of social networks, the reachability distances can be used in order to determine the *degrees of separation* between different nodes in the network. This is closely related to the problem of determining the statistics of large graphs such as the underlying density and the diameters. This can be used to measure the spread of influence in social networks. Recently broad measures of graph diameters and reachability have been studied (Chakrabarti et. al. 2008), (Leskovec et. al. 2007) for the case of large social and information networks. It has been shown that graph diameters are typically of modest size in real applications such as instant messaging networks (Leskovec et. al. 2007).

Viral Influence in Networks

Viral spread is an important problem in the context of communication and social networks. The problem is defined as ne in which a given node can infect another node across an edge, and we would like to determine the rate of spread over the network, as well as the origin of the infection. This kind of problem is very useful in modeling viral spread (Wu et. al. 2004), (Zhou et. al. 2004) in communication networks. Numerous epidemic models for measuring epidemic spread in networks are proposed in (Chakrabarti et. al. 2008), (Leskovec et. al. 2007), (Newman, Forrest and Balthrop 2002) (Newman 2002) (Wu et. al. 2004), (Zhou et. al. 2004). The work on viral spread was first studied in the context of viral spread (Newman, Forrest and Balthrop 2002), (Wu et. al. 2004), (Zhou et. al. 2004) in computer networks. These studies have been used to model both the rate of spread and the influence of any particular node in the process of transmission and dissemination.

Subsequently, it was observed (Chakrabarti et. al. 2008), (Newman 2002), (Wu et. al. 2004) (Leskovec et. al. 2007) that the techniques discussed above can be used in the context of a variety of other social networking related applications, which show similar behavior models as viral spread. These applications are: (1) We would like to determine the most influential members of the social network; i.e. members who cause the most flow of information outwards. (2) Information in the social behavior often cascades through it in the same way as an epidemic. We would like to measure the information cascade rate through the social network, and determine the effect of different sources of information. The idea is that monitoring promotes the early detection of information flows, and is beneficial to the person who can detect it. The cascading behavior is particularly visible in the case of blog graphs, in which the cascading behavior is reflected in the form of added links over time. Since it is not possible to monitor all blogs simultaneously, it is desirable to minimize the monitoring cost over the different blogs, by assuming a fixed monitoring cost per node. This problem is NP-hard (Leskovec et. al. 2007), since the vertex-cover problem can be reduced to it. The main idea in (Leskoven, McGlohon et. al. 2007) is to use an approximation heuristic in order to minimize the monitoring cost. Such an approach is not restricted to the blog scenario, but it is also applicable to other scenarios such as monitoring information exchange in social networks, and monitoring outages in communication networks. (3) We would like to determine the conditions which lead to the critical mass necessary for uncontrolled information transmission. Some techniques for characterizing these conditions are discussed in (Chakrabarti et. al. 2008) and (Wang et. al. 2003). The work in (Wang et. al. 2003) relates the structure of the adjacency matrix to the transmissibility rate in order to measure the threshold for an epidemic. Thus, the connectivity structure of the underlying graph is critical in measuring the rate of information dissemination in the underlying network. It has been shown in (Wang et. al. 2003) that the eigenstructure of the adjacency matrix can be directly related to the threshold for an epidemic.

Graph Streaming Applications

A number of important applications for networks are designed for the streaming scenario. The streaming scenario arises because many networking applications such as telecommunication networks and social networks have continuously arriving data in the form of *edge streams*. For example, the phone calls in a telecommunication network can be modeled as edges between different user nodes. Similarly, the messages between users in a social network can be modeled as edges between user nodes.

Graph streaming applications pose special challenges to mining algorithms, because the entire graph is not available on disk at one time. Therefore, a natural approach is to design methods for *synopsis construction* of the underlying graph. An interesting synopsis method has been designed in (Cormode and Muthukrishnan 2005), which summarizes the degree distribution of edge streams in the context of telecommunication networks. The key challenge is that the *repetition* of the edges over time disallows the use of traditional data structures such as sketches (Alon et. al. 1996) for counting purposes. Therefore, the technique in (Cormode and Muthukrishnan 2005) uses a min-wise hash structure in order to estimate the degree distributions of the underlying nodes.

A natural application in graph streams is that of determining the *page rank* in a graph stream. As defined by (Brin and Page 1998), the page rank is a useful definition of importance of nodes (web pages, entities) in a graph (web, social network), by utilizing the linkage of nodes to one another. Typically nodes of greater importance are linked more heavily to other nodes. The well known search engine *Google* uses page rank in order to determine the relative importance of different web pages. Since the web is a (relatively) slowly changing entity, as compared to many other recent applications such as the blogosphere and social networks, it suffices to use static algorithms for determining page rank. Such a static algorithm (Brin and Page 1998), and its subsequent improvements have been used for the *Google* search engine.

Many of the more recent network applications are highly dynamic. For example, the messaging and linkage structure of social networks continues to evolve rapidly over time. Similarly, the blogosphere is a portion of the web, which evolves far more rapidly than most of the stable portions of the world wide web. Such applications require more dynamic page rank algorithms which can adjust rapidly with the evolving network structure. Many of these recent network applications continue to be relevant for search, though in a slightly different context. For example, entity-search can be enhanced in social networks by using page-rank computations. This is because individuals with highly relevant linkage structure are more useful to a search than isolated nodes. A method for dynamic determination of page rank has been proposed in (Sama et. al. 2008). The method in (Sama et. al. 2008) uses a probabilistic random walk model in order to estimate the page rank of different nodes in the stream.

Graph Spectrum Analysis and Statistics

Many forementioned properties concerning graphs and operations on them have equivalent formulations based on spectra-decomposition of the graph adjacency matrix (and their variants). In Physics and Mathematics, complex network studies consider applications as far as Internet to social and biological networks (Vendruscolo et. al. 1999), (Newman 2006), (Newman 2001), (Newman et. al. 2002), (Lazer et. al. 2009), (Kleinberg 2000), (Cheng et. al. 2008), (Albert and Barabasi 2000), to name a few. These studies give both phenomenal and generative views concerning properties of the system in question. For example, (Kleinberg 2000) considered routing and topology problems in small-world networks and (Newman 2001) characterized community structures in social networks through products of adjacency matrix. Thus, the graph spectrum becomes the major role in determining properties of graphs and operations such as random walk on top of them.

Fundamental results of graph spectrum analysis are mostly discussed in the field of algebraic and spectral graph theory (Biggs 1994) (Chung 1997) which studies the relation between the spectrum of a graph adjacency matrix and its graph theoretic properties. For example, the second smallest eigenvalue (Λ_2) in the spectrum provides bounds on the connectedness of the graph and the number of zeroes give number of disconnected components in the network. Below, we first give the formalisms from spectral graph theory so as to give some flavors of what the results are like.

Fact 1 (Chung 1997). *Let $G=(V,E)$ be a graph with vertices V and $E:V\times V$ the edges. The degree of connectivity is defined as $\phi(G)$ =* $\min_{S\subset V} E(S,S')/\min\{|S|, |S'|\}$, the isoperimetric number of graph *G*, where S' is the complement of S.

The relation between Λ and ϕ *is* as below:

$$2\phi \geq \Lambda_2 \geq \phi^2/2$$

For distance between k-subgroups $\{X_1, X_2, \ldots, X_k: X V\}$ *in the system, their pair-wise distance is upper-bounded by:*

$$\min dis\left(X_i, X_j\right) \leq \max\Gamma \frac{\log \frac{vol\left(G\right)}{\sqrt{vol\left(X_i\right) vol\left(X_j\right)}}}{\log \frac{\lambda_{n-1} + \lambda_k}{\lambda_{n-1} - \lambda_k}}$$

where dis(· , ·) denote the number of edges connecting two sets of vertices.

Consider a random walk matrix $P = \left[p_{i\,j}\right]$ over G such that

$$P_{i\,j} = \begin{cases} p_{i\,j} = \frac{1}{d_i} if\left(i; j\right) \in E \\ 0 \end{cases}$$

There exists an equilibrium state π *in which* $P^t P = \pi$, provided that P is aperiodic [1]. *The rate at which a random walk converges to* π *is measured in terms of* relative pointwise distance

$$\Delta\left(t\right) = \max \frac{\left|P_{ij}^{\,t} - \pi_i\right|}{\pi_i}$$

It is upper-bounded by

$$\Delta\left(t\right) \leq exp\left(t\left(\max\left|1 - \Lambda_i\right| - 1\right)\right) \frac{vol\left(G\right)}{\min_x d_x}$$

Having established the general relation between graph spectrum and its properties, applications such as stream-query processing and network routing require further characterizations about the *in situ* or *ad hoc* configurations. These applications work under arbitrary and stringent conditions and hence one often seeks statistical, rather than absolute, guarantees. It turns out that spectrum of random matrices of finite independently, identically distributed (iid) entries obeys a limiting distribution, an equivalent behavior to the central limit theorem. For cases where iid condition must be relaxed, the geometric random graphs often used by wireless networks for example, their spectrum still have limiting behaviors (Mézard et. al. 1999). Through these results, a graph with stochastic properties becomes amenable to statistical treatments. Below, we give an introduction to the fascinating results in the field of Random Matrix Theory (RMT).

The limiting behavior of spectrums due to random matrices are first noticed in the 1950s. It was observed in many nuclei energy level experiments that the energy level distribution follows a distribution irrespective of the particular system state or individual particle charateristics. It was later demonstrated that the energy level (spectra) distribution depends only on the structure of the nuclei system (represented as a matrix) in question. Exact distributions for eigenvalues of matrices classes such as Gaussian and circular ensembles were derived and the conjecture was that all distributions converge to the empirical distribution (Mehta 1991).

A little more than a decade after these results are experimentally and numerically verified, Wigner derived the exact distribution—the semi-circular distribution—for random hermitian matrices with Gaussian distributed entries. Exact solutions also exist for other matrix ensembles such as circular ensembles (Mehta 1991). Below, we state the exact form of the semi-circular law below:

Fact 2 (The Semi-Circular Law). *Let* $W = \left[w_{ij}\right]$ be a random real matrix of order n such that

- Random variables $\{w_{ij} : 1 \le l \le j \le n\}$ are i.i.d.
- $w_{ij} = w_{ij}$ (Symmetric)
- $Ew_{ij} = 0$ and $\mathrm{var}(w_{ij}) = 1$

That is, if W is hermitian with upper-diagonal entries i.i.d with 0 mean and unit variance, then the eigenvalue distribution of $\frac{1}{\sqrt{n}} W$ converges to the empirical spectrum distribution (ESD):

$$\Pr[x] = \frac{1}{2\pi}\sqrt{4 - x^2} \qquad asn \to \infty$$

Inspired by various numerical experiments later, this result was then generalised to complex entries requiring only finite mean and variance (Tao et. al. 2008) for the seminal work of (Girko 2005).

A natural question to ask is then whether this applies to systems with a prefined state. In the standard queueing network terms, this amounts to the distribution of the equilibrium states (Kleinrock 1975). The applicability of Semi-Circular Law is greatly reduced since it does not give results concerning random matrices with a given starting point. (Tao et. al. 2008) have shown universality principles which extends to matrices due to a random operator and an predefined state. This theorem provides the necessary foundation on which we may safely ignore initial conditions of the system and analyse the equilibrium state directly. Also, it indicates that given minor perturbation (*M*), the spectrum is not easily changed. We note that this theorem generalizes to non-symmetric, complex random matrices as shown in (Tao et. al. 2008).

Fact 3. Universality Principal (Tao et. al. 2008)

Let X_n and Y_n be n×n matrices with entries from i.i.d. sequences { $x_{i,j}$} and { $y_{i,j}$} *such that EX* = 0, *and var(X)* = 1. *For each n, let M_n be a random nxn matrix independent of X_n, Y_n such that* $\Pr\left[\lim \frac{1}{n^2} |M_n|_2^2 = F\right] = 1$ there F is a constant. Consider the random base matrics $A_n = M_n + X_n$ and $B_n = M_n + Y_n$. Their empirical spectrum distribution $\Pr\left[ESD(A_n) - ESD(B_n) \to 0\right] = 1$.

CONCLUSION

In this paper, we presented a review of the different graph applications which are useful in the context of information networks. We present an overview of the different graph management and mining problems such as indexing, clustering, classification, frequent-pattern mining, and graph matching. In each case, we provide an overview of the different methods as well as the applications of these methods to information networking applications. The broad applications for which graph mining algorithms can be used are as follows:

- Graph mining algorithms are useful for determining closely related entities in information networks. They can also be used to determine clusters of well connected nodes (communities) in social and communication networks. This can typically be achieved by frequent pattern mining and node-clustering algorithms.
- Recently, the network structure of social and communication networks has been explored in the context of spread of information (or computer viruses) in a network. The spread of information in a social network has tremendous similarity to the spread of viruses in a communication network. It has been shown (Chakrabarti et. al. 2008) that the analysis of the structural behavior of the underlying network is critical information in determining the threshold for an epidemic in a given network.
- In cases in which the relationships in the

information network are stored in the form of ontologies, it is sometimes useful to design query processing applications in which particular structural regions of these ontologies are determined. In such cases, graph indexing algorithms can be used in order to determine regions of a graph which are similar to a particular target.

- When the data is derived from different distributed sources, it is useful to apply schema-matching techniques in order to determine redundant pieces of information. Schema-matching is naturally defined as a graph matching problem, in which the structural behavior of two different schema are matched for the purposes of similarity.
- The shortest reachability problem is particularly useful in communication networks from the perspective of routing applications. In an information network, the reachability problem provides an understanding of the relationship between different entities in the network. Recently developed reachability algorithms can be used for this purpose. The analysis of reachability behavior can also be used to facilitate effective network design.
- Recently, the scenario of streaming edges has been explored in the context of graph data. This is a particularly challenging scenario, but it is extremely important for networking applications such as social networks or the blogosphere which are highly dynamic.

Many of the graphs for the applications discussed above are massive in nature. Typically, such graphs may have millions of nodes, and an even greater number of edges. For example, a graph with nodes may have as many as edges. This makes such problems even more difficult from a practical perspective, because the graph cannot be held in main-memory and must be held in disk. In some cases, even the disk space is not sufficient to store and process such graphs. Most of the currently proposed algorithms are designed for the memory-resident case. This is clearly not sufficient for the case of information networks.

In future research, it would be worthwhile to examine a variety of techniques which can perform the mining techniques necessary for very large graphs. This can be achieved by designing *structural synopsis structures* which can maintain the summary behavior of the graph. This summary behavior of the graph can be used in order to respond to the underlying queries. An example of such a synopsis structure could be a structural generalization of the sketch technique, which is often used in the context of data streams (Aggarwal and Yu 2007). The stream scenario for graph data (Cormode and Muthukrishnan 2005), (Sarma et. al. 2008) has barely been explored, and presents a fertile area for future research.

REFERENCES

Aggarwal, C., Ta, N., Feng, J., Wang, J., & Zaki, M. J. (2007). XProj: A Framework for Projected Structural Clustering of XML Documents. In *KDD Conference*.

Aggarwal, C., Xie, Y., & Yu, P. (2009). GConnect: A Connectivity Index for Massive Disk-Resident Graphs. In *VLDB Conference*.

Aggarwal, C., & Yu, P. (2007). *A Survey of Synopsis Construction in Data Streams. Data Streams: Models and Algorithms*. Berlin: Springer.

Agrawal, R., Borgida, A., & Jagadish, H. V. (1989). Efficient Maintenance of transitive relationships in large data and knowledge bases. In *ACM SIGMOD Conference*.

Ahuja, R., Orlin, J., & Magnanti, T. (1992). *Network Flows: Theory, Algorithms, and Applications*. Englewood Cliffs, NJ: Prentice Hall.

Alon, N., Matias, Y., & Szegedy, M. (1996). The Space Complexity of Approximating the Frequency Moments. In *ACM Symposium on Theory of Computing*.

Biggs, N. (1994). *Algebraic Graph Theory (Cambridge Mathematical Library)*. Cambridge, UK: Cambridge University Press.

Brin, S., & Page, L. (1998). The anatomy of a large-scale hypertextual Web search engine. In *WWW Conference*.

Broder, A. Z., Charikar, M., Frieze, A., & Mitzenmacher, M. (1997). Syntactic clustering of the web. *WWW Conference, Computer Networks, 29*(8–13), 1157–1166.

Brugger, A., Bunke, H., Dickinson, P. J., & Riesen, K. (2008). Generalized Graph Matching for Data Mining and Information Retrieval. In *ICDM Conference* (pp. 298-312).

Chakrabarti, D., Wang, Y., Wang, C., Leskovec, J., & Faloutsos, C. (2008). Epidemic thresholds in real networks. *ACM Transactions on Information and System Security, 10*(4). doi:10.1145/1284680.1284681

Chakrabarti, D., Zhan, Y., & Faloutsos, C. (2004). R-MAT: A Recursive Model for Graph Mining. In *SDM Conference*.

Chawathe, S. S. (1999). Comparing Hierachical data in external memory. *Very Large Data Bases Conference*.

Cheng, M.-K., Lu, Y.-E., Vendruscolo, M., Lió, P., & Blundell, T. L. (2008). Prediction by graph theoretic measures of structural effects in proteins arising from non-synonymous single nucleotide polymorphisms. *PLoS Computational Biology, 4*(7). doi:10.1371/journal.pcbi.1000135

Cheriyan, J., Hagerup, T., & Melhorn, K. (1996). An -time maximum-flow algorithm. *SIAM Journal on Computing, 25*(6), 1144–1170. doi:10.1137/S0097539791278376

Chung, F. (1997). *Spectral Graph Theory* (No. 92: CBMS Regional Conference Series in Mathematics). Providence, RI: American Mathematical Society.

Cormode, G., & Muthukrishnan, S. (2005). Space-Efficient Mining of Multigraph Streams. In *ACM PODS Conference*.

Dalamagas, T., Cheng, T., Winkel, K., & Sellis, T. (2005). Clustering XML Documents Using Structural Summaries . In *Information Systems*. Thousand Oaks, CA: Elsevier.

Dijkstra, E. W. (1959). A note on two problems in connection with graphs. *Numerische Mathematik, 1*, 269–271. doi:10.1007/BF01386390

Réka, A., & Albert-László, B. (2000). Dynamics of complex systems: Scaling laws for the period of boolean networks. *Physical Review Letters, 84*(24), 5660. doi:10.1103/PhysRevLett.84.5660

Cheng, J., Xu Yu, J., Lin, X., Wang, H., & Yu, P. S. (2008). Fast Computing Reachability Labelings for Large Graphs with High Compression Rate. In *EDBT Conference*.

Cheng, J., Xu Yu, J., Lin, X., Wang, H., & Yu, P. S. (2006). Fast Computation of Reachability Labelings in Large Graphs. In *EDBT Conference*.

Cohen, E. (1997). Size-estimation framework with applications to transitive closure and reachability. *Journal of Computer and System Sciences, 55*(3), 441–453.

Cohen, E., Halperin, E., Kaplan, H., & Zwick, U. (2002). Reachability and distance queries via 2-hop labels. In *ACM Symposium on Discrete Algorithms*.

Cook, D., & Holder, L. (2007). *Mining Graph Data*. New York: John Wiley & Sons, Inc.

Dijkstra, E. W. (1959). A note on two problems in connection with graphs. *Numerische Mathematik, 1*, 269–271.

Faloutsos, M., Faloutsos, P., & Faloutsos, C. (1999). On Power Law Relationships of the Internet Topology. In *SIGCOMM Conference*.

Feigenbaum, J., Kannan, S., McGregor, A., Suri, S., & Zhang, J. (2005). On Graph Problems in a Semi-streaming Model. *Theoretical Computer Science, 348*(2).

Fjallstrom, P.-O. (1998). Algorithms for Graph Partitioning: A Survey. *Linkoping Electronic Articles in Computer and Information Science, 3*(10).

Flake, G., Tarjan, R., & Tsioutsiouliklis, M. (2003). Graph Clustering and Minimum Cut Trees. *Internet Mathematics, 1*(4), 385–408.

Garey, M. S., & Johnson, D. S. (1979). *Computers and Intractability: A Guide to the Theory of NP-completeness*. New York: W. H. Freeman.

Gibson, D., Kumar, R., & Tomkins, A. (2005). Discovering Large Dense Subgraphs in Massive Graphs. In *VLDB Conference*.

Girko, V. L. (2005). The circular law. twenty years later. part iii. *Random Operators and Stochastic Equations, 13*(1), 53–109.

He, H., & Singh, A. K. (2005). Efficient Algorithms for Mining Significant Substructures in Large Graphs with Quality Guarantees. In *ICDM Conference*.

Inokuchi, A., Washio, T., & Motoda, H. (2000). An apriori-based algorithm for mining frequent substructures from graph data. In *PKDD Conference*, (pp. 13–23).

Kuramochi, M., & Karypis, G. (2001). Frequent subgraph discovery. In *ICDM Conference*, (pp. 313–320).

Jain, A., & Dubes, R. (1998). *Algorithms for Clustering Data*. Upper Saddle River, NJ: Prentice Hall.

Kashima, H., Tsuda, K., & Inokuchi, A. (2003). *Marginalized Kernels between Labeled Graphs*. ICML.

Kleinberg, J. (2000). The small-world phenomenon: an algorithm perspective. In *Proceedings of the thirty-second annual ACM Symposium on Theory of Computing*, (pp. 163–170). New York: ACM Press.

Kleinrock, L. (1975). Queueing Systems: *Vol. 1. Theory*. New York: John Wiley & Sons.

Jiang, X., & Bunke, H. (2008). Graph Matching . In *Case based Reasoning in Images and Signals*. Berlin: Springer.

Kernighan, B. W., & Lin, S. (1970). An efficient heuristic procedure for partitioning graphs. *The Bell System Technical Journal, 49*, 291–307.

Kudo, T., Maeda, E., & Matsumoto, Y. (2004). An Application of Boosting to Graph Classification. In NIPS Conf.

Kumar, R., Raghavan, P., Rajagopalan, S., Sivakumar, D., Tomkins, A., & Upfal, E. (2000). The Web as a Graph. In *ACM PODS Conference*.

Lazer, D., Pentland, A., Adamic, L., Aral, S., Barabasi, A.-L., & Brewer, D. (2009). Social Science: Computational Social Science. *Science, 323*(5915), 721–723.

Lee, M., Hsu, W., Yang, L., & Yang, X. (2002). XClust: Clustering XML Schemas for Effective Integration. In *ACM Conference on Information and Knowledge Management*.

Leskovec, J., Krause, A., Guestrin, C., Faloutsos, C., VanBriesen, J., & Glance, N. S. (2007). Cost-effective outbreak detection in networks. In *KDD Conference*, (pp. 420–429).

Leskovec, J., Kleinberg, J. & Faloutsos, C. (2007). Graph Evolution: Densification and Shrinking Diameters. *ACM Transactions on Knowledge Discovery from Data (ACM TKDD), 1*(1).

Leskovec, J., McGlohon, M., Faloutsos, C., Glance, N., & Hurst, M. (2007). Cascading Behavior in Large Blog Graphs. In *SIAM International Conference on Data Mining (SDM)*.

Leskovec, J., & Faloutsos, C. (2007). Scalable modeling of real graphs using Kronecker multiplication. *ICML Conference*, (pp. 497–504).

Leskovec, J., Lang, K., Dasgupta, A., & Mahoney, M. (2008). Statistical Properties of Community Structure in Large Social and Information Networks . In *World Wide Web*. WWW.

Lian, W., Cheung, D. W., Mamoulis, N., & Yiu, S. (2004). An Efficient and Scalable Algorithm for Clustering XML Documents by Structure. *IEEE Transactions on Knowledge and Data Engineering, 16*(1).

Lal Mehta, M. (1991). *Random Matrices* (2nd ed.). San Diego, CA: Academic Press.

Melnik, S., Garcia-Molina, H., & Rahm, E. (2002). Similarity Flooding: A Versatile Graph Matching Algorithm and its Application to Scheme Matching. In *ICDE Conference*.

Mézard, M., Parisi, G., & Zee, A. (1999). Spectra of euclidean random matrices. *Nuclear Physics B, 559*, 689–701.

Newman, M. E. J., Forrest, S., & Balthrop, J. (2002). Email networks and the spread of computer viruses . *Physical Review E: Statistical, Nonlinear, and Soft Matter Physics, 66*, 035101.

Newman, M. E. J. (2002). The spread of epidemic disease on networks . *Physical Review E: Statistical, Nonlinear, and Soft Matter Physics, 66*, 016128.

Newman, M. E., Watts, D. J., & Strogatz, S. H. (2002). Random graph models of social networks. *Proceedings of the National Academy of Sciences of the United States of America, 99*(Suppl 1), 2566–2572.

Newman, M. E. J. (2001). Clustering and preferential attachment in growing networks. *Physical Review E: Statistical, Nonlinear, and Soft Matter Physics, 64*(2), 025102.

Newman, M. E. J. (2006). Finding community structure in networks using the eigenvectors of matrices. *Physical Review E: Statistical, Nonlinear, and Soft Matter Physics, 74*(3), 036104.

Papadias, D., Zhang, J., Mamoulis, N., & Tao, Y. (2003). Query Processing in Spatial Network Databases. In *VLDB Conference*, (pp. 802-813).

Pei, J., Jiang, D., & Zhang, A. (2005). On Mining Cross-Graph Quasi-Cliques. In *ACM KDD Conference*.

Pei, J., Jiang, D., & Zhang, A. (2005). Mining Cross-Graph Quasi-Cliques in Gene Expression and Protein Interaction Data. In *ICDE Conference*.

Pei, J., Han, J., Mortazavi-Asl, B., Pinto, H., Chen, Q., Dayal, U., & Hsu, M.-C. (2001). PrefixSpan: Mining sequential patterns efficiently by prefix-projected pattern growth. In *ICDE Conference*, (pp. 215-224).

Raghavan, S., & Garcia-Molina, H. (2003). Representing web graphs. In *ICDE Conference*, (pp. 405–416).

Ranu, S., & Singh, A. K. (2009). GraphSig: A Scalable Approach to Mining Significant Subgraphs in Large Graph Databases. In *ICDE Conference*.

Rattigan, M., Maier, M., & Jensen, D. (2007). *Graph Clustering with Network Sructure Indices*. ICML.

Samet, H., Sankaranarayanan, J., & Alborzi, H. (2008), Scalable Network Distance Browsing in Spatial Databases. In *ACM SIGMOD Conference*.

Sarma, A. D., Gollapudi, S., & Panigrahy, R. (2008). Estimating Page Rank on graph streams. In *ACM PODS Conference*.

Tao, T., Vu, V., & Krishnapur, M. (2008). *Random matrices: Universality of esds and the circular law.*

Tsay, A. A., Lovejoy, W. S., & Karger, D. R. (1999). Random Sampling in Cut, Flow, and Network Design Problems. *Mathematics of Operations Research, 24*(2), 383–413.

Wang, H., He, H., Yang, J., Xu-Yu, J., & Yu, P. (2006). Dual Labeling: Answering Graph Reachability Queries in Constant Time. In *ICDE Conference*.

Vendruscolo, M., Najmanovich, R., & Domany, E. (1999). Protein folding in contact map space. *Physical Review Letters, 82*(3), 656–659.

Wang, Y., Chakrabarti, D., Wang, C., & Faloutsos, C. (2003). *Epidemic Spreading in Real Networks: An Eigenvalue Viewpoint* (pp. 25–34). SRDS.

Wu, F., Huberman, B., Adamic, L., & Tyler, J. (2004). Information Flow in Social Groups. *Physica A . Statistical and Theoretical Physics, 337*(1-2), 327–335.

Yan, X., & Han, J. (2002). gSpan: Graph-Based Substructure Pattern Mining. In *ICDM Conference*.

Yan, X., & Han, J. (2003). CloseGraph: Mining Closed Frequent Graph Patterns. In *ACM KDD Conference*.

Yan, X., Cheng, H., Han, J., & Yu, P. S. (2008). Mining Significant Graph Patterns by Scalable Leap Search. In *SIGMOD Conference*.

Yan, X., Yu, P. S., & Han, J. (2005). Substructure Similarity Search in Graph Databases. In *SIGMOD Conference*, (pp. 766-777).

Yan, X., Yu, P. S., & Han, J. (2004). Graph Indexing: A Frequent Structure-based Approach. In *SIGMOD Conference*.

Zaki, M. J., & Aggarwal, C. C. (2003). XRules: An Effective Structural Classifier for XML Data, *KDD Conference*.

Zhou, C., Towsley, D., & Gong, W. (2004). *Email Virus Propagation Modeling and Analysis*. UMass Technical Report TR-CSE-03-04.

ENDNOTE

1 For full details, see (Kleinrock 1975)

Chapter 4
Region Coverage and Protection with Sensors: A Survey

Ted Brown
City University of New York, USA

Peter Brass
City University of New York, USA

Matthew P. Johnson
City University of New York, USA

Simon Shamoun
City University of New York, USA

ABSTRACT

Covering an area with sensors has been an active research area in recent years. Coverage problems for sensors include the positioning of the sensors in order to cover much or all of a region, once or many times, and using sensors whose coverage abilities vary. Certain problem extensions arise in security applications and when sensors are deployed in hostile environments: it may not be possible to safely enter the area, in which case sensors may be distributed randomly from a distance; even if the positions can be chosen, there may be some minimal placement error which must be compensated for; it may not be possible to provide complete coverage, in which case we may settle for partial coverage or only barrier coverage and position sensors for improved intrusion detection. Another factor to consider when parties are acting in a coalition is that differing types of sensors may be deployed by the different parties, which must be taken into account when choosing positions. This short survey deals with some recent results that are especially applicable to such settings.

DOI: 10.4018/978-1-61520-855-5.ch004

INTRODUCTION

A major objective in the deployment of a network of sensors is effective coverage of a chosen region. This takes the form of monitoring for signs of danger, the presence of intruders, and other events. The coverage provided by a sensor network depends on the capabilities of the sensors, the environment in which they are deployed, their locations within that environment, and interactions between sensors and multiple parties sharing coverage of the region. This chapter presents results from select papers on the coverage provided by various deployment strategies, with emphasis on the kinds of situations that may apply to security and military contexts.

Sensor networks have many applications in military, environmental, and other contexts, which were not feasible before the advent of inexpensive, self-contained sensor nodes with enough capacity to store and relay data over wireless channels. For example, besides the obvious uses within military surveillance applications, sensor networks have been used in environmental studies to provide early alerts of forest fires and to monitor habitat, volcanic, and glacial activity.

There has been an explosion of research on the topic of ad hoc sensor networks in the last decade. For example, a popular article by Meguerdichian et al. (2001a) on best and worst case coverage had over 800 citations on Google Scholar as of August, 2009. A prominent survey paper published by Akylidiz et al. (2002) had more than 3,000 citations. Interest continues to mushroom. In 2009 alone, more than 30 conferences included sensor networks as a topic. Within the very large body of literature on ad hoc sensor networks, much effort has been expended examining placement issues. Since it is not feasible to create a comprehensive survey on the topic of sensor coverage, our goal is to provide broad outlines of some of the research that is especially of interest for ground-based coverage in security and military applications.

The sensors considered here are small, and perhaps capable of performing multiple tasks. As sensors, they are able to sense some aspect of the environment local to their physical location, and in addition are able to receive information and programming instructions, transmit information, execute a program, store a small amount of past data, and perform calculations. They can reconfigure their connectivity on the fly as needed (i.e., create and recreate an ad hoc network). They can be programmed to measure and report data upon request, periodically, or upon being triggered. In all cases, the information is forwarded to a base station for analysis.

This chapter focuses mostly examining those how sensors may be placed in large, continuous areas, with a special interest in coalition environments that arise in security and military environments. By a coalition environment, we mean one in which multiple partners are cooperatively interested in the coverage of an area. We assume, for the most part, that all the sensors of a coalition partner are of the same type with identical characteristics, e.g., the same fixed radius of coverage from their placement positions, although different partners in the coalition may have sensors with different characteristics. Two general categories of sensor deployment are random placement, e.g., by dropping the sensors from an airplane, as may be done in a hostile environment; and deliberate placement in specified locations, which can be accomplished when the region is small and accessible. A special case of deterministic placement is when sensors are placed on or approximately near grid points spaced equally apart in a regular pattern, such as a square or hexagonal lattice. We restrict ourselves here to coverage with statically placed sensors. That is, we do not consider the literature on mobile sensors (MANETs) and we only consider communications issues in passing. We leave other aspects of ad hoc sensor networks, including communications and network configuration, and the scheduling of sensor sleep cycles, to other chapters and other books. Two recent surveys

[Younis & Akkaya (2008), Ghosh & Das] delve into some areas not covered in this chapter and so are worthy of examining. Younis and Akkaya (2008) have a valuable discussion of placement of sensors for improved longevity of the network. They survey papers that look at static placement and papers that look at moving sensors to improve the lifetime of the network. Ghosh and Das (2006) in a longer survey paper spend much of the survey on exposure paths (see below for a definition) again for static and mobile situations.

We shall discuss grid and random placements and within these two we shall discuss placement of sensors to create blanket (complete) coverage of a region, coverage that is not complete so gaps between sensing regions exist, edge effects in the case of small regions, and what is called barrier coverage, which is covering only a narrow area at the edge of the region to be protected against intrusions. In many military situations, an exact placement cannot be assumed. For example, although it is well known that (under certain conditions) a hexagonal grid pattern requires the fewest sensors, in some situations precise placement of nodes is not feasible. The consequences of inexact placement are considered. For both placement types (random and in a grid pattern), we shall examine research that discusses protecting a given region (in the plane) completely or partially by binary sensors with constant coverage radius r. If a region can only be partially covered, then the selection of covered and uncovered areas can be very important. This too is discussed.

We begin by discussing different notions of coverage and by introducing some terminology. We then address blanket (complete) coverage by random and grid placement. We continue with weaker guarantees such as partial coverage and intrusion detection, and the stronger guarantee of k-coverage. In many coalition situations, precise placement of nodes is not feasible. The consequences of inexact placement are considered next. Finally, we take up boundary conditions; barrier coverage, which is the coverage of a narrow area at the edge of the region to protect against intrusions; and other factors that influence coverage in non-idealized models.

COVERAGE OF A REGION

The region within a sensor node's sensing range is commonly referred to as its *sensing area*. The region to be covered is the *region of interest* (when it is clear from the context, we will simply say the *region*). The coverage problems we have in mind here regard positioning and configuring nodes so that their sensing areas together cover the region of interest, according to some notion of coverage. Four important problem settings arise from two sets of choices: placement of sensors can be deterministic or random, and the region of interest may be a large region or shaped such as a narrow belt-like region.

The simplest, most commonly used model for the coverage provided by an individual sensor is the binary, or Boolean, sensing model. In this model, a sensor detects all events occurring within its sensing range and only those events. Sensors are most commonly assumed to be isotropic, i.e., directionless, in which case a binary sensor's coverage area is simply a disc centered on the sensor node. Most of the references below use this model. More complicated isotropic models reduce the likelihood of coverage and increase the likelihood of noise as a function of distance. In such models, the coverage guarantee at a given point is probabilistic rather than absolute, motivated by the signal attenuation that typically occurs with distance between sensor and event (Clouqueur et al., 2002; Megerian et al., 2002, 2005; Liu & Towsley, 2004). In a detection context, for example, the coverage probability might decrease with distance d according to a function such as $e^{-\alpha d}$, α being a parameter (Dhillon et al., 2002; Dhillon & Chakrabarty, 2003; Zou & Chakrabarty, 2003), or simply with distance to a power greater than one. A more complex isotropic probability model might assume detection with probability one up to a certain fixed distance then have a monotonically decreasing function up a further point after which the probability of coverage is zero (Wu et al., 2006, 2007; Zou & Chakrabarty, 2003a, 2004a, 2004b). Noise factors and other extensions have also been considered by these authors. For most of the remainder of this chapter, we restrict ourselves to binary coverage, although non-disc coverage ranges, motivated by non-isotropic sensors like cameras or by topography, including obstructions or non-flat terrain studied, among others, by Dhillon & Chakrabarty (2002, 2003); Huang & Tseng (2005); Wang et al. (2006); Zou & Chakrabarty (2003, 2004a, 2004b) will be touched upon.

When one considers coverage of an area by sensors, *covering* can have multiple meanings and various terms have been used for the same concept. Some of the literature uses the terms *full coverage*, *area coverage*, or *blanket coverage*, meaning that every point within the region has at least one active sensor capable of sensing activity at that point. We shall use the term *blanket covering*. More precisely, let C_i indicate the sensing area of sensor *i*. Then the region of interest *A* is *blanket-covered* if *A* is contained within the union the C_i coverage areas. There may be a need to have multiple sensors track points, for instance, for triangulation. Hence there may be a need for some points in *A* to be more than *blanket-covered*: *k-covering* means that every point in the region is within the range of at least by *k* sensors, i.e., covered by at least *k* sensors. *1-covered* is equivalent to *blanket-covered* under this definition.

Unless the region of interest is small, blanket coverage, much less *k*-coverage, may often be too demanding. A less stringent requirement is to have some portion of the region guaranteed covered. In a security or intrusion contexts, knowing only that a certain fraction of a region is covered is insufficient for protection; *which* portions are covered is important. Where the gaps in the protection of the region are is an important consideration. For example, the gaps' location could relate to the probability of finding intruders taking a particular path when trying to breach the region. Intrusion detection is important when a narrow strip of sensors is used as a barrier between the enemy forces and the coalition forces (this is called *barrier coverage* in the literature), or it might be a narrow strip of sensors used to belt an area to protect it from intrusions. A barrier may have gaps because of the random placement of sensors or the size of the barrier zone and the number of sensors available, and so intrusion detection is an important consideration here too.

RANDOM PLACEMENT OF SENSORS TO COVER A REGION

If the positions of sensors in a region *A* are generated by some random process, then the probability of a point *x* in *A* being covered by a sensor with sensing radius *r* is the probability that at least one sensor falls within the disc of radius *r* around *x*. This local detection probability can be integrated over *A* to get the probability that a random event in *A* will be detected. If the sensors are placed independently and uniformly at random and *A* is a sufficiently large region, then their locations can be modeled by a stationary two-dimensional Poisson point process with density λ, the number of sensors per unit area. The probability that any size A_i part of the region contains *n* sensors follows a Poisson distribution with parameter λA_i. The detection probability of any event that is at least distance *r* from the border is equivalent to the probability that at least one sensor falls with the disc of radius *r* around that event, which is $1 - e^{-\lambda\pi r^2}$ (Lui & Towsley, 2004). Although the detection probability rapidly approaches 1 for increasing density λ and sensing radius *r*, under random distribution there are never enough sensors in the sense that detection can be guaranteed. See Koskinen (2004) for a discussion on the probability of full coverage by a given set of sensors.

Suppose there are k parties in the coalition who jointly try to observe a region A (a large enough region so we can disregard boundary effects). If all sensing radii are the same, we get the classical single-party situation just described. This situation has been studied in numerous papers; see Brass (2007) for a summary. The more interesting case is when the parties use sensors with different characteristics such as sensing radii. There are two cases to consider. First, suppose that the parties place its sensors throughout the region. If each party i distributes their sensors according to an independent Poisson process of density λ_i, then the probability of a point x (sufficiently far away from the boundary) to be missed by the ith of these processes is $e^{-\lambda_i \pi r_i^2}$. The probability of a point to be missed by all k independently placed types of sensors is the product of these k probabilities, which is $e^{-\pi \sum \lambda_i r_i^2}$. If $\lambda r^2 = \sum \lambda_i r_i^2$, λ is the overall density, and r is the radius of a single sensor type, then the probability of the total region being covered by the sensors is exactly the same, whether they be of one type or many types.

An alternative approach to calculating the coverage by sensors with different sensing radii comes from Lazos & Poovendran (2006). Their formula more accurately accounts for the size and shape of the field and the sensing areas and the effects of sensors intersecting the border of the field, whether they are placed inside or outside the field. The formula is restricted, however, to convex continuous fields and sensing areas. If the area of the field is A_0 and the perimeter is P_0, the coverage by a single sensor with area A_1 and perimeter P_1 can be characterized by the formula

$$\frac{2\pi A_1}{2\pi\left(A_{0+}A_1\right) + P_0 P_1}$$

This formula generalizes as follows to characterize the coverage by N heterogeneous binary sensors whose sensing areas intersect the region:

$$1 - \prod_{i=1}^{N}\left(\frac{2\pi A_0 + P_0 P_i}{2\pi\left(A_{0+}A_i\right) + P_0 P_i}\right)$$

RANDOM PARTIAL COVERAGE

If the entire region of interest cannot be blanket covered by the coalition for whatever reason, then necessarily there will be some areas in the region uncovered. One deployment objective, when full coverage is not possible, is to maintain a fairly uniform distribution of coverage throughout the region (Liu & Liang, 2005). A more important type of objective is to minimize the probability of non-detection, or equivalently, to maximize the probability of detection, of an object or person moving through the region.

One approach to measuring the probability of non-detection is discussed by Chakrabarty and collaborators in a series of papers (Dhillon et al., 2002; Zou & Chakrabarty, 2003a, 2003b, 2004a, 2004b). They assign a probability value to the grid points within the sensor field that represent the minimum detection probability of that point. The fineness of the grid is adjusted to tradeoff the quality of approximation for the run time of its calculation. Preferential areas of coverage are assigned higher minimum confidence levels. To determine if the coverage requirements are met, they model the sensor field as an $(m \times n)$ grid. (They recommend representing a non-rectangular field using several rectangular grids.) Since the detection and miss probabilities are complementary of one another, it is sufficient to set a maximum miss probability for each point. The miss probability of any point is the product of the miss probabilities of all sensors covering that point. They guarantee minimum coverage of non-grid points by making sure that all distances in the solution are two times the distance between sensors and grid points. An $(m \times n)$ coverage threshold matrix is maintained for the entire grid, as well as an $(m \times n)$ matrix of the miss probabilities according to the sensor locations. The measure of performance is the fraction of grid points covered according to their minimum thresholds. Another measure, the DT-Score, uses this model as well (Wu et al., 2006, 2007), except that detection probability of a point is the maximum detection probability of any sensor covering that point. This is effectively the binary sensor model, since to count as being covered, a point must be within a certain minimum distance of at least one sensor. The only distinction is that preferential areas of coverage will require a higher sensor density.

If a region is only partially covered, then a natural question to ask regarding intrusion is how easily an adversary can move through the region without being detected. Two notions have been proposed: *maximal breach weight* and *maximal support weight,* corresponding to worst-case coverage and best-case coverage, respectively (Meguerdichian et al. 2001a; Megerian et al., 2005). A *maximal breach path,* between a given starting point and end point, is a path that keeps as far away from the sensors as possible, i.e., maximizing the minimum distance between it and any sensor node. Such a path would arguably be best for a knowledgeable intruder. It turns out that a maximal breach path can be computed efficiently using Voronoi diagrams. Given a finite set of sites (here sensor nodes), a Voronoi diagram partitions a region into convex pologons corresponding to the nodes; each polygon consists of all those points closer its corresponding node than to any other. Since the points on each edge of a Voronoi diagram are equidistant between two sites, a maximal breach path through the interior of the Voronoi diagram will run along the Voronoi edges. If each edge is labeled with the minimum distance between it and its two neighboring sensors nodes, then the breach weight of a path will be the minimum among these values assigned to its edges. Once the Voronoi diagram is computed, a maximal breach path can be efficiently computed from it using standard graph algorithms.

Conversely, a maximal support path between a given starting point and end point is one that contains the smallest maximum distance from the nearest sensor, i.e., minimizing the maximum distance between it and any sensor node. The distance is called the *support weight.* We can compute this path and weight analogously to what was done above, using Delunay triangulation. The Delaunay triangulation of a set of sites is dual to the Voronoi diagram: edges are drawn between two sensors if their Voronoi faces share an edge. It turns out that a maximal support path will follow Delaunay edges, traveling from sensor node to sensor node.

An alternative measure of detection ability is *path exposure* (Meguerdichian et al., 2001b, Megerian et al., 2002), which is the integral of either the signal strength of a single sensor or the combined signal strength of all sensors along a path through the field. The exposure of a path is approximated by imposing a grid over the sensor field. The path with the minimum exposure is a measure of the ability to detect a moving abject. A different approach to measuring exposure is taken by Clouqueur et al. (2002a, 2002b). They use the more general model of sensor coverage including random noise to the signal strength to evaluate point coverage using both value fusion and decision fusion. Detection occurs when the fused values are above a certain threshold, which occurs with a certain probability. Exposure is the net probability of detection along a path. They superimpose a Manhattan grid on the sensor field, set the weight of each edge to the *log* of the miss probability along the edge, connect a vertex on each side of the field to all vertices on the respective side, and use Dijkstra's algorithm to find the path of minimal exposure between both sides.

More information on different approaches to calculating paths and their coverage measures is contained in the survey article by Ghosh, and Das. (2006).

GRID PLACEMENT COVERAGE

One can always achieve blanket coverage by placing sensors purposefully within the region of interest. It is known that if one ignores boundary conditions, the most efficient arrangement of sensors for blanket coverage is a hexagonal lattice (also sometimes called a triangular lattice) (Kershner, 1939). With this arrangement, placing sensors as far apart as possible while maintaining full coverage puts the minimum pair-wise distance between sensors as $\sqrt{3}r$, which gives rise to an approximately 20.9% overlap overall. Consequently the fewest sensors required to blanket cover an area A is, as is easily calculated to be approximately $0.383A/r^2$ [Bar-Noy et al. (2007)]. In practice, it might be easier to place sensors in a rectangular grid pattern. Blanket coverage with this arrangement has approximately 57% of the area doubly-covered and so is much less efficient.

Alternatively, if we can choose the position for each sensor type by some means, then we can do with several types of sensors at least as well as with a single type. We can divide the region of interest into k pieces, proportional to the area provided by each sensor type, and arrange in each piece the sensors in the properly scaled hexagon lattice arrangement. If we disregard the boundary effects, both of the original boundary and of the new internal boundaries, then this is on each piece as good as the combined total sensor area would be with a single type of sensor. But at least for some combinations of r_i and λ_i it is known that we can achieve a coverage better than the coverage that can be achieved with a single type of discs (Fejes Tóth, 1972). For example, we can pack big discs, and then use very small discs in the remaining gaps; that would give a better density than packing big discs in one part of A, and small discs in another part of A.

Another interesting approach is to consider a fine-grain grid and only place sensors at some of these grid points taking into account other factors such as line-of-sight. Chakrabarty et al. have a series of paper addressing "coverage optimization under the constraints of imprecise sensor detection and terrain properties" using this approach. The papers co-authored with Dhillon & Iyengar (Dhillon et al., 2002; Dhillon & Chakrabarty, 2003) primarily address variations in the terrain, including preferential areas of coverage in the sensor region and partial occlusion by obstacles. The methodology they use assigns to each grid i two matrices of values. One is the probability that a sensor at this point would detect a target at grid point j and a second, the probability that a target at this grid point i would be seen by a sensor at j. This very general formulation can take into account obstacles and other extensions. The papers co-authored with Zou (Zou & Chakrabarty, 2003, 2004) additionally address uncertainty of the final sensor locations subsequent to deployment. The survey article by Ghosh and Das (2006) has more details on these ideas.

GRID PLACEMENT *K*-COVERAGE

A natural extension of the blanket coverage goal is k-coverage, under which every point in the region is covered by at least k sensors. K-coverage may be required for robustness reasons or in order to perform triangulation. Another desirable goal may include covering the region with k different type of sensors for multimodal data input. A possible consideration is ensuring that all sensors are positioned as far apart as possible in the sense of maximizing the minimum distance between sensors. One motivation is reduce the likelihood of exposure to unfriendly events. In this section, we discuss sensor placement for k-coverage, both exactly and inexactly and the stronger goal of k-coverage with different sensor types.

Given the known optimality of placing sensors in a hexagon lattice pattern for 1-coverage on a plane (i.e., without boundary conditions), grid (lattice) configurations are a natural starting point for k-coverage. A naive k-coverage configuration is simply to superimpose k copies of the hexagon grid one on top of another, by which is meant having k sensors (almost) on top of one another. There is an infinite number of ways this superposition can be done; think of the k grids as being on k separate transparent sheets that are able to slide over one another. If each grid pattern corresponds to a different sensor type, then this configuration naturally provides a reasonable feature of distinct-type coverage.

Another approach is taken in Johnson et al. (2009). Consider an optimal hexagonal grid configuration for blanket coverage with a sensor that has a sensing radius of r. All of the region will be at least 1-covered and some portion will be 2-covered (approximately 20.9% with the hexagonal grid). If the radii of all sensors are sufficiently increased, then some, and eventually all, of the region will become 3-covered. As r is increased, keeping the grid size fixed (or equivalently, if the grid size is shrunk, keeping r fixed), the multiplicity k of coverage will continue to grow. For many values of k (such as 3) the multiplicity of coverage will equal the factor by which the number of sensors has increased. Interestingly, for certain values of r, the coverage multiplicity can be greater than this. For example, when r is increased enough to increase the number of sensors 11-fold, it turns out that 12-coverage is obtained. This occurs at other higher values of k. See Johnson et al. (2009) for details.

As an aside, Pach & Tóth (2007) study the problem of decomposing a k-cover into a maximum number of disjoint 1-covers. They note that when the region of interest is a disc, the only positive result that has been claimed is that any 33-cover may be decomposed into two disjoint 1-covers.

Focusing on probabilistic sensors, Zhang & Hou (2006) give bounds on density for guaranteeing full k-coverage with high probability with various arrangements. This non-local guarantee requires increasing sensor density by a logarithmic factor. Wang & Tseng (2008) provide k-coverage arrangements for binary and probabilistic sensors. They note the presence of overlap among discs but take advantage of it only in the setting of small coverage radius r. For large r they simply duplicate the 1-covering k times.

Zhou et al. (2004) give approximation algorithms extended by Gupta et al. (2006) for selecting a connected k-covering. Kumar (2004) considered sensors on a regular grid which are randomly turned on. Distributed algorithms for assembling (moving) sensors into a hexagon lattice are in Bartolini et al. (2008). The formula for blanket coverage by Lazos & Poovendran (2006) is generalized for k-coverage in their paper.

One approach to simplify coverage problems is to replace the goal of full, continuous coverage with the relaxed goal of covering a discrete set of points. An example of this approach is Chakrabarty et al. (2002) where they provide k-coverage only to the grid points of a fixed-precision square grid lattice using disc sensors of different sizes. An integer programming approach allows them to solve small instances, and for larger instances they suggest a divide-and-conquer heuristic. Abrams et al. (2004) solved a related but relaxed problem in which a set of sensors is partitioned into subsets with the objective of maximizing the sum of the (partial) covers of the sets.

A related problem is recognizing the amount of coverage provided by an existing arrangement. Huang & Tseng (2003, 2005) introduce a distributed method for determining if the sensor field is k-covered. They prove that if all sensors have k-coverage on their perimeters, then the entire region is k-covered. They use the following technique to determine if the perimeter of a sensor is k-covered, assuming all sensors have equal radius. First, label the points of intersection with all other sensors as either the left or right point when inspecting in a counterclockwise motion along the perimeter. Next, sort all the points according to their angle along the x-axis. Finally, starting with the level of coverage at angle 0°, traverse the list, adding 1 when a left point is en-

Figure 1. Central disk can be moved (B and C) to another position and still fully cover the area

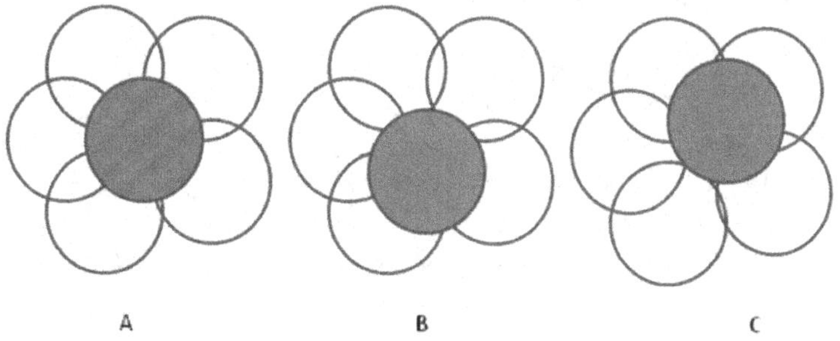

countered and subtracting 1 when a right point is encountered. The lowest number is the *k*-coverage of the perimeter. For differently sized sensor radii, they use a similar approach except that they deal with different cases depending on the radii and distances between sensors.

COVERAGE OF AN AREA WITH APPROIMATE POSITIONING OF SENSORS

In many situations, perhaps especially military ones, sensors cannot be positioned *exactly* where we would like. The effect on coverage of a local area with imprecisely positioned sensors is the topic of this section. The discussion below applies to both random and grid-based placements and refers to Figure 1. Consider the coverage of a particular sensor and of those surrounding it, as shown in Figure 1(a). The four sensors overlap each other *and* the center sensor, providing 2- and 3-coverage in the overlapped areas. If the positions of the sensors around the centrally placed sensor are fixed, then the centrally located sensor can be in a slightly different position (in all directions in the figure) from where it is now and still have full area coverage as before. This is true as long as the areas that were 2- and 3-covered remain at least 1-covered. Figures 1(a), 1(b) and 1(c) illustrate this. Clearly, it is necessary and sufficient that these areas are maintained at least 1-covered. Specifying the area in which the central sensor can be placed and still maintain full coverage is more complicated (see Brown et al (2007)).

In practice, it may be more useful to use a circle around the given central sensor to characterize its imprecise positioning, even though the resulting area is smaller. Let $\{x_i: i=1,2,,\ldots\}$ be the set of the closest points to the center of the central sensor for each of the 3-covered areas. Let r_a be the radius of a circle centered on the intended location in which the sensor could have been placed and still maintaining full coverage of the area. Clearly $r_a = \min\{\mathbf{dist}(r, \boldsymbol{x}_i), i=1,2,\ldots\}$, where **dist** is the Euclidean distance between points. It is possible to then move or position the central sensor in the direction of any of these $\boldsymbol{x}_i$ or any vector combination of them up to a distance r_a, while maintaining full coverage. The distance r_a is sufficient but not necessary to maintain full coverage of the area.

If *all* of the sensors in the figure are allowed to be repositioned from their current position, then the central sensor cannot be allowed move as much as above. However, there is a lot of flexibility with respect to how much allowance to provide each of these sensors while guaranteeing coverage. The mechanism is to choose any point within each at least 3-covers. Each sensor can move in away

Figure 2. Placement of sensors in a hexagon lattice showing boundaries with some sensors overhanging the edge of region and consequently not being useful

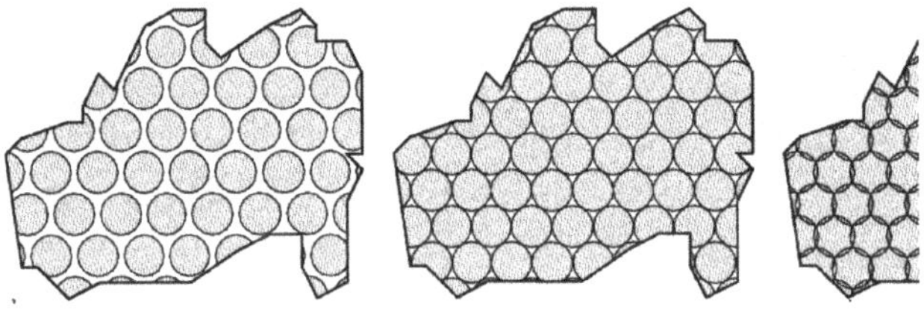

from (or towards) this point with the restriction of not moving beyond the point. There is further information on this technique in Brown et al. (2007) and in Bar-Noy et al. (2007).

BOUNDED REGIONS

So far we have not considered the boundary of the region, assuming the region of interest is large enough its effects are insignificant, i.e., that the number of additional sensors required to cover the boundary will be small compared to the number covering the interior. If not, boundary effects may produce specific problems. If the region's shape is sufficiently irregular, then this effect could be significant; see for example Figures 2 and 3. In another context, if multiple parties are responsible for covering different portions of the area, as in the parties to a military coalition, then there may be challenges in connecting the two covers to create a blanket cover throughout, particularly if the parties use different types of sensors.

One possible way to handle boundary issues is simply by assuming the region to be covered is larger than it really is. If a sensor is positioned at a point closer than r from the boundary, only some of its sensing range will be useful. Therefore, if the sensors are being placed randomly, it might make sense to distribute the sensors within a larger region B that contains both the region of interest A and all points within distance r to it. The same can be done if the sensors are distributed on a grid: assume the grid size is sufficiently larger so as to include sensors whose discs lie partially inside the actual region.

One recent example explicitly dealing with boundary is from Brass (2007) relying upon Fejes Tóth (1972): to cover a region A by sensors with sensing radius r, a sufficient number of sensors is

$$\frac{2}{3\sqrt{3}}\frac{area(A)}{r^2}+\frac{2}{\pi\sqrt{3}}\frac{perimeter(A)}{r}+1$$

Figure 3. Placement of sensors in which each coalition member covers its area with sensors of a specific sensing range. This creates additional boundary considerations

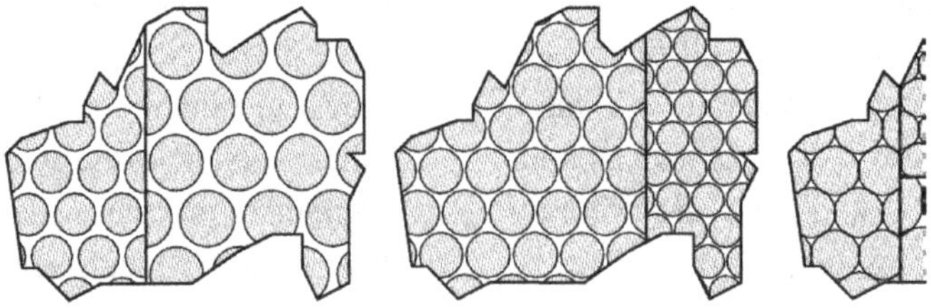

However, the original formula assumes that a point can be covered by a sensor in any position around it. This is clearly not the case when considering points along the border, in which case they can only be covered by sensors inside the region.

The coverage by a number of heterogeneous sensors intersecting a bounded region was derived by Lazos & Poovendran (2006) and provided in the section on random coverage. There is also work on random placement in a bounded circular region. The coverage of a large circular region of radius R by n sensors with radius r placed within the disc is the inverse probability that an arbitrary point is not covered by any sensor. When $n=1$, this probability depends on the area of intersection between the sensor and the disc and is conditioned on the distance ρ of the point from the center of the disc to be covered. If $\rho \leq \left|R - r\right|$, then the area of intersection is $\min\left\{\pi r^2, \pi r R^2\right\}$. If $\rho > \left|R - r\right|$, then the area of intersection is (Koskinen, 2004):

$$A\left(\rho, R, r\right) = \frac{-\sqrt{(-\rho + r + R)(\rho + r - R)(\rho - r + R)(\rho + r + R)}}{2} + r^2 \arccos\left(\frac{(\rho^2 + r^2 - R^2)}{2\rho r}\right) + R^2 \arccos\left(\frac{(\rho^2 - r^2 + R^2)}{2\rho R}\right)$$

OTHER FACTORS IN COVERAGE FOR A REGION

There are many factors ignored by an idealized Boolean sensing model. Obstacles are considered, for instance, by Bottino & Laurentini (2006) in work making use of the art gallery problem. The art gallery problem is formulated as minimizing the number of guards needed to see all areas within an art gallery with multiple rooms. Walls between the rooms interfere with the view guards have, so they need to be positioned so they can see a maximum distance. See O'Rourke (1987) for many applications. If we allow a "guard" to see only up to a fixed distance even if there is no obstacle in the way, the art gallery problem is transformed into a sensor coverage-like problem. Howard et al. (2002) address sensor deployment in indoor environments with doorways and passageways that allow and obstruct sensing.

Different modalities affect the parameters of a sensor's sensing ability. An ordinary camera, for example, is far from being iso-tropic. There are sensors that measure temperature, infrared, acoustic, optical, seismic, etc; each will differ in their engineering specifications. And in real-world situations, there are additional considerations such as weather, wind direction, view angle, terrain, soil type (for acoustic sensors), etc., that could or should be taken into account in an actual deployment.

To maintain tractability, these compounding factors are most often ignored, although a few papers have made some contributions. See e.g., Kaplan & Cevher (2007). These authors adapt a model that includes a fixed budget. A budget is the constraint: different fixed cameras with differing fields of vision and different costs are spread out randomly in a region. Under some assumptions they are able to formalize a linear programming model to minimize the cost of surveillance of multiple targets.

An altogether different approach has been taken by Wilson, Vecherin and Pettit in a series of papers. Vecherin et al. (2009) consider the factors above based on energy dissipation effects from the ground geometry obstacles, noise, etc., between the sensor and the point to be sensed. In Pettit et al. (2009), they experimentally examine the heuristics-based grid ideas of Dhillon et al. (2003) (see above), looking at how well acoustic sensors and cameras would work if there are obstacles in the way. Furthermore they make the solution more realistic by assuming that an intruder has non-zero dimensions, i.e., any part of him would set off a sensor alarm.

USING SENSORS TO CREATE A BARRIER

Rather than considering covering a region for protection, another possibility is to use sensors to cover a narrow area at the edge or edges of a region to create a protection boundary. A barrier is simply a region in which the length is much larger than the width. Thus, many of the ideas for both grid and random type placement hold. If the barrier region is blanket-covered, the barrier could be thought of as impenetrable, but this is often too ambitious a goal, as the length of barrier region is often long. To create an impenetrable barrier, one row of sensors just touching one to the next would suffice. A shortcoming of this approach is that the width of this narrowest sensor field is a point, where two sensors touch each other. A solution to this is to have the sensors overlap enough to create a wider sensor field. Verma et al. (2008) provide additional examples of one row of sensors used to create a complete perimeter sensor ring.

This is a stringent requirement. Thus, as in the partial coverage literature, the goal in most of the papers on this topic is to detect attempted penetration through the boundary with sufficiently high probability. The terminology is slightly different in these papers since the focus is on penetration. *Full coverage* means that an intruder going through the zone (through the width) would be sensed by at least one sensor; *k*-coverage is used to mean that at least *k* sensors would detect an intruder going across the zone. If partial coverage is all that is possible, one has to also consider the possibility of the intruder choosing a path that tries to avoid detection. The earlier definitions of maximal breach weight and maximal support weight measures are useful in this context, except here the interest is in any starting point on one side of the width and any ending point on the other side, which has been called *strong barrier coverage*.

A lesser requirement, defined by Kumar et al. (2007), is *k-weakly barrier covered*, by which they mean that all orthogonal paths through the zone are *k*-covered. They prove that for *k*-coverage the optimal configuration is to have rows of sensors. Each sensor in a row is configured in a hexagonal lattice.

Liu et al. (2008) have some important asymptotic results when sensors are deployed randomly in the zone. They show that for a rectangular region with width w, asymptotically smaller than the length L, i.e., $w=o(\log L)$, asymptotically no sensor density is enough to guarantee detection. On the other hand, if $w=\Omega(\log L)$, then with probability of one there is no perpendicular crossing path an intruder can follow and avoid detection.

CONCLUSION

In this chapter we have touched upon some of the major issues in sensor coverage that are of importance in a coalition environment. We discussed blanket coverage of a region and partial coverage of a region, with deterministic placement and random placement and barrier coverage. The chapter does not claim to be complete; the number of research articles is at least in the hundreds. In doing so, we have only scratched the surface of what continues to be a very active research area, with much exciting new work on topics such as mobility, optimizing network lifetime, and multimodal sensors, and provided starting points for the reader interested in learning more about these areas.

ACKNOWLEDGMENT

Research was partially sponsored by US Army Research laboratory and the UK Ministry of Defence and was accomplished under Agreement Number W911NF-06-3-0001. The views and conclusions contained in this document are those of the authors and should not be interpreted as representing the official policies, either expressed or implied, of the US Army Research Laboratory, the U.S. Government, the UK Ministry of Defense, or the UK Government. The US and UK Governments are authorized to reproduce and distribute reprints for Government purposes notwithstanding any copyright notation hereon.

REFERENCES

Abrams, Z., Goel, A., & Plotkin, S. A. (2004, April 26-27). Set k-cover algorithms for energy efficient monitoring in wireless sensor networks. In ISPN 04, Berkeley, CA, (pp. 424-432).

Akylidiz, I., Su, W., Sankarasubramariam, Y., & Cayirci, E. (2002, Aug). A survey on sensor networks. *IEEE Communications Magazine*, *40*(8), 102–113. doi:10.1109/MCOM.2002.1024422

Bar-Noy, A., & Brown, T. M, Johnson, P., Sarïoz, D., Verma, D. & Wu, C. W. (2008). More is More: the Benefits of Dense Sensor Deployment. In MASS 08.

Bar-Noy, A., Brown, T., Johnson, M. P., & Liu, O. (2007, November 9-10), Covering With Inexactly Placed Sensors. In *17th Fall Workshop on Computational and Combinatorial Geometry*. Yorktown Heights, NY: IBM T.J. Watson Research Center.

Bar-Noy, A., Brown, T., Johnson, M. P., & Liu, O. (2009, June). Cheap or Flexible Sensor Coverage, Intl Conference on Distributed Computing in Sensor Systems. In *DCOSS*.

Bartolini, N., Calamoneri, T., Fusco, E. G., Massini, A., & Silvestri, S. (2008). A Self-deployment Algorithm for Mobile Sensor Networks . In *DCOSS 2008*. Snap and Spread.

Bottino, A., & Laurentini, A. (2006, December 17 - 20). Experimental results show near-optimality of a Sensor location algorithm. In *Proc. 2006 IEEE International Conference on Robotics and Biomimetics*, Kunming, China.

Brass, P. (2007). Bounds on coverage capabilities for models of networks of mobile sensors. *ACM Transactions on Sensor Networks, 3*(2).

Brown, T., Sarïoz, D., Bar-Noy, A., La Porta, T., & Verma, D. (2007, September). Full Coverage of a Region Allowing Inexact Placement of Sensors. In *Annual Conference of ITA*.

Brown, T., Sarïoz, D., Bar-Noy, A., La Porta, T., Verma, D., Johnson, M. P., & Rowaihy, H. (2007). Geometric considerations for distribution of sensors in ad-hoc sensor networks. *Proc. SPIE.*

Chakrabarty, K., Iyengar, S. S., Qi, H., & Cho, E. (2002). Grid coverage for surveillance and target location in distributed sensor networks. *IEEE Transactions on Computers*, *12*(51), 1448–1453. doi:10.1109/TC.2002.1146711

Chen, A., Lai, T., & Xuan, D. (2008). *Measuring and guaranteeing quality of barrier-coverage in wireless sensor networks*. MobiHoc.

Clouqueur, T., Phipatanasuphorn, V., Ramanathan, P., & Saluja, K. K. (2002a). Sensor deployment strategy for target detection . In Raghavendra, C. S., & Sivalingam, K. M. (Eds.), *WSNA* (pp. 42–48).

Clouqueur, T., Phipatanasuphorn, V., Ramanathan, P., & Saluja, K. K. (2002b). Sensor deployment strategy for detection of targets traversing a region. *MONET*, *8*(4), 453–461.

Dhillon, S. S., & Chakrabarty, K. (2002). Sensor placement for grid coverage under imprecise detections . In *FUSION* (pp. 1581–1587). Iyengar.

Dhillon, S. S., & Chakrabarty, K. (2003). *Sensor placement for effective coverage and surveillance in distributed sensor networks* (pp. 1609–1614). WCNC.

Fejes Tóth, L. (1972). *Lagerungen in der Ebene, auf der Kugel, und im Raum (2. Auflage)*. Berlin: Springer Verlag.

Ghosh, A., & Das, S. (2006). Coverage and Connectivity Issues in Wireless Sensor Networks. In Shorey, Ananda, Chan, & Ooi (Eds.), Mobile, Wireless and Sensor Networks: Technology, Applications and Future Directions. Chichester, UK: John Wiley & Sons

Gupta, H., Zhou, Z., Das, S. R., Gu, Q. (2006). Connected sensor cover: self-organization of sensor networks for efficient query execution. *IEEE/ACM Transactions on Networks, 1*(14), 55-67.

Howard, A., Mataric, M. J., & Sukhatme, G. S. (2002a). An incremental self-deployment algorithm for mobile sensor networks. *Autonomous Robots, 13*(2), 113–126. doi:10.1023/A:1019625207705

Howard, A., Mataric, M. J., & Sukhatme, G. S. (2002b). *Mobile sensor network deployment using potential fields: A distributed, scalable solution to the area coverage problem* (pp. 299–308). DARS.

Huang, C.-F., & Tseng, Y.-C. (2003). The coverage problem in a wireless sensor network . In Raghavendra, C. S., Sivalingam, K. M., Govindan, R., & Ramanathan, P. (Eds.), *Wireless Sensor Networks and Applications* (pp. 115–121). New York: ACM.

Huang, C.-F., & Tseng, Y.-C. (2005). The coverage problem in a wireless sensor network. *Mobile Networks and Applications, 10*(4), 519–528. doi:10.1007/s11036-005-1564-y

Kaplan, L., & Cevher, V. (2007). Design considerations for a heterogeneous network of bearing-only sensors using sensor management. In IEEE Aerospace Conf, 2007, (pp. 1-14).

Kershner, R. (1939). The number of circles covering a set. *American Journal of Mathematics, 61*, 665–671. doi:10.2307/2371320

Koskinen, H. (2004). On the Coverage of a Random Sensor Network in a Bounded Domain. In *Proceedings of 16th ITC Specialist Seminar*, (pp. 11-18).

Kumar, S., Lai, T., & Arora, A. (2007). Barrier Coverage with wireless sensors. *Wireless Networks, 13*, 817–834. doi:10.1007/s11276-006-9856-0

Lazos, L., & Poovendran, R. (2006). Stochastic coverage in heterogeneous sensor networks. *ACM Transactions on Sensor Networks, 2*(3), 325–358. doi:10.1145/1167935.1167937

Liu, B., Dousse, O., Wang, J., & Saipulla, A. (2008). *Strong barrier coverage of wireless sensor networks*. MobiHoc.

Liu, B., & Towsley, D. (2004). *A study of the coverage of large-scale sensor networks*. MASS.

Liu, Y., & Liang, W. (2005). *Approximate coverage in wireless sensor networks* (pp. 68–75). LCN.

Megerian, S., Koushanfar, F., Qu, G., Veltri, G., & Potkonjak, M. (2002). Exposure in wireless sensor networks: Theory and practical solutions. *Wireless Networks, 8*(5), 443–454. doi:10.1023/A:1016586011473

Meguerdichian, S., Koushanfar, F., Potkonjak, M., & Srivastava, M. B. (2001a). *Coverage problems in wireless ad hoc sensor networks* (pp. 1380–1387). INFOCOM.

Meguerdichian, S., Koushanfar, F., Qu, G., & Potkonjak, M. (2001b). *Exposure in wireless ad hoc sensor networks* (pp. 139–150). MOBICOM.

Meguerian, S., Koushanfar, F., Potkonjak, M., & Srivastava, M. B. (2005). Worst and best-case coverage in sensor networks. *IEEE Transactions on Mobile Computing, 4*(1), 84–92. doi:10.1109/TMC.2005.15

Pach, J., & Agarwal, P. K. (1995). *Combinatorial Geometry* (3rd ed.). New York: Wiley-Interscience, New York.

Pach, J., & Tóth, G. (2007). Decomposition of multiple coverings into many parts. *SoCG, 2007*, 133–137.

Pettit, C., Vecherin, S., & Wilson, D. K. (2009). *On the influence of problem definition on sensor placement optimization. Defense Transformation and Net-Centric Systems*. SPIE.

Thul, A. (1910). On the densest packing of congruent circles in the plane. (in Norwegian), *Skr. Vidensk-Selsk . Christiania, 1*, 3–9.

Vecherin, S., Wilson, D. K., & Pettit, C. (2009). *Optimal sensor placement with terrain-based constraints and signal propagation effects. Unattended ground, sea, and air sensor technologies and applications*. SPIE.

Verma, D., Brown, T., Bar-Noy, A., Toce, A., Kaplan, L., & Nixon, M. (2008, December). The Base Zone Protection Problem. ASC.

Verma, D., Brown, T., Kaplan, L., & Nixon, M. (2008, September).The Green Zone Protection Problem. *ACITA 2008*.

Verma, D., Wu, C., Brown, T., Bar-Noy, A., Shamoun, S., & Nixon, M. (2008). Location dependent heuristics for sensor coverage planning. *ACITA, 2008*, 346–347.

Wang, G., Cao, G., & La Porta, T. F. (2006). Movement-assisted sensor deployment. *IEEE Transactions on Mobile Computing, 5*(6), 640–652. doi:10.1109/TMC.2006.80

Wang, Y.-C., & Tseng, Y.-C. (2008, September 9-19). Distributed Deployment Schemes for Mobile Wireless Sensor Networks to Ensure Multilevel Coverage. In IEEE Transactions on Parallel and Distributed. Systems.

Wu, C., & Verma, D. (2007). *A sensor placement algorithm for redundant covering based on riesz energy minimization*. ISCAS.

Wu, C.-H., Lee, K.-C., & Chung, Y.-C. (2006). A delaunay triangulation based method for wireless sensor network deployment. In ICPADS, (1, pp. 253–260).

Wu, C.-H., Lee, K.-C., & Chung, Y.-C. (2007). A delaunay triangulation based method for wireless sensor network deployment. *Computer Communications, 30*(14-15), 2744–2752. doi:10.1016/j.comcom.2007.05.017

Younis, M., & Akkaya, K. (2008, June). Strategies and Techniques for Node Placement in Wireless Sensor Networks: A Survey. *Ad Hoc Networks, 6*(4), 621–655. doi:10.1016/j.adhoc.2007.05.003

Zhang, H., & Hou, J. (2004). *Maintaining sensing coverage and connectivity in large sensor networks*. WTASA.

Zhang, H., & Hou, J. (2005) Maintaining Sensing coverage and connectivity in large sensor networks. In Ad Hoc and Sensor Networks, (pp.89-124).

Zhang, H., & Hou, J. (2006). *Is deterministic deployment worse than random deployment for wireless sensor networks?*INFOCOM.

Zhou, Z., Das, S., & Gupta, H. (2004). Connected {K}-coverage problem in sensor networks. In ICCCN 2004, (pp. 373--378).

Zou, Y., & Chakrabarty, K. (2003a). Sensor deployment and target localization based on virtual forces. In INFOCOM 200.

Zou, Y., & Chakrabarty, K. (2003b). *Uncertainty-aware sensor deployment algorithms for surveillance applications* (pp. 2972–2976). GLOBECOM.

Zou, Y., & Chakrabarty, K. (2004a). Sensor deployment and target localization in distributed sensor networks. *ACM Trans. Embedded Comput. Systems, 3*(1), 61–91. doi:10.1145/972627.972631

Zou, Y., & Chakrabarty, K. (2004b). Uncertainty-aware and coverage-oriented deployment for sensor networks. *Journal of Parallel and Distributed Computing, 64*(7), 788–798. doi:10.1016/j.jpdc.2004.03.019

Section 2

Security and Policy Technologies for Coalitions

Chapter 5
Security across Disparate Management Domains in Coalition MANETs

Mudhakar Srivatsa
IBM T.J. Watson Research Center, USA

Dakshi Agrawal
IBM T.J. Watson Research Center, USA

Andrew D. McDonald
Roke Manor Research Ltd, UK

ABSTRACT

Designing a coalition network for chaotic environments (e.g., responding to a large catastrophe) is challenging because such systems cannot rely on availability of a fixed communication or a security infrastructure. In such situations, a coalition may use Mobile Ad-hoc NETworks (MANETs) to communicate and to extend its operational reach and tempo. In this scenario, bootstrapping security and networking protocols requires that networking protocols cannot assume full existence of operational security protocols and vice-versa. In this chapter, the authors outline a realistic bounded resource adversary model and examine bootstrapping problems in the physical & link layer and the routing layer with the goal of identifying new research challenges and novel solution methodologies. In particular, (i) the authors examine secure link key set up protocols at the physical & link layer that neither use computationally intensive PKI mechanisms nor assume pre-configured shared keys between nodes that belong to different coalition partners, (ii) identify new security issues owing to power saving intra-domain routing protocols that use sophisticated packet matching and forwarding mechanisms; in a coalition setting they also examine inter-domain routing protocols that preserve domain autonomy and yet permits scalable network monitoring and misbehavior detection, (iii) examine identity management issues in MANETs and outline a wireless fingerprinting approach to condone a malicious node from spoofing and forging one or more identities on the network.

DOI: 10.4018/978-1-61520-855-5.ch005

INTRODUCTION

Large corporations are slowly being transformed from monolithic, vertically integrated entities, into globally disaggregated value networks, where each member focuses on its core competencies and relies on partners and suppliers to develop and deliver goods and service. The ability of multiple partners to come together, share sensitive business information and coordinate activities to rapidly respond to business opportunities, is fast becoming a key driver for success.

The defense sector too has similar dynamic information sharing needs. The decentralized, dynamic and distributed threat of global terrorism has created a need for information sharing between intelligence agencies of different countries and between multiple security and law-enforcement agencies within a country. Furthermore, traditional wars between armies of nation-states are being replaced by highly dynamic missions where teams of soldiers, strategists, logisticians, and support staff, drawn from a coalition of military organizations as well as local (military and civilian) authorities, fight against elusive enemies that easily blend into the civilian population (Roberts et. al., 2007). Securely disseminating mission critical tactical intelligence to the pertinent people in a timely manner will be a critical factor in a mission's success.

Mobile ad hoc networks (MANETs) have been developed to support communication in tactical missions wherein, the availability of a fixed communication infrastructure cannot be assumed. Many such situations require resources from a coalition wherein multiple groups and organizations come together, communicate, and collaborate, all within a short period of time; for example, in a disaster recovery operation, the local police force, fire-fighters, military forces, medical crews, and other organizations may all coordinate their activities. Such situations call for a *coalition MANET* – interconnect of several MANETs governed by different administrative domains – to enable end-to-end communication. However, most approaches to date have considered MANETs from an intra-domain perspective with a flat network under a single administrative entity. In order to allow inter-operation among heterogeneous network domains operated by different organizations, inter-domain routing approaches such as IDRM (Inter-Domain Routing Protocol for MANETs) (Chau et. al., 2008) are now being considered.

This chapter explores issues in bootstrapping networking and security protocols in a coalition MANET. The key challenge in bootstrapping a coalition MANET is the establishment a secure and reliable end-to-end packet delivery service starting from a point where nodes belonging to different organizations are placed in a field *without* any pre-configured coordination of either the networking or the security protocols amongst the members of a coalition. In this scenario, bootstrapping security and networking protocols requires that networking protocols cannot assume full existence of operational security protocols and vice-versa.

In this chapter, we examine how to enable secure communication at the lower layers (physical and data link layer) as well as at the network layer leading to secure end-to-end packet delivery service in a coalition MANET. We start with the challenges in establishing physical layer communication including discovery of other nodes and secure configuration of physical & link layer parameters. We illustrate bootstrapping issues by considering two cases: in the first case, nodes use a wireless communication protocol similar to the family of IEEE 802.11 standards that are vulnerable to jamming attacks, and in the second case, nodes use a low probability of interception and detection (LPI/LPD) wireless communication protocol from the military grade wideband waveform that are resilient to jamming attacks, but require pre-configuration (e.g., shared spreading code, hopping pattern, etc.). Using these considerations, we abstract out security implications for the networking layer.

We then examine the networking (routing) layer and specifically examine routing protocols used to set up end-to-end paths for packet delivery. Unlike fixed networks which use simple IP-address prefix based matching and routing, MANETs use more sophisticated routing protocols such as flexible end point based routing, network coding based cooperative routing, in network semantic aggregation, etc. with the goal of minimizing the overall power consumption (and thus, the longevity of the nodes and the network itself). We argue while such novel protocols offer performance benefits, they bring forward new security challenges in key management algorithms, data confidentiality & integrity and denial of service (DoS) mitigation; issues such as key management are exacerbated in coalition settings wherein nodes belong to different administrative domains that do not entirely trust each other. To simplify routing over coalition MANETs, several authors have argued that in most circumstances, the nodes belonging to each organization will organize themselves into one or more routing domains (Chau et. al., 2008). Any communication between two nodes belonging to different organizations will be done through *inter-domain routing gateways* using BGP-like protocol to discover end-to-end path through routing domains. We will investigate security issues such as route authentication, cooperative fraud detection, etc. in inter-domain routing algorithms. As a result of our analysis, we present a catalogue of different security functions that need to be implemented and enabled for building secure coalition MANETs.

The rest of this chapter is organized as follows. Section II outlines a space-time delimited adversary model. Sections III and IV discuss security issues in physical layer and routing layer in coalition MANETs. Section V presents trust models and solutions for inter-domain routing protocols for MANETs followed by a summary in Section VI.

ADVERSARIAL MODELS AND SECURITY METRICS

Space-Time Delimited Adversary

A MANET environment is typically characterized by susceptibility of nodes to physical attacks and capture and by resource limitations (e.g., battery power) on the nodes. Prior work on security mechanisms for MANETs, have enhanced traditional adversarial models (that were originally designed for fixed networks) by taking such aspects of the MANET environment into account. For example, some of these models assume that a certain fraction of nodes can be physically captured and compromised anywhere in the network; others assume that corrupted nodes may be Byzantine, and yet others assume that the ``man-in-the-middle'' (MITM) attacks are possible everywhere. These models assume that arbitrary nodes can get corrupted, corrupted nodes are capable of injecting false and misleading information anywhere into the network, and that the corrupted nodes are able to clone other nodes, and collude with one other.

While each of these adversarial models is applicable in some scenarios, it is unlikely that all modes of adversarial operation would be active at once in a network. Without systematically limiting adversarial power, it will be hard to derive meaningful conclusions for designing highly reliable and trusted coalition MANETs. To that end, we propose to examine *space and time delimited* adversarial models which have not been considered thus far explicitly by the prior work. An adversary which is space delimited can only cause "local" damage, e.g., a corrupted node can only corrupt nodes or information in its immediate vicinity. An adversary which is time delimited can only cause damage "slowly". It is our thesis that such models are more realistic in a MANET environment, where an adversary has to incur costs and risks for performing an attack (e.g., node capture, node substitution, etc.); and the costs and risks are unlikely to be uniform within a theatre of operations. For example, certain portion of the MANET may exist in a difficult terrain or a well guarded terrain. Typically, attacks that involve physically capturing a node may be difficult to mount in heavily protected areas.

Even a compromised node has to work in a resource limited environment with limited range wireless links to communicate with other nodes. This put a limit on the "reach in space" of an adversary. A restriction on resource consumption will limit how fast an adversary can compromise more nodes in different parts of the network. Let us consider a scenario wherein a compromised node jams physical layer communication in a coalition network. In a space-time delimited adversary model one may quantify a jammer in terms of its spatial-temporal reach as follows. For example, a jammer may be capable of jamming 2 square miles for one hour; the same jammer may also be capable of jamming one square mile for four hours. We note that space and time delimited adversarial model does permit multiple points of attack on the network. It is only that even if multiple points of attack are successful, further damage to the security mechanisms will have limits on its propagation both in space and time.

Physical Layer Metrics

In the light of above model, we will now define some new security metrics for MANETs. At the physical layer, we categorize fundamental resources under four dimensions: space, time, energy, and radio spectrum. The goal of an adversary is to deny the usage of this resource along any of these four dimensions or a combination thereof to a MANET. The efficiency of the attack is measured by the ratio of resources the adversary has to spend to deny a certain amount of resources to the MANET. For example, using a 802.11x protocol, an adversary can send an unauthenticated *deauth* frame causing the handset to disconnect from an access point (AP), wait for a time-out period (say, one minute) and seek a new channel to reconnect to the AP. Let us suppose that sending a *deauth* frame (by forging the AP's MAC address) takes one unit of time, one unit of energy, and one unit of radio spectrum, and one unit of space. The result is wastage of 1000 units of time, 16 units of energy, one unit of radio spectrum, and one unit of space.

Towards this goal, (Xu et. al., 2005) define two metrics to measure the effectiveness of a jammer at the data link layer: (a) Packet Send Ratio (PSR) which is the ratio of packets that are successfully sent out by a legitimate traffic source compare to the number of packets it intends to send out at the data link layer. This metric is geared towards MAC layers that employ some form of carrier-sensing multiple access control before transmission may be performed. Such protocols require that the channel be sensed idle before a node can transmit a packet. A radio interference attack in this case may cause the channel to be sensed as busy causing the legitimate node to wait before it can transmit packets. (b) Packet Delivery Ratio (PDR) which is the ratio of the packets that are successfully delivered to a destination compared to the number of packets that have been sent out by the sender. This captures the scenario when attacks target corruption of packets in transit.

Another interesting set of metrics has been proposed by (Law et. al., 2005) where they measure effectiveness of an attack by consider *attrition rate* R_a, the fraction of additional energy a node has to spend in presence of a jamming attack. In addition, they consider the *effort ratio* R_e, defined as the ratio of the attacker's per node energy expenditure to mount an attack to the network node's energy consumption when not under attack. These two numbers are then used to calculate the *lifetime advantage* R_l of a jammer node over a sensor node, that is, how long a jammer node can live compared to a sensor node. (Law et. al., 2005) use these metrics to compute lifetime advantage for various classes of MAC layer protocols (including, S-MAC, A-MAC and L-MAC protocols).

Routing Layer Metrics

The effectiveness of routing layer attacks on MANET routing protocols depends on the number of malicious nodes in the network and the rate of benign failures in the network (e.g., node crash failure, link failure due to interference, mobility, etc.). Consequently, routing layer metrics are defined as a function of the fraction of malicious nodes and the rate of benign failures in the network. Routing layer metrics are defined as *multipliers* on performance metrics. For instance, *bit multiplier* denotes the ratio of the number of bits required to send (reliably) a unit size message in the presence of an adversary to that in the absence of the adversary. Similar metrics include *latency multiplier*, *jitter multiplier*, *control data multiplier*, *storage cost multiplier*, *computation cost multiplier*. An adversary may introduce routing black holes, grey holes, loops, sub-optimal routes, etc. and adversely affect the latency and jitter of end-to-end communication. Control data refers to the average amount of computation and communication cost expended by a node on handling control traffic; an adversary may trigger heavy control traffic by inducing heavy network flux, exploiting multicast tree construction and maintenance protocols, etc. Storage cost is measured by the average case size (and the worst case size) of the routing state maintained by each node in the network; the storage cost can be significantly affected by poisoning routing caches in source-based routing protocols. Computation cost is measured by the number of CPU cycles expended by a node on forwarding a packet in the network.

While some of the above metrics equally apply to fixed networks, there are additional security considerations for MANETs. Routing layer in fixed (wired) networks typically rely on simple address prefix based packet header matching and forwarding. This allows efficient hardware based implementation of routers and facilitates the network to reliably sustain large data transfer rates. However, in MANETs, energy and battery life considerations drive routing protocols to implement more sophisticated matching and forwarding algorithms such as content based routing protocols, network coding based cooperative routing protocols, in network semantic aggregation based routing protocols (discussed in section IV). Essentially all these protocols require additional functionality from the routing nodes (as against simple address based prefix matching and forwarding in IP networks), thereby creating new failure modes that did not exist in fixed network routing protocols. On one hand, such novel routing protocols present an opportunity to improve the performance of the network, whilst carrying additional security risks and new challenges for designing coalition MANETs.

PHYSICAL LAYER COMMUNICATION IN COALITION MANETS

The broadcast nature of wireless radio transmissions makes them particularly vulnerable to *jamming* based Denial-of-Service (DoS) attacks. The aim of these attacks is to prevent devices from exchanging any useful information by interfering with their communication. Possible communication jamming attacks include signal annihilation and modification (bit-flipping, overshadowing, etc.). Secure communication at the physical layer must satisfy the following properties:

- *Physical Layer Integrity*: An adversary J must not be able to jam all communication between two nodes T and R. We note that detecting the compromise of the integrity of transmitted data is handled by higher layers in the networking stack. The goal at the physical layer is to ensure that most of the communication between T and R can neither be intercepted nor modified by J.

- *Physical Layer Confidentiality*: An adversary J must be not able to overhear communication between two nodes T and R. We note that protecting the confidentiality of transmitted data is handled by higher layers in the networking stack. The goal at the physical layer is to hide (or at least obscure) the fact that two nodes T and R are engaged in a conversation from an adversary J. This protects all communication between T and R from side-channel attacks such as traffic analysis, timing analysis, power analysis, etc.
- *Physical Layer Identity*: An adversary J must not be able to spoof multiple identities to a node T. We note that the logical identity of a node is handled by higher layers in the networking stack. The goal at the physical layer is to associate an opaque but immutable identity to the adversary J such that any attempt by J to spoof a large number of identities (e.g., Sybil attack (Douceur, 2002)) is futile.

A class of well-known countermeasures against physical layer interception and detection attacks is spread-spectrum based techniques such as frequency hopping, direct-sequence spread spectrum, chirp spread spectrum, etc. Common to all these techniques is a secret (spreading) code that is shared between the communicating entities. These secret codes enable the sender to spread the signal (in time and/or frequency) such that its transmission becomes undetectable for a third party, thus reducing the probability of interference. However, for these schemes to work, the secret code must be shared between the partners prior to communication, generally precluding unanticipated transmissions between unpaired devices. The requirement of a shared code has so far been fulfilled by pre-configuring the devices. This approach has scalability disadvantages in environments where a large number of nodes potentially take part in a pair-wise communication. In particular, in a coalition MANET, it may be infeasible to assume that any node T in domain A may have a shared secret with some node R in domain B.

There are several cryptographic protocols (e.g., Diffie-Hellman key exchange protocol) that allow two nodes with no preconfigured secrets to construct a shared secret. However, operating a key exchange protocol over an unprotected physical channel renders it vulnerable to physical layer jamming attacks. Hence, establishing physical layer communication between two nodes in a coalition MANET suffers from the following bootstrapping problem. Securing physical layer communication between T and R using spread spectrum based techniques requires a shared secret. Establishing a shared secret between T and R (without preconfiguration) requires a secure physical layer communication between the nodes.

In this section we examine the problem of bootstrapping coalition MANETs at the physical layer. We first describe an adversary model for physical layer attacks and then describe candidate solutions that address the bootstrapping problem. In particular, we describe various classes of physical layers and discuss two approaches to jamming resistant key establishment in the physical layer: (i) uncoordinated frequency hopping approach to cryptographic key exchange protocol, and (ii) information theoretic secure key exchange protocol.

Jamming Attacks

Let us consider a setting wherein two nodes T and R that belong to different domains A and B wander into the radio range of each other. The nodes T and R are interested in establishing a communication channel that is resilient to physical layer attacks. We assume that the domains A and B have a priori agreed upon a band of c channel frequencies which is also known to the adversary

J. However, the nodes *T* and *R* do not possess any shared secret to start with. Nodes *T* and *R* first attempt to establish a shared secret and use fast frequency hopping or spread spectrum codes for subsequent communication. An adversary *J* is present in the vicinity of *T* and *R* whose goal is to disrupt any communication between *T* and *R*. We note that once the nodes establish a shared secret, it is harder for an adversary *J* to launch physical layer attacks (Karkatzounis, 2004). Hence, the adversary *J* invests its resources on thwarting all attempts by *T* and *R* to establish a shared secret.

We assume that the adversary *J* is active and is subject to a maximum power constraint of p_j. We remakr that if the adversary *J* were passive, then the nodes *T* and *R* could use a standard Diffie-Hellman key exchange protocol to establish a pair-wise secret. The jamming capabilities of the attacker can then be expressed as the probability with which a packet *m* is jammed over a fixed time horizon. This probability depends on the length of the packet and on the strength of the attacker: the longer the packet is and the stronger the attacker is, the more likely it is that the packet will get jammed. More precisely, let t_p be the time to transmit a packet *m* on frequency channel f_i and *t*p be the minimum jamming period during which the attacker has to interfere with the transmission of packet *m* such that it cannot be decoded. The portion of a message the attacker has to interfere with in such a manner depends on the coding scheme used to encode packets and can be as high as 15% of the message size (Lin and Noubir, 2005).

Following previous classifications (Poisel, 2006), we distinguish between static, sweep, random, and follower jammers. Static, sweep, and random jammers do not sense for ongoing transmissions but permanently jam on c_j channels. However, whereas sweep and random jammers switch the output channels after a duration of *t*p, static jammers remain on the same channels for a time $t >> t_p$. Also, sweep jammers systematically update the output channels in a way that after c/c_j jamming cycles all channels have been jammed once (but do not have to follow a particular order). Random jammers, on the other hand, always choose c_j channels at random and might thus jam the same channels more than once in subsequent jamming cycles.

The *follower jammer* is a sophisticated jammer that attempts to intercept the instantaneous frequency of the frequency hopping (FH) system, and then it can generate an appropriate jamming in a narrow range about this frequency. Follower jammers may be capable of *wide band sensing*. We use c_s to denote the number of channels that the jammer can sense in parallel and by t_s (t_j) the required time to switch the frequency of the input (output) channels. Additionally, the responsiveness of a follower jammer is limited by the laws of physics. Let d_{TJ}, d_{RJ} and d_{TR} to denote pair-wise distances between nodes (*T*, *J*), (*R*, *J*) and (*T*, *R*) respectively. Assuming the frequency f_i used for communication between *T* and *R* is unknown to *J*, then *J* cannot jam the first any communication between *T* and *R* in the first $(d_{TJ}+d_{RJ}-d_{TR})/v_c$ time units, where v_c denotes the speed of light. Indeed, if $d_{TJ} + d_{RJ} >> d_{TR}$, then the nodes *T* and *R* might have sufficient time to establish a shared secret before *J* may attempt to disrupt communication between *T* and *R*.

On a real battlefield, the jammer's distance from the receiver is of vital importance for the jammer's survival. The closer the jammer is to the link between the transmitter and the receiver, the more vulnerable it is to the enemy's fire power. (Karkatzounis, 2004) showed that if the frequency hopping rate is 20 khops/second, the distance between the follower jammer and the link can be at most 14.36 km, compared to the distance of 7.5 km for the higher rate 60 khops/sec. Distances under 8 km is often considered physically infeasible for a jammer on the real battlefield (Karkatzounis, 2004).

Smart (Energy Efficient) Jamming Attacks over 802.11x Networks

We have so far discussed jamming attacks over LPI/LPD (low probability interception/detection) networks. However, we note that detectable networks (such as 802.11x networks) are vulnerable to smart and targeted jamming attacks. A resource constrained adversary may choose strategic locations in a MANET to inject physical layer jamming attacks. Additionally, an adversary may partition the network by jamming the links along a minimal cut in the network. In particular, for a random power-law topology, the size of a min-cut (number of links in the min-cut) is a constant that is independent of the network size. Hence, an adversary would require only constant amount of resources to partition an arbitrary network.

Further, an energy constrained adversary can launch smart jamming attacks by targeting data link layer protocols with the goal of degrading a link's availability and reliability (Kyasanur & Vaidya, 2003; Radosavac et. al., 2005; Cardenes et. al., 2005). Here the adversary exploits two facts: first, wireless medium is a shared medium and therefore data link layer will have a medium access control (MAC) protocol that detects when a node can use the wireless medium. Attacks on the MAC protocols can make the channel seem busy for a long duration, thus preventing the packets from getting transmitted. Other attack involves corrupting the transmission of a packet by creating short bursts of noise that are sufficient to overcome forward error correction codes. Jamming attacks are attractive at the data link layer since they require only protocol specific information, they do not presume the knowledge of any instance specific parameters. Along these lines, (Law et. al., 2005), categorize MAC layer protocols into three types (S-MAC, A-MAC and L-MAC) and present low power jamming attacks on each of these classes of link layer protocols.

Jamming Resistant Physical Layer Key Exchange

Uncoordinated Frequency Hopping

A recent paper in IEEE Security and Privacy Symposium proposed using uncoordinated frequency hopping (Strasser et. al., 2008) for jamming resistant key establishment in wireless networks. Uncoordinated Frequency Hopping (UFH) enables jamming-resistant communication between two nodes in the presence of a jammer without a pre-shared code. The nodes can then use this key to create a secret hopping sequence and communicate using coordinated frequency hopping, thereby abandoning the use of the UFH scheme for subsequent communication.

UFH is closely related to coordinated frequency hopping: each message is split into multiple parts and then sent across the air on random hopping frequencies chosen from a fixed frequency band. Like coordinated frequency hopping, UFH is based on the assumption that the attacker cannot jam all frequency channels on which the nodes communicate at the same time so that the sender and the receiver can still communicate through the remaining channels. However, in UFH, the sender and the receiver do not agree on a secret channel sequence a priori but instead transmit and listen on randomly selected channels. Hence, all communication in UFH underlies the observation that, with sufficient transmission attempts, the sender and receiver will send and listen on the same channels in a number of time slots, even if they did not agree on them beforehand. Intuitively, given 500 channels and given a sender hopping among the channels at a high rate of, for instance, 1500 Hz and a receiver hopping at a low rate (e.g., 100 Hz), the receiver will be listening on the frequency where the sender is transmitting in average 1500/500 = 3 times per second. Building on this observation, the UFH scheme constructs a key establishment protocol that is highly resistant to packet losses and active interference by an attacker.

Unfortunately, the adversary model used in (Strasser et. al., 2008) appears to be flawed. They assume that the number of channels sensed by adversary c_s is significantly smaller than the total number of communication channels c. However, there are several efficient techniques to sense signal over wide bands (Tian & Giannakis, 2007) and commercially available follower jammer kits that are capable of jamming 10 khops/second (Jammer-Kit). Although (Strasser et. al., 2008) includes follower jammers in its adversary model, they seem to have incorrectly assumed that the jammer is limited in its sensing capability. In reality, wideband sensing is a highly parallelizable operation and may be implemented in hardware. Indeed in the presence of a follower jammer neither c_s (number of channels sensed concurrently) nor t_s (time to sense) is the key limiting factor. As pointed out in (Karkatzounis, 2004) the key limiting factor in this case arises from the laws of physics. Assuming d_{TJ}, d_{RJ} and d_{TR} denote distance between nodes (T, J), (R, J) and (T, R) respectively, an adversary J cannot jam the first any communication between T and R in the first $(d_{TJ}+d_{RJ}-d_{TR})/v_c$ time units, where v_c denotes the speed of light. Indeed if transmitter T and receiver R use fast frequency hopping (hopping frequency $> v_c/(d_{TJ}+d_{RJ}-d_{TR})$ then all communication between T and R is guaranteed to be jamming resistant. Hence, UFH (Strasser et. al., 2008) fails to offer protection against an active jammer who is in the vicinity of T and/or R.

Information Theoretic Secure Key Exchange

In coalition scenarios, neither PKI based mechanisms (no pre-configuration; but vulnerable to jamming attacks) nor pre-configured shared symmetric keys may work. Here a third approach offers to extract a secret key from a common source of physical randomness that is shared between a transmitter and a receiver, namely, fading coefficients. This approach essentially exploits the fact that if an intended recipient R and an eavesdropper E are a few wavelengths apart, then the fading experienced by R and E from a given transmitter T will be independent. This difference in physical channels can be exploited to achieve perfect information-theoretic secrecy (Hero, 2003; Khisti et. al., 2006; Khisti & Wornell, 2007; Barros & Rodrigues, 2006; Maurer, 1993). (Koorpaty et. al., 2000) have proposed a technique that uses the short-term reciprocity of the radio channels and rapid de-correlation of radio channels in time, space, and frequency domains to provide a means for secret cryptographic key agreement between two users. Their fundamental idea can be further extended to design space-time transmission schemes that exploit MIMO antenna configurations for a secret key agreement (Li et. al., 2005; Li et. al. 2006). (Li et. al., 2007) have done further experimental verification of similar techniques. We note that a price paid for secrecy using physical layer approaches is the reduced rate of information transfer between transmitter and legitimate receiver. However, the reduced information rate is not a concern for the initial handshake used to exchange secret keys for subsequent communication as the number of bits transferred during this initial handshake is very small (less than 100 bytes). However these approaches fail to provide a key exchange protocol in the presence of an active adversary, and are thus not directly amenable to deployment in coalition networks.

Physical Layer Identity

We have so far examined physical layer integrity and confidentiality in coalition MANETs. We will briefly examine identity management issues at the physical layer before examining the routing layer.

Naming service (e.g., domain name service (DNS)) is one of the core services offered by any network. In an open ad hoc network wherein arbitrary nodes can join the network, it is hard to verify the identity and the credentials associated with a new node in the absence of a common certification authority (CA). The worst case scenario manifests itself as a Sybil attack. In a Sybil (named after the subject of the book *Sybil*, a case study of a woman with multiple personality disorder) attack (Douceur, 2002), a single malicious entity presents multiple identities and uses them to gain a disproportionately large influence, thereby undermining the outcome of de facto election algorithms, redundancy control algorithms and trust management systems.

"One can have, some claim, as many electronic personas as one has time and energy to create."
-Judith S. Donath (Douceur, 2002)

Theoretically, in the absence of a trusted authority, an arbitrarily powerful adversary can forge infinitely many identities without being detected by the network. However, in practice, a network's vulnerability to a identity attacks depends on how cheaply identities can be generated, the degree to which the network accepts inputs from entities that do not have a chain of trust linking them to a trusted authority, and whether the network treats all entities identically. There are two prominent approaches used to mitigate Sybil attacks – however, the absence of a logical identity at the physical layer makes this problem very challenging.

In a *Resource Challenge* based approach the amount of resources available to an adversary is challenged. An example of a CPU resource challenge is a cryptographic puzzle (Juels & Brainard, 1999; Wang & Reiter, 2003). A cryptographic puzzle ensures that if an adversary has ρ times as computationally powerful as the network nodes, then the adversary can spoof no more than ρ identities. Other challenge mechanisms focus on memory size (using matrix inversion test) and the number of radios available at a malicious entity (number of simultaneous conversations) and limit the number of spoofed identities.

In a *Radio Frequency (RF) Localization* based approach builds upon a space-time delimited adversary model. The key idea is to have the neighbors of a malicious entity use RF localization techniques (Nicelescu & Nath, 2003) to determine the physical coordinates of a malicious entity. In the event of a Sybil attack, the purported locations of all fake identities reported by a malicious entity would appear geographically clustered (within a small radius). Standard detection theoretic approaches (e.g., hypothesis testing) can be used to analyze cluster size and cluster radius to detect spoofed identities.

A new and a promising approach to handle identity attacks is *radiometric fingerprinting* (Brik et. al., 2008). The key idea here is to identify a node by its intrinsic hardware properties (such as the physical characteristics of its antenna including: frequency drift, modulation, phase, etc.). Such intrinsic properties must be hard to spoof and forge; in particular, the properties should be chosen such that an adversary should not be able to spoof more than ρ identities. Such an approach solves the *resurrecting duckling* problem (Stajano & Anderson, 1999) wherein a malicious node vanishes and reappears with a new identity elsewhere in the network. For instance, one can bind the reputation of a node to its wireless fingerprint. The network nodes may use a gossip based protocol (or a trust management system) to propagate this information on the network, thereby, curbing the resurrecting duckling problem.

COOPERATIVE ROUTING IN COALITION MANETS

In this section, we describe security issues at the routing layer in MANETs. First, we explore intra-domain routing protocols in MANETs and identify new security challenges in view of sophisticated packet routing and forwarding protocols used in MANETs. Given that the link layer protocols support mutual authentication and facilitate secure

Figure 1. Simple packet forwarding

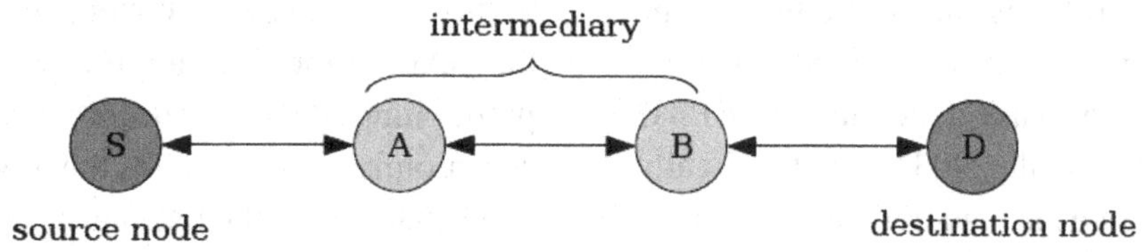

exchange of link keys, an adversary can launch routing layer attacks only by compromising one or more nodes in the network. Hence, the routing layer protocols must be designed such that they can tolerate a small fraction of malicious nodes in the network; also network management infrastructure must be capable of detecting compromised nodes and either repair them or expel them from the routing network. As we point out in the previous section, expelling a compromised node is a challenging problem in itself. For instance, a malicious node may reappear elsewhere in the network and forge one or more new identities. Hence, security at the routing layer should be implemented as a combination of robust routing protocols and a network management system (NMS) that quickly detects malicious or compromised nodes. For a system to be stable the rate at which NMS detects compromised nodes must be faster than the rate at which an adversary may compromise nodes; this ensures that in steady state, only a small fraction of network nodes are in a compromised state, thereby allowing the routing protocols to operate efficiently.

Second, several authors have argued that coalition MANETs are most likely to use inter-domain routing protocols (IDRP). We argue that IDRP offers autonomy for each coalition partner while providing certain routing guarantees (e.g., throughput, latency, reliability, etc.) to the end points. It also allows one to device novel root cause analysis and fraud detection techniques that can detect misbehaving routing domains (and gateways) assuming a threshold number of coalition partners are honest. In the rest of this section, we first examine intra-domain routing issues followed by inter-domain routing issues with the goal of identifying new challenges and unsolved research problems.

Simple Packet Forwarding Network

In a traditional packet forwarding network, a node *S* may send packets to a destination *D* through a path *S, A, B, D*; here, nodes *A* and *B* perform packet forwarding function to deliver packets (Figure 1: Simple Packet Forwarding). Node *S* requires cooperation from the routing nodes to discover a route to node *D* (route set up and discovery). A malicious node can create a *black hole* by advertising low cost routes and drop all packets routed to it, destabilize routes, and create routing loops. A detailed survey of routing layer attacks on ad hoc networks is presented in (Hu & Perrig, 2004; Karlof & Wagner, 2002).

Several authors have proposed solutions to verify the integrity of route construction protocols. The division between *proactive* and *reactive* protocols is reflected in the security mechanisms. SAODV (Zapata, 2006; Zapata & N. Asokan, 2002) provides an example of reactive routing protocol security. It uses hash chains to avoid manipulation of hop counts in route discovery messages, and digital signatures are used for the immutable parts of these messages, to provide end-to-end confirmation that the request reached the owner of the address. SLSP (Papadimitratos & Haas, 2003) is an example of a security mechanism for a proactive routing protocol. It uses signatures on link state update messages to avoid manipulation of the topology information. The SAODV solution is focused on verifying the validity of the path, whereas the SLSP approach is based around determining the correctness of the network topology. Recent research has also explored the possibility of using Identity-based PKC (ID-PKC) (Khalili et. al., 2003), which radically simplifies certificate management. In both cases, the existence of a simple Public Key Infrastructure (PKI) is assumed, but a fuller exploration of the key management and key distribution problem has not been undertaken.

Resource constraints of the typical MANET environment have played a role in shaping these proposed security solutions. Low processing requirements and minimal packet overhead are seen as common requirements. Recently, several research proposals have used cooperative network monitoring based on root cause analysis techniques to detect malicious and faulty nodes in networks. Cooperative monitoring techniques range from physical layer power estimation for detecting jamming attacks (Ho et. al., 2008; Zafer et. al., 2008), MAC layer misbehavior detection (Law et. al., 2005; Radosavac et. al., 2003) to routing layer faults and anomaly detection (Srivatsa et. al., 2008). However, to date, all cooperative root cause analysis techniques assume that the monitors are honest. While this is a reasonable assumption for an intra-domain setting wherein all monitors belong to a single domain, an inter-domain setting is faced with the challenge of handling faulty monitors that may be malicious, rational-selfish or Byzantine.

All past work on MANET routing protocols have alluded to an inherent tradeoff between the robustness and the communicating cost of packet routing and forwarding protocol. For instance, if each node uses a broadcast (flooding) protocol, it may maximize the probability that a packet from *S* reaches *D*; however, its packet forwarding cost is linear in the size of the network. In any event, if an adversary can partition the network (by compromising a *vertex cut*) then it can completely control all communications between the partitions. Fortunately, in a mobile network the topology changes dynamically and arbitrarily; and thus the vertex cut is not static. On one hand, the network may exploit mobility to defend against network partitioning attacks, while forcing the adversary to compromise new nodes as the topology dynamically changes. In the limiting case, if the network had unlimited mobility, then it can completely avoid the use of multi-hop routing protocols; the communicating parties can simply move close to one another and correspond over a direct link. Most papers in literature have failed to explicitly include mobility as a resource (available to both the network and the adversary) while analyzing the security properties of routing protocols.

Content-Based Routing

First, several routing protocols in MANETs use flexible end point based routing, wherein, the end point is not defined by an identity (such as an IP-address); instead the end point could be based on: (i) fixed attributes of an entity (e.g., role): deliver this packet to any entity who is authorized to play a role *R*, (ii) dynamic attributes of an entity (e.g., geographical routing): deliver this packet to any entity within a geographical bounding box, or to the entity closest to geographical coordinates (*lat*, *lng*), (iii) content-based routing (e.g., publish/subscribe networks): deliver this packet with attributes *e* (e.g., e_1 = {<mission, A>, <secrecyLevel, Secret>}) to all entities who are authorized to listen to a filer *f* (e.g., f_1 = {<mission, =, A>, <secrecyLevel, ≤, topSecret>}, f_2 = {<mission, =, A>, <secrecyLevel, ≤, classified>}) such that match(*e*, *f*) = *true* (e.g., match(e_1, f_1) = *true*, match(e_1, f_2) = *false*). Unlike (i) and (ii) wherein an end point is defined by a predicate over the entity's fixed or dynamic attributes, (iii) defines an end point using a predicate over the packet payload's attributes.

While research in the network security community has explored identity based end point routing (e.g., multicast address), flexible end point routing appears to be largely untapped. First, flexible end point based routing protocols rely on efficient *multicast overlay trees to match and forward packets on the network.* Such routing protocols have not been properly inspected in the presence of compromised routing nodes that attempt to actively subvert the protocol. Improperly constructed routing trees may significantly increase the transmission cost. Further, with flexible end points, the leaves in the routing tree may change dynamically over time. Malicious nodes may artificially increase the churn rate for a routing tree, thereby, significantly increasing the cost of constructing and maintaining routing trees.

Second, flexible end point based routing *requires novel key management algorithms to ensure that the packet payload are intelligible only to authorized recipients.* For example, in geographical routing, a packet intended for a node in a geographical bounding box *B* must be intelligible to only some node within the box *B*. However, the key challenge here is that the sender may not a priori know the identity of the recipient. *This precludes the possibility of encrypting the packet either using a symmetric key or using a public key that is bound to the identity of the recipient.* A naive approach is to first identify a node in the box *B* and use the identity of the node to appropriate encrypt the packet (with a shared symmetric key or a public key). However, this defeats the performance and scalability benefits of geographical routing; also, when the node moves out of the box *B*, the sender has to redo the task of identifying a new node in box *B*. One alternative approach is to leverage identity based (attribute based) public key management protocols (ID-PKC), wherein, the sender encrypts the packet using the attributes associated with the bounding box *B*. In this case, the problem reduces that of key distribution: when a node is in box *B*, the network must distribute keying material to the node that allows it to generate an appropriate private key; also, the key must be revoked when the node leaves the box *B*.

Network Coding Based Cooperative Routing

Recent research on reducing transmission power in MANETs has proposed network coding based cooperating routing approach. Let us suppose there are two sources s_1 and s_2 each of which generate data streams x_i and y_i at the rate of one bit per second. All wireless channels can carry at most one bit per second. There are two destination nodes d_1 and d_2 both of which need the bit streams x_i and y_i. One can show that no packet routing and forwarding scheme can achieve the goal of delivering both x_i and y_i to both d_1 and d_2. On the other hand, this goal can be met if n_1 uses packet coding and transmits x_i XOR y_i (Figure 2: Network Coding Example).

In addition to packet forwarding and routing attacks, the routing nodes may improperly execute packet coding, thereby, compromising the integrity of a packet. While it may be possible for the receiver to ultimately detect a corrupted packet (and subsequently drop it), the malicious node has succeeded in wasting the transmission power at every node on the route from the sender to the receiver (this problem becomes worse when we use multicast routing trees). In a simple packet forwarding scenario a source *S* and destination

Figure 2. Network coding example

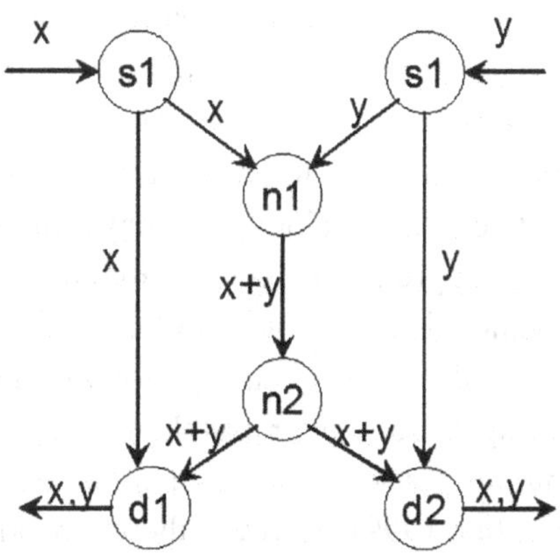

D can protect packet data confidentiality using payload encryption; every intermediary node (including the receiver D) on the routing path can verify the integrity of the packet by checking the sender's signature. In network coding schemes, confidentiality from packet coding and forwarding nodes can be achieved using payload encryption (most network coding schemes operate on the entire payload as if they were bit strings); however, it may not be possible to generate sig(x_i XOR y_i) from sig(x_i) and sig(y_i); one solution is to attach both sig(x_i) and sig(y_i); however, this increases packet size and consequently the transmission cost. One possible approach is to use a computation intensive (Goldwasser & Micali, 1982) digital signature that is homomorphic on the XOR operator. However, using expensive computational operations exposes a node to DoS attacks (similar to SSL handshake based DoS attacks that exploit the cost of public key signature generation and verification). Hence, it is very crucial to couple network coding schemes with cryptographic schemes that allow all intermediary nodes to efficiently verify the integrity of a packet before expending its energy to forward the packet.

In-Network Data Aggregation

In a sensor network, the base station may not be interested in collecting all the raw data; but may require only some aggregates on the raw data such as *sum*, *average*, *count*, *variance*, *min*, *max*, etc. Hence, sensor nodes may perform in-network packet data aggregation to reduce transmission costs (Figure 3: In-Network Aggregation Example). In addition to packet forwarding and routing attacks, a malicious node may improperly execute the aggregation operator. In the aggregation schemes, it is hard to maintain both confidentiality and integrity data from forwarding and aggregating nodes. One possible approach is to use homomorphic encryption schemes (Benolah, 1999; Paillier, 1999); however, there is no known cryptosystem which preserves the ring structure of the plaintexts, that is, allows both addition and multiplication on a group. Some authors have suggested using statistical properties of the raw data (e.g., spatial-temporal locality in temperature readings) to build probabilistic integrity check mechanisms over a limited class of aggregation operators (Chan et. al., 2006, Chan et. al., 2008). In the absence of efficient general purpose security primitives, the network is faced with the challenge of effectively trading off efficiency (in the worst case by turning off in-network aggregation) with the goal of improving its resilience to routing layer attacks.

INTER-DOMAIN ROUTING IN COALITION MANETS

In this section, we will examine security issues in inter-domain routing protocols for MANETs. We will first develop a trust model of actors in the network, their relationships, and the information that needs to be protected. Previous research on inter-domain routing on the Internet provides us with a useful starting point to model entities and information items that are critical to developing security solutions across disparate management domains.

Figure 3. In-network aggregation example

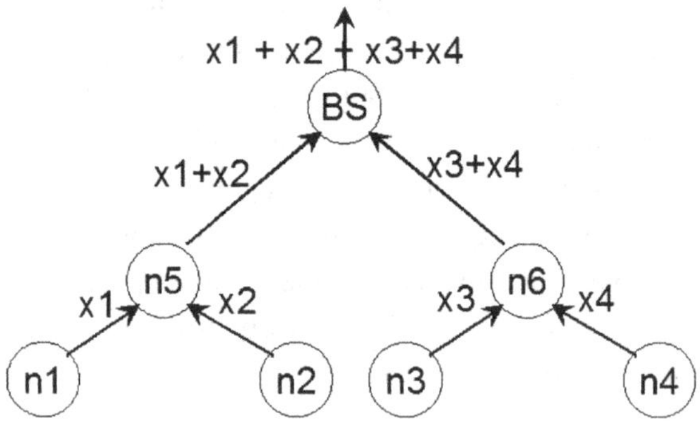

Inter-Domain Routing Security on the Internet

Inter-domain routing on the Internet is managed using BGP4 (Border Gateway Protocol) (Rekhter et. al., 2006). This was originally developed for use in a trusted environment, and so provides little security against attackers or misconfiguration. The need for additional security mechanisms has been recognized in recent times, and demonstrated by the AS7007 incident (Mahajan et. al., 2002; Misel, 1997) and more recent "hijacking" of a part of the YouTube address space (RIPE NCC, 2008). Both incidents are believed to have been due to misconfiguration, rather than malicious intent.

Current BGP operations depend completely on peers trusting one another not to inject bad information into the routing updates. This is coupled with limited filtering (e.g., to filter out advertisements of unallocated address space, and to ensure that downstream customers only advertise their own address prefixes). In addition to such filtering, there is some use of TCP-MD5 (Heffernan, 1998) to provide integrity protection for the protocol between peer routers. Recent work has been undertaken in the IETF to understand the requirements for securing BGP. Three general classes of threat against routing protocols have been identified (Barbir, 2002):

- Attacks on the protocol itself
- Falsification of information carried in the protocol
- Forwarding traffic along a different path to that identified by the routing protocol

Threats against the protocol itself include attempts to spoof a network identity for a peer network. These attacks are the simplest to address, since it is essentially a matter of creating a secure channel between two cooperating entities (neighboring BGP routers). Falsification is identified as forming a significant set of threats through sending false routing information in the routing protocol:

- Overclaiming: a malicious router or an outsider claims ownership of some prefix it does not own, or is not authorized to advertise.
- Misclaiming: an attacker advertises network resources it owns, but in a way different from that intended by the authoritative network administrator. e.g., advertising inappropriate link costs or path lengths.
- Falsification by Forwarders: a router modifies information in the routing protocol that is meant to be immutable.
- Misstatement: an attacker modifies mutable route attributes in an invalid manner. e.g., deleting, inserting or substituting elements of the AS PATH in BGP.

Protecting the routing information against such falsification attacks is a challenging multi-party security problem, and the core focus of research on inter-domain routing security. Verifying correct behavior of the forwarding plane requires active monitoring, and again relies on multi-party cooperation to achieve the security goal. There have been a number of different proposals for adding security to BGP, such as S-BGP (Kent et. al., 2000) Secure Origin BGP (so-BGP) (White, 2003), and Pretty Secure BGP (psBGP) (Wan et. al., 2005). These competing proposals, embody different views on the appropriate model for authenticating ownership of identifiers (such as AS numbers and prefixes), even in the relatively well understood Internet case.

These solutions tend to rely heavily on public key signatures, although some attempts are made to ensure that results of signature verification can be cached. Both the computational burden and the key and certificate storage requirements are significant for a protocol operating on an Internet scale. To address this, other proposals have been made where such signature use is minimized, e.g., Secure path vector (SPV) (Raghavan et. al., 2007; Hu et. al., 2004). Approaches to BGP security which avoid the use of cryptographic components by relying on BGP policy tools have also been proposed. One solution, pgBGP (Pretty Good BGP) (Karlin et. al., 2006), simply adjusts BGP policies to provide some additional cautiousness in accepting new routes, thereby protecting the Internet from (short-lived) prefix or sub-prefix hijacking when used in conjunction with appropriate monitoring systems. RPSL (Route Policy Specification Language) (Alaettinoglu et. al., 1999) provides a way for ISPs to describe their routing policies. For example, it will indicate what routes they accept from a particular neighboring AS, and what routes they advertise to them. This information is stored in one of a number of central databases, and can be automatically extracted to perform filtering on a router. However, deployment is limited and in practice this information tends to be more useful for diagnostics than filtering.

Inter-Domain Routing in MANETs

Although the Internet inter-domain scenario provides a useful basis for considering the MANET inter-domain issues, the topology is more dynamic in MANET environments, and there is a need to avoid reliance on centralized services. Figure 5: Military Address Assignment illustrates sample address allocation architecture, based on organizational structure in a military context. As compared to the Internet model (Figure 4: Internet Address Assignment), the organization is less closely related to the network topology. In particular, a single regiment may provide resources to more than one battle group, and so have its address space partitioned across diverse locations. This implies that address space is 'PI' rather than 'PA'. In the Internet context this is undesirable due to scalability issues. This is less significant in military MANET deployments where fewer domains and prefixes are expected. A further difference from the Internet scenario arises since, if any domain can potentially act as a transit network, the distinction between stub and transit networks is less relevant. However, some domains may still be restricted from providing transit through policy.

Address Ownership

A single approach to the question of address and identity ownership is unlikely to be feasible, since this is strongly driven by the deployment model. Therefore, it is appropriate to consider both hierarchical and peer-to-peer (web-of-trust) models of managing identity and address space ownership. Real deployments may actually need to use a combination of these at different parts

Figure 4. Internet address assignment

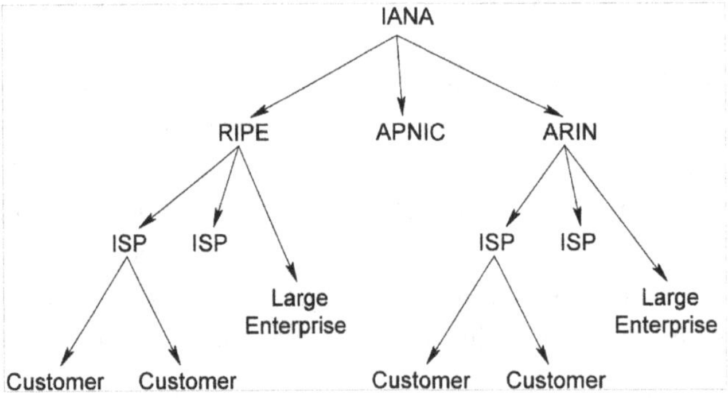

of the allocation hierarchy (e.g., through cross signing of CAs).

A standard PKI approach offers a route to providing hierarchical address and network identity assignment. In the Internet context, this is the approach used by S-BGP (Kent et. al., 2000). It makes use of two hierarchical PKIs using X.509 certificates for assignment of AS numbers and prefixes to organizations. Signatures are used to confirm that a router is authorized to act on behalf of an AS, that an AS is authorized to advertise itself as the origin of a prefix (address attestations). In addition, certificates are provided to routers to confirm that they are authorized to speak on behalf of an AS, and so secure communications between peer routers using IPsec. Certificate distribution is performed out-of-band. However, in the MANET context reliance on a central repository would need to be avoided.

Secure Origin BGP (so-BGP) (White, 2003)) and Pretty Secure BGP (psBGP) (Wan et. al., 2005) mix the use of hierarchical and "web of trust" models for address ownership and AS identities. so-BGP uses a hierarchical X.509 PKI for AS number assignment, with a small number of root CAs, and the possibility of multiple levels in the PKI hierarchy. The use of multiple levels actually moves this more towards a "web of trust" model, where ASs confirm the ownership of public keys by other ASs. Signatures by ASs are then used to authorize another AS to advertise a particular prefix. This removes the division between the organization to which a prefix is assigned and the AS that can originate it that exists in S-BGP.

psBGP uses a centralized trust model for AS numbers, with a PKI using RIRs acting as certificate authorities to confirm AS bindings to public keys. A decentralized trust model is used to confirm prefix ownership, where a small number of the peers of an AS endorse its ownership of a prefix. In addition, signatures from ASs are used to confirm that BGP speakers (routers) are authorized to act on its behalf. As an alternative to traditional PKI, Identity-based Public Key Cryptography may offer some advantages. In particular, the need to store and exchange certificates is avoided. Some initial proposals have been made on applying ID-PKC to inter-domain routing (Boldyreva et. al., 2007).

Route Correctness

Attacks on routing can target both the control plane (route set up) and the data plane (packet

Figure 5. Military address assignment

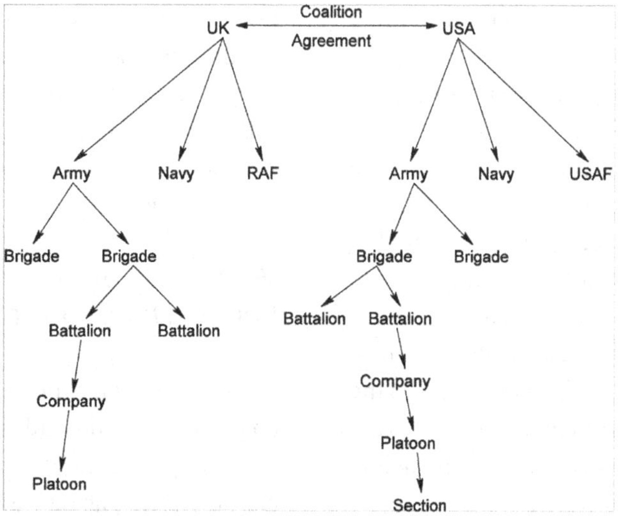

forwarding). The security goal of control plane protocols is to ensure that only valid path updates are propagated on the network. Validity of a path update (e.g., <*A, B, C, D, dst*>) is defined according to four primary criteria:

- The router that sent the update was authorized to act on behalf of the domain (*A*) it claims to represent (by virtue of placing that domain ID in the path).
- The domain from which the update emanates (*A*) was authorized by the preceding domain (*B*) in the path to advertise the destination (*dst*) contained within the update.
- The first domain in the path (*D*) owns the set of destinations *dst* (e.g., a set of prefixes), and is thus authorized to advertise them.
- If the update withdraws one or more routes, then the sender must have advertised the route(s) prior to withdrawing it (them).

S-BGP addresses these requirements using the concept of address and route attestation. Address attestation binds a set of destination nodes (*dst*) to a domain (*D*). It ensures that a destination (address or prefix) is advertised by a domain if and only if the domain owns the destination. Route attestation created by one domain authorizes a neighbor domain to advertise paths to a set of destinations. Route attestation ensures that a malicious node cannot spoof paths, shrink a path (e.g., create black holes) or alter a path (other than appending itself to the tail of a valid path).

so-BGP provides a weaker form of route attestation, providing only assurance of *path plausibility*. Rather than requiring evidence that *B* was authorized to advertise the destination, only evidence of the existence of a peering relationship between *A* and *B* is needed, which in turn provides the ability to verify the correctness of the AS topology map. In combination with address attestation, this identifies that there is a plausible path to the given prefix.

psBGP uses a model of *path consistency* checks, where ASs endorse one another's ownership of address space, and peering assertions (i.e., that *A* is a peer of *B*) are signed. The combination of the two components allows routes to be verified, based on a seeing a consistent set of assertions from the possible paths available. This approach does, however, provide opportunities for collusion.

S-BGP, so-BGP and psBGP rely on a PKI based infrastructures for route attestation. Secure path vector (SPV) (Raghavan et. al., 2007; Hu et. al., 2004) provides an alternative using only efficient symmetric key cryptographic primitives. One of the key drawbacks of these approaches is that the signature size in the update message grows with path length. More recently, Identity-based Sequential Aggregate Signatures (IBSAS) (Boldyreva et. al., 2007) have been proposed to generate compact signatures that allow signers to attest a path update message as well as the order in which they signed (albeit, incurring higher computation costs than S-BGP and SPV), making it more suitable for a MANET setting.

While attestations ensure that all advertised paths are valid, it does not prevent a malicious node from selectively propagating only those advertisements that are of strategic interest (say, with the goal of attracting more paths for traffic analysis or repelling paths to throttle transit traffic, etc.). Hence, attestations can only guarantee validity, but not the optimality of path updates. However, if the network is tightly connected (offering multiple alternate paths between any two nodes), then path optimality may not be a serious concern.

Verification of Correct Forwarding Behavior

Route correctness ensures that all good nodes commit to a valid path in the control plane. On the data plane, however, nodes may misbehave and forward packets along paths that deviate from their path advertisements. The security goal in the data plane protocols is to ensure that all nodes forward packets in accordance to the paths advertised. This may be achieved through active mechanisms (e.g., probing) or passive mechanisms (e.g., monitoring).

One recently proposed active approach, Ordered Multi-Signatures (OMS) (Boldyreva et. al., 2007), allows multiple (ordered) signers to sign the same packet to allow verification of the path it followed through the network. However, OMS is computationally very expensive, making it nearly infeasible to operate at high speeds. One can use secure implicit sampling (SIS) techniques in which: (i) only a small (pseudo-randomly selected) set of packets are signed, and (ii) the selected packets are signed by a (pseudo-randomly selected) sub-set of nodes in the advertised path. Implicit sampling allows the sender (and receiver) to pseudo-randomly determine the subset of nodes in an advertised path that are required to sign a packet. Secure implicit sampling ensures that the choice of nodes that are required to sign a packet is a priori known only to the selected nodes and the sender/receiver. Nonetheless, SIS techniques incur significantly high key management overhead.

In a MANET setting one can use cooperative monitoring techniques to verify a node's forwarding behavior. Cooperative monitoring can be achieved at two levels: neighbor monitoring and end-to-end monitoring. Neighbor monitoring exploits the broadcast nature of wireless channels. A node's neighbor (acting as a *witness*) can observe packet transmissions by that node. Even if the packet payloads are encrypted, a node can use packet headers to decide if the node is behaving consistently with its advertised path. We remark that a node may exploit MAC-layer misbehavior to appease a witness while ensuring that the transmitted packet does not reach the next node in the network. Fortunately, such denial of service attacks can be easily detected using end-to-end monitoring.

End-to-end monitoring techniques rely upon special nodes (*monitors*) in the network. Monitors may not explicitly participate in the packet forwarding task; however, their goal is to collect end-to-end performance metrics between pairs of monitors. The collected monitoring information is pooled and root cause analysis techniques are employed to detect misbehaving nodes in the network. For instance, let us suppose that two end-to-end tests $<M_1, A, C, M_2>$ and $<M_3, B, C, M_4>$ show poor performance (e.g., heavy packet loss) then the monitors may use this *evidence* to conclude that *C* is the most *likely suspect*. Indeed, the monitors may adaptively run more tests to confirm their hypothesis. However, the monitors have to camouflage their test (or *probe*) packets with normal traffic; otherwise, a malicious router may offer differential treatment to the probe packets, thereby defeating all probe based end-to-end monitoring techniques.

In an inter-domain setting wherein monitors may belong to different administrative domains (with different trust levels), misbehavior detection becomes a more challenging problem. For example, a colluding subset of monitors may attempt to actively manipulate the monitored data with at least one of the following two goals: *shilling* – make a bad node appear good and *bad mouthing* – make a good node appear bad. Trust based mechanisms need to be deployed here to incentivise monitors whose data matches the consensus (e.g., the monitor's vote matches the majority vote), while punishing those monitors whose data deviate significantly from the consensus. Finally, the results of these monitoring observations may be used to build trust based routing policies.

SUMMARY

In this chapter, we have examined security issues in coalition MANETs. Primary contribution of

Figure 6. Coalition MANETs - physical and link layer issues

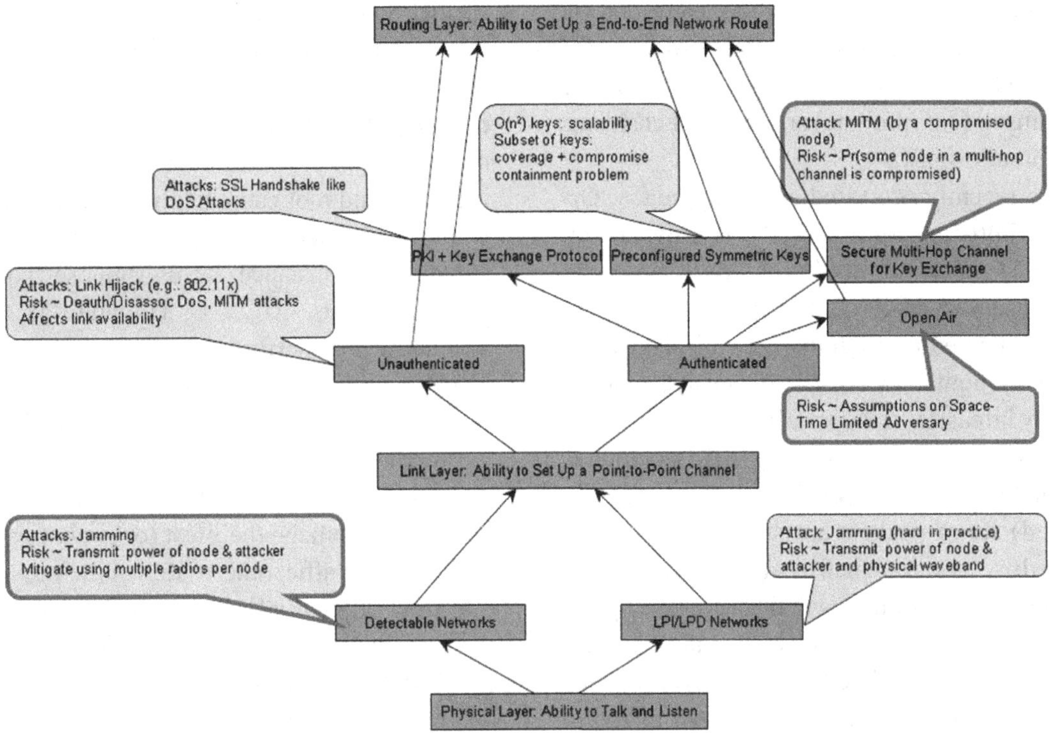

this chapter is three fold. First, we have outlined a realistic space and time delimited adversary model for MANETs. The model allows us to quantify security protocols as a function of the amount of resources (e.g., space, radio spectrum, energy, mobility, etc.) available to the network nodes and the adversary. Based on this model we have examined security requirements at two networking layers: physical & link layer (see Figure 6: Coalition MANETs - Physical and Link Layer Issues) and the routing layer (see Figure 7: Coalition MANETs - Routing Layer Issues).

At the physical and link layer, we have argued that traditional key set up protocols (PKI and a priori shared symmetric keys) may not be applicable to coalition MANETs. We have examined the feasibility of two alternate approaches: uncoordinated frequency hopping (UHF) and information theoretic key exchange. In particular, we have argued that the UHF based approach can be used as long as the jammer is sufficiently far away from the transmitter and/or the receiver; the information theoretic key exchange mechanisms offer unconditional (perfect) secrecy but significantly decays in performance with the jamming power of the adversary. We have also examined identity management issues in the physical layer and examined a radiometric fingerprinting based approach to condone Sybil attacks in coalition MANETs.

At the routing layer, we have argued that sophisticated power saving routing protocols used by MANETs opens up security vulnerabilities in the network layer. In particular, we have argued for: novel key management algorithms for flexible end point based routing, new cryptographic schemes to support confidentiality and integrity in network coding based cooperative routing schemes and in-network aggregation schemes, and robust routing protocols to mitigate DoS attacks on multicast routing tree construction and maintenance protocols. We have also examined

Figure 7. Coalition MANETs - routing layer issues

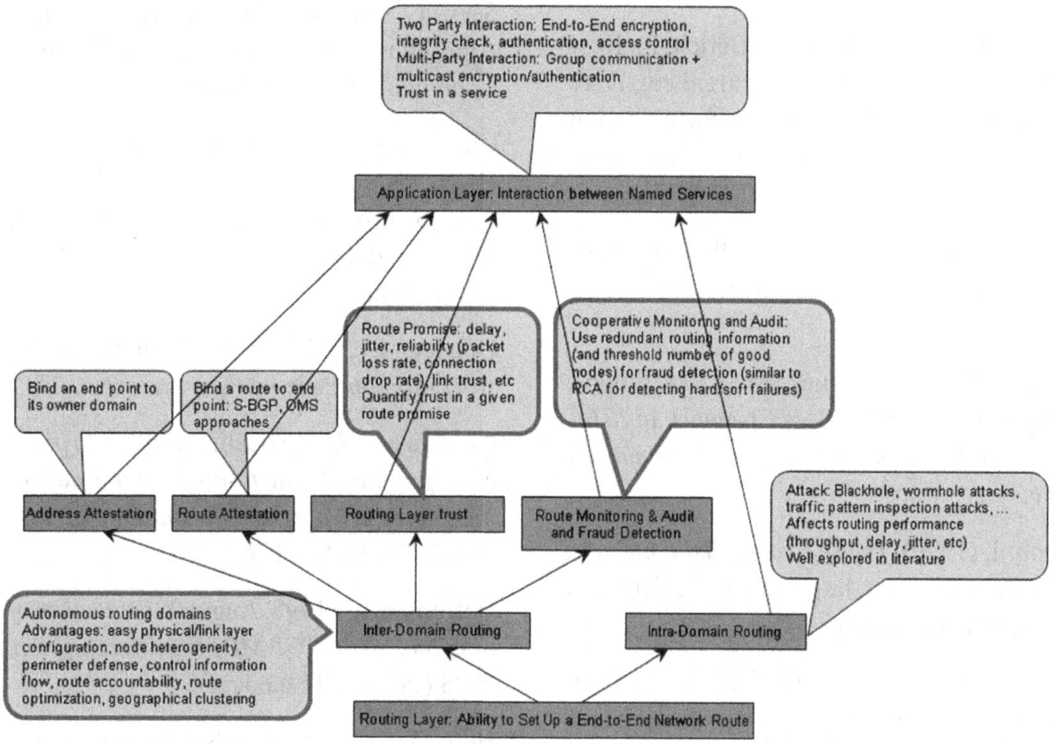

security issues arising from inter-domain routing protocols in MANETs. We have emphasized on the need to provide strong authentication of the ownership of identifiers, which then supports the ability to verify routes advertised by the routing protocol. Existing research shows the prospect for developing techniques that offer a balance between security and processing or bandwidth overhead. In addition to identity and path protection, techniques have been identified to provide monitoring functionality which can be used to confirm whether there is misbehavior on the forwarding path.

ACKNOWLEDGMENT

This work immensely benefited from many discussions held in various meetings and workshops organized by "International Technology Alliance" and "Collaborative Technology Alliance". In particular, the authors would like to thank Jon Crowcroft, Robert Hancock, Pankaj Rohatgi, Steffen Reidt, Dinesh Verma and anonymous reviewers for their helpful comments and insights. This research was sponsored by the U.S. Army Research Laboratory and the U.K. Ministry of Defence and was accomplished under Agreement Number W911NF-06-3-0001. The views and conclusions contained in this document are those of the author(s) and should not be interpreted as representing the official policies, either expressed or implied, of the U.S. Army Research Laboratory, the U.S. Government, the U.K. Ministry of Defence or the U.K. Government. The U.S. and U.K. Governments are authorized to reproduce and distribute reprints for Government purposes notwithstanding any copyright notation hereon.

REFERENCES

Alaettinoglu, C., Villamizar, C., Gerich, E., Kessens, D., Meyer, D., Bates, T., Karrenberg, D. & Terpstra, M. (1999). Routing Policy Specification Language (RPSL). *RFC 2622 (Proposed Standard)*, June 1999. *Updated by RFC 4012*.

Barbir, A., Murphy, S. & Yang, Y. (2006, October). Generic Threats to Routing Protocols. *RFC 4593 (Informational)*.

Barros, J. H., & Rodrigues, M. R. D. (2006). Secrecy Capacity of Wireless Channels. In *IEEE International Symposium on the Information Theory*.

Benaloh, J. (1999). *Dense Probabilistic Encryption*. Retrieved from http://research.microsoft.com/crypto/papers/dpe.ps

Boldyreva, A., Gentry, C., O'Neill, A., & Yum, D. H. (2007). *Ordered multi-signatures and identity-based sequential aggregate signatures, with applications to secure routing. In 14th ACM conference on Computer and communications security* (pp. 276–285). New York.

Brik, V., Banerjee, S., Gruteser, M., & Oh, S. (2008, September). Wireless Device Identification using Radiometric Signatures. In ACM Mobicom, Burlingame, CA.

Camtepe, S. A., & Yener, B. (2007). *Key Distribution Mechanisms for Wireless Sensor Networks: Survey*. RPI Technical Report TR-05-07.

Cardenes, A., Radosavac, S., & Baras, J. H. (2005). Detection and Prevention of MAC Layer Misbehavior in Ad-Hoc Networks. In *ACM Workshop on Security of Ad-Hoc and Sensor Networks (SASN)*.

Chan, H., Perrig, A., & Song, D. (2006). Secure Hierarchical In-Network Aggregation in Sensor Networks. In *ACM Conference on Computer and Communication Security (CCS)*.

Chan, H., Perrig, A., & Song, D. (2008). Efficient Security Primitives from a Secure Aggregation Algorithm. In *ACM Conference on Computer and Communication Security (CCS)*.

Chau, C. K., Crowcroft, J., Lee, K.-W., & Wong, H. Y. (2008, January). *IDRM: Inter-Domain Routing Protocol for Mobile Ad Hoc Networks*. Technical Report UCAM-CL-TR-708, University of Cambridge, Computer Laboratory.

Douceur, J. (2002). The Sybil Attack. In *2nd IEEE Workshop on Peer-to-Peer Systems (IPTPS)*.

Goldwasser, S., & Micali, S. (1982). *Probabilistic encryption and how to play mental poker keeping secret all partial information. In 14th* (pp. 365–377). ACM STOC.

Heffernan, A. (1998, August). *Protection of BGP Sessions via the TCP MD5 Signature Option*. RFC 2385 (Proposed Standard).

Hero, A. O. III. (2003). Secure space-time communication. *IEEE Transactions on Information Theory*, *49*(12). doi:10.1109/TIT.2003.820010

Ho, I. W., Ko, B.-J., Zafer, M., Bisdikian, C., & Leung, K. (2008). Cooperative Transmit-Power Estimation in MANETs. In *IEEE Wireless Communication and Networking Conference (WCNC)*.

Hu, Y. C., & Perrig, A. (2004). A Survey of Secure Wireless Ad Hoc Routing. In IEEE Security and Privacy Magazine.

Hu, Y-C., Perrig, A. & Sirbu, M. (2004, October). SPV: secure path vector routing for securing BGP. *ACM SIGCOMM Computer Communications Review*, *34*(4).

Jammer-Kit. (n.d.). *Follower Jammer Kit*. Retrieved from http://www.polarizone.com/milzone.html

Juels, A., & Brainard, J. (1999). A Cryptographic Defense Against Connection Depletion Attacks. In *Network and Distributed Systems Security (NDSS)*. Client Puzzle.

Karkatzounis, K. (2004). *Performance evaluation of different jamming strategies over uncoded non-coherent fast frequency hopping mfsk communication systems*. Naval Postgraduate School Thesis.

Karlin, J., Forrest, S., & Rexford, J. (2006, November). Pretty Good BGP: Improving BGP by Cautiously Adopting Routes. In *The 14th IEEE International Conference on Network Protocols (ICNP)*.

Karlof, C., & Wagner, D. (2002). Attacks and Countermeasures. In *Elsevier's Ad-Hoc Networks Journal, Special Issue on Sensor Network Applications and Protocols*. Secure Routing in Wireless Sensor Networks.

Kent, S., Lynn, C., & Seo, K. (2000, April). Secure Border Gateway Protocol (Secure-BGP). *IEEE Journal on Selected Areas in Communications*, *18*(4), 582–592. doi:10.1109/49.839934

Khalili, A., Katz, J., & Arbaugh, W. A. (2003, January). Toward secure key distribution in truly ad-hoc networks. In *Symposium on Applications and the Internet Workshops*, (pp. 342–346).

Khisti, A., Tchamkerten, A., & Wornell, G. (2006). Secure Broadcasting with Multiuser Diversity. In *44th Allerton Conference on Communication, Control and Computing*.

Khisti, A., & Wornell, G. (2007). The MIMOME Channel. In *45th Allerton Conference on Communication, Control and Computing*.

Koorapaty, H., Hassan, A. A., & Chennakeshu, S. (2000). Secure information transmission for mobile radio. *IEEE Communications Letters*, *4*(2). doi:10.1109/4234.824754

Kyasanur, P., & Vaidya, N. (2003). Detection and Handling {MAC} Layer Misbehavior in Wireless Networks. In *IEEE Symposium on Dependable Systems and Networks (DSN)*.

Law, Y. W., Hoesel, L. V., Doumen, J., Hartel, P., & Havinga, P. (2005). Energy-efficient link-layer jamming attacks against wireless sensor network MAC protocols. In *SASN '05: Proceedings of the 3rd ACM workshop on Security of ad hoc and sensor networks*, (pp. 76–88), New York.

Li, X., Chen, M., & Ratazzi, E. P. (2005). A Randomized Space-Time Transmission Scheme for Secret-Key Agreement. In *39st Annual Conference on Information Sciences and System* (*CISS*).

Li, X., Hwu, J., & Ratazzi, E. P. (2006). Array Redundancy and Diversity for Wireless Transmissions with Low Probability of Interception. In *IEEE International Conference on Acoustics, Speech and Signal Processing (ICASSP)*.

Li, Z., Trappe, W., & Yates, R. (2007). Secret Communication via Multi-antenna Transmission. In *41st Annual Conference on Information Sciences and System* (*CISS*).

Lin, G. & Noubir, G. (2005). On link layer denial of service in data wireless LANs. *Wireless Communications and Mobile Computing, 5*(3).

Mahajan, R., Wetherall, D., & Anderson, T. (2002). Understanding BGP misconfiguration. In *SIGCOMM '02: Proceedings of the 2002 conference on Applications, technologies, architectures, and protocols for computer communications*, (pp. 3–16), New York.

Maurer, U. M. (1993). Secret key agreement by public discussion from common information. *IEEE Transactions on Information Theory*, *39*(3). doi:10.1109/18.256484

Misel, S. A. (1997). *Wow, AS7007!* Retrieved from http://www.merit.edu/mail.archives/nanog/1997-04/msg00340.html

Nicelescu, D., & Nath, B. (2003). Ad Hoc Positioning (APS) using AOA. In *IEEE International Conference on Computer Communications (Infocom),* (pp: 1734-1743).

Paillier, P. (1999). *Public-Key Cryptosystems Based on Composite Degree Residuosity Classes* (pp. 223–238). EUROCRYPT.

Papadimitratos, P., & Haas, Z. J. (2003, January). Secure link state routing for mobile ad hoc networks. In *IEEE Workshop on Security and Assurance in Ad Hoc Networks*, (pp. 379–383).

Poisel, R. A. (2006). *Modern communications jamming principles and techniques*. Boston: Artech House Publishers.

Radosavac, S., Baras, J. H., & Koutsopoulos, I. (2005). A framework for MAC misbehavior detection in wireless networks. In *ACM Workshop on Wireless Security (WiSe)*.

Raghavan, B., Panjwani, S., & Mityagin, A. (2007, April). Analysis of the SPV Secure Routing Protocol: Weaknesses and Lessons. In *ACM SIGCOMM. Computer Communication Review*, *37*(2).

Rekhter, Y., Li, T. & Hares, S. (2006, January). *A Border Gateway Protocol 4 (BGP-4)*. RFC 4271, Draft Standard.

RIPE NCC. (2008, February). *YouTube Hijacking: A RIPE NCC RIS case study*. Retrieved from http://www.ripe.net/news/study-youtube-hijacking.html

Roberts, D., Lock, G., & Verma, D. C. (2007). Holistan: A Futuristic Scenario for International Coalition Operations. In *4th Conference on Knowledge Systems for Coalition Operations (KSCO)*.

Srivatsa, M., Ko, B.-J., Beygelzimer, A., & Madduri, V. (2008). Topology Discovery and Link State Detection using Routing Events. In *IEEE Symposium on Reliable Distributed Systems (SRDS)*.

Stajano, F., & Anderson, R. (1999). The Resurrecting Duckling: Security Issues for Ad-Hoc Wireless Networks. In *Workshop on Security Protocols,* (LNCS 1796, pp. 172-194). Berlin: Springer-Verlag.

Strasser, M., Popper, C., Capkun, S., & Cagalj, M. (2008). Jamming resistant key establishment using uncoordinated frequency hopping. In *IEEE Symposium on Security and Privacy*.

Tian, Z., & Giannakis, G. (2007). Compressed sensing for wideband cognitive radios. In *IEEE Conference on Acoustics, Speech and Signal Processing (ICASSP)*.

Wan, T., Kranakis, E., & Oorschot, P. C. (2005 February). Pretty Secure BGP (psBGP). In *The 12th Annual Network and Distributed System Security Symposium*.

Wang, X., & Reiter, M. K. (2003). Defending Against Denial-of-Service Attacks with Puzzle Auctions. In *IEEE Symposium on Security and Privacy*.

White, R. (2003, September). Securing BGP Through Secure Origin BGP. *Internet Protocol Journal, 6*(3).

Xu, W., Trappe, W., Zhang, Y., & Wood, T. (2005). The feasibility of launching and detecting jamming attacks in wireless networks. In *ACM Symposium on Mobile Ad-Hoc Networks (MobiHoc)*.

Zafer, M., Ko, B.-J., & Ho, I. W. (2008, May). Cooperative Transmit-Power Estimation under Wireless Fading. In *ACM Symposium on Mobile Ad-Hoc Networks (Mobihoc)*, Hong Kong.

Zapata, M. G. (2006, September). *Secure ad hoc on-demand distance vector (SAODV) routing*. Internet Draft. Retrieved from http://personals.ac.upc.edu/guerrero/saodv.html

Zapata, M. G., & Asokan, N. (2002, September). Securing Ad hoc Routing Protocols. In *ACM Workshop on Wireless Security (WiSe)*, (pp. 1–10).

Chapter 6
Policy Technologies for Federation and Interoperation of Coalition Networks

Mandis S. Beigi
IBM Research, USA

Seraphin B. Calo
IBM Research, USA

David A. Wood
IBM Research, USA

Petros Zerfos
IBM Research, USA

ABSTRACT

The salient characteristic of a coalition network is that two networks that may have evolved independently are brought together for a common operation. Dynamic communities of interests need to be established among the two networks, and establishing such communities requiring federating the assets and resources available in the disparate networks that make up the coalition. Each coalition partner may have guidelines restricting the operating conditions for working with other partners, and sharing information with the other partners. This chapter presents a life-cycle for policy management in the context of coalition operations which can be used to federate different information assets in a coalition, and allow improved interoperability among different constituents of the network.

INTRODUCTION

A coalition operation entails an ad hoc arrangement between organizations that act together to pursue a common objective. Such a coalition will involve two or more organizations with their own inherent restrictions on how they are allowed to operate. These restrictions are usually stated as a set of policies that govern information security and fusion, and the sharing/dissemination of information. Within a coalition, ad hoc Communities of Interest (CoIs) come together, perhaps for only a short time, with different assets such as sensors, sensor platforms,

DOI: 10.4018/978-1-61520-855-5.ch006

data fusion elements, and networks, to conduct a task (or set of tasks) with different coalition members taking different roles. The environment is therefore very dynamic, and the policy support for it must provide for the rapid assembly and synthesis of disparate elements while maintaining system security requirements.

A key capability that is typically needed in the context of a coalition operation is to deliver the right information to the right coalition partner, in conformance with the policies of all. The assumption is that each member of the coalition has control of its own network of resources, some of which will also be provided for use by coalition partners.

The flow of information from the organizations' networks must thus be controlled by a set of coalition policies. All data sent between the coalition members' networks should be routed through a network of information filtering elements (e.g. secure gateways) that federate data according to the policies of the organizations involved. This network of gateways would enforce policies concerning: what information about each coalition member's assets will be exchanged with other coalition members; and, who, with what authentication, and under what conditions can access sensor information, and in what form it can be provided. These gateways are thus used to extend current communications frameworks into the tactical coalition environment.

This chapter presents policy technologies for federation and interoperation of coalition networks. To guide the discussion and assist in deriving a set of key requirements for such technologies based on the unique characteristics and functional needs of the target operational environment, a use case scenario is employed that involves a contemporary Peace Support Operation. According to this scenario, UK and US Coalition forces have been deployed into a region to assist the indigenous Government forces. The issues related to information flow interoperability across coalition operations can be divided into three distinct stages: the mission planning stage, the operational planning stage, and the tactical operations stage. Sample policies that address a variety of information requirements such as distribution, information sharing and context awareness as well as various approaches to enforce them in all these phase are elaborated in the context of this scenario.

Over the course of a coalition operations scenario, the policy management lifecycle takes place. It is a multi-staged process that starts with policy authoring and extends to policy deployment and enforcement. In this process, various capabilities and software components are needed to provide the framework that enables policy management. While the same concepts can be applied to manage policies in many different domains, our continuing frame of reference is that of ad hoc sensor networking in support of coalition operations. In such environments, policy management must support the rapid assembly/dynamic control of a network of sensors, platforms, and networks to support multiple concurrent coalition missions. Two of the key capabilities required at multiple stages in the policy lifecycle are that of policy authoring and analysis.

Policy authoring can be done in a number of ways. Policies can be stated in a natural language and then be converted into policy languages (e.g., SPARCLE, JK2005). There are also alternative approaches that follow a template paradigm, as a natural language interface may not always be required or desirable. When the user completes the specification of a policy, it is saved in a computer-interpretable policy language, e.g., the Simple Policy Language (SPL) from the Distributed Management Task Force (DMTF). Many different types of analysis may be useful in policy management environments. These include conflict checking, which detects whether any two pairs of policies have overlapping condition regions and, at the same time, specify conflicting actions; coverage checking for determining whether policies have been defined for all the intended ranges of input parameters; and dominance checking that detects whether a policy might never be applicable due to the existence of other policies of higher priority.

Once the policies have been authored, negotiated, analyzed, and transformed into a deployable format, they can be used within the managed system. Two exemplary technologies that can provide policy-enabled information filtering in federated networks are elaborated in the latter part of this chapter: a policy-enabled distributed, federated database (called GaianDB), which controls information flows that follow the relational model. GaianDB is augmented with software components for performing decisions based on policies (Policy Decision Points –PDPs) and enforcing their actions (Policy Enforcement Points –PEPs); and a publish/subscribe messaging middleware (called the ITA Sensor Fabric) that is similarly extended with PDP and PEP components for performing message filtering on the MQTT protocol (MQ Telemetry Transport, n.d.).

The rest of this chapter is structured as follows: scenario and requirements for information federation and interoperation in coalition networks are described in Section 2. The policy lifecycle, system architecture for managing it, as well as the library and tools that implement this architecture are presented in Section 3. Section 4 describes the two policy-enabled federation and interoperation technologies based on the relational and publish/subscribe paradigms respectively. The chapter concludes in Section 5 with a summary and several insights gained from the development of these technologies.

SCENARIO AND REQUIREMENTS

As a reference point for further guiding the discussion, consider a contemporary warfare scenario that involves multiple nations forming a coalition to conduct a Peace Support Operation. In such a scenario, we assume that UK and US Coalition forces have been deployed into a region to assist the indigenous Government forces in encounters with an active insurgency and reassuring/supporting the local population. The coalition nations (for example the UK, US and possibly the indigenous Government) must operate together to protect the forces in the region and control the region (to protect and support the local population, and deter/defeat the insurgency). This requires the use of ISR (Intelligence, Surveillance and Reconnaissance) networks to provide suitable and actionable intelligence information.

Each coalition partner operates and maintains an ISR network that interconnects the various assets belonging to that partner. These ISR networks consist of a multitude of devices with heterogeneous energy, processing, storage and communication capabilities (Future Combat Systems, n.d.). Nodes can be anything from sensor platforms to unmanned aerial and ground vehicles equipped with wireless communications. As it is central to the Future Combat Systems vision (Future Combat Systems, n.d.), these networks are mobile and formed in an ad hoc manner (henceforth called mobile ad hoc networks – MANETs), offering unprecedented capabilities over conventional networks in terms of deployment efficiency and flexibility, as they require no infrastructure support in the battlefield (Chiang, Chadha, Newman, Lo & Bauer, 2007). Any technologies for federated, controlled data access for network-centric coalition operations should be flexible enough for deployment in such networks.

Each of the ISR networks that are operated by the coalition partners needs the ability to support access from other members of the coalition who may not all be equally trusted. This requires that policy and access control restrictions be put in place to control the flow of ISR data and control operations throughout the network environment. For example, consider the following policies that dictate how information is shared amongst the forces:

1. US can only view UK assets that are assigned to the mission.
2. UK can only view US assets that are located in the mission's perimeter.
3. The munitions state of UK tanks assigned to the mission are not exposed to non-UK coalition members.
4. Only share the location of US or UK assets with trusted parties.
5. US asset information may not be routed through non-US network elements.

The traditional approach for enforcing policies is in a centralized fashion, by requiring that all data flowing between the coalition networks be routed through a well known gateway. This may be the simplest solution to deploy and it decreases processing overhead at the other nodes of the network, but it has a number of undesirable characteristics: if only one gateway exists, it becomes a single point-of-failure and a potential traffic bottleneck. Due to the dynamic nature of the networks under consideration, connectivity to the gateway may be lost and then no traffic can be securely transmitted between affected nodes. Alternatively, every node in the network could be responsible for enforcement of all policies applicable to it and the data that flows through it. The system becomes more robust, but the awareness of every node to more global conditions needs to be increased, with large sets of redundant policies being deployed throughout the coalition networks. An approach that is positioned between the above two extremes is to have network traffic flow through a network of federation gateways and have all relevant policies be accessible from a distributed policy repository. Policies can then be deployed selectively, with more global federation-related controls being applied at the gateways. Based on the above scenario, several key requirements emerge for policy technologies that enable federation and interoperation of mobile ad hoc networks in a coalition setting:

- *Distributed, ad hoc policy deployment*: the ability to define policies locally and yet have them be enforced globally in the ISR network is highly desirable. Network elements will join and leave, together with the policies they would like enforced. As they join, their policies need to be visible and applied across the network.
- *Data-centric policies for information sharing*: agile control of information flow between two or more coalition members of a joint mission should apply at the application-layer protocol used to carry the data that is shared. This allows for a finer grained approach to information sharing than simple allow/deny rules, which typically consider only the origin and destination of information requests. For example, if limited access to data regarding artillery capacity is allowed to some members of the coalition, data sharing policies might specify transformation rules that generate only aggregate summaries, instead of detailed reports on deployed artillery.
- *Transparent deployment*: as coalition networks are managed by disparate administrative domains, it would be unrealistic to expect requestors of information as well as information sources to modify their heterogeneous systems to accommodate controlled access by allied members. Instead, it is desirable to interpose federation technology that is easy to deploy and requires no modifications to existing end-systems, their schemas or data models.
- *Context-aware interoperability policies*: coalition sensor networks consist of heterogeneous nodes with varying degrees of processing, communication, and battery capabilities, operating in highly dynamic environments due to increased mobility and under a variety of unpredictable and adversarial conditions in the battlefield. For this reason, information sharing has to

take into account the context in which it takes place, such as spatio-temporal constraints, device capabilities, level of security clearance, etc.

POLICY LIFECYCLE

As has been indicated, policies form a key underlying decision component in coalition operations. The policy management lifecycle, as depicted in Figure 1, illustrates the manner in which policy technologies are employed in a military environment from the time of policy authoring to that of policy deployment. As communities of interest are formed to conduct coalition operations, the essential characteristics of each mission must be determined and supported, which include the required information technologies, policies, roles, levels of trust, etc. Policies need to be conveniently defined at a high level of abstraction and refined into the rules, constraints, and decision constructs that can be executed by the security mechanisms in the coalition computing and networking systems. They are thus specified using an authoring tool that captures high level goals and requirements in an intuitive manner. This can be done in terms of natural language or predefined templates, either free form or structured. Once authored, the sets of policies need to be analyzed to ensure that they are meaningful, consistent, and non-conflicting. A number of different kinds of analysis may be performed on the policy sets depending upon the model of the system for which they have been defined. The system model at the specification level will deal with appropriate system concepts and their relationships.

The policy management lifecycle can be divided into three distinct stages: the mission planning stage, the operational planning stage, and the tactical operations stage. In the mission planning stage, the forces from different coalition partners come together to develop the common operating principles and policies that would apply to coalition information flows, and any Intelligence, Surveillance and Reconnaissance (ISR) assets that may be shared during coalition operations. Software components supporting policy negotiation would be employed along with those that maintain asset inventories. Policies would be agreed to by

Figure 1. Policy lifecycle – from authoring to deployment

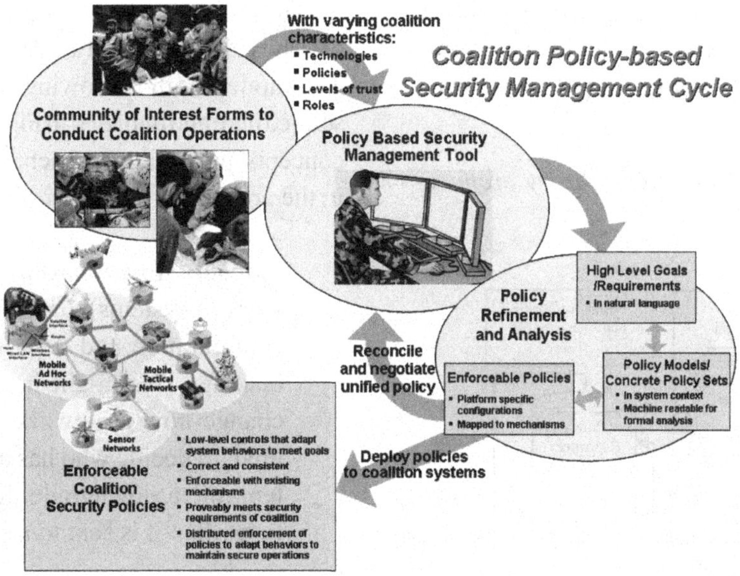

the various members of the coalition at a high level. These would need to be further refined at subsequent stages of the lifecycle. The overall process thus requires the use of components for policy authoring, policy analysis, policy transformation, and policy deployment. Analysis needs to be performed at each stage to ensure that the intent of the original policies has been preserved and that no unwanted artifacts have been introduced in the transformations.

In the operational planning stage, the coalition team has been formed and its operational policies have been defined. The coalition team would be tasked to undertake one or more operations, and the appropriate networking support would need to be determined. The team would also need to determine the necessary set of ISR assets and their network configuration. The high level policies agreed to at the previous stage would now be refined into lower level policies that are specific to the needs of the mission and the equipment that will be used.

In the tactical operations stage, the networks are deployed and available on the ground. They would be interconnected through policy-enabled gateways that enforce any policy constraints that are applicable as information flows between them. Given the dynamics of the environment, the deployed policies may be changed or overridden as the mission progresses. Such changes would need to be reflected back to the policy management system so that their overall effect upon the original intent and the other policies active in the system can be determined.

Figure 2. Policy management library architecture

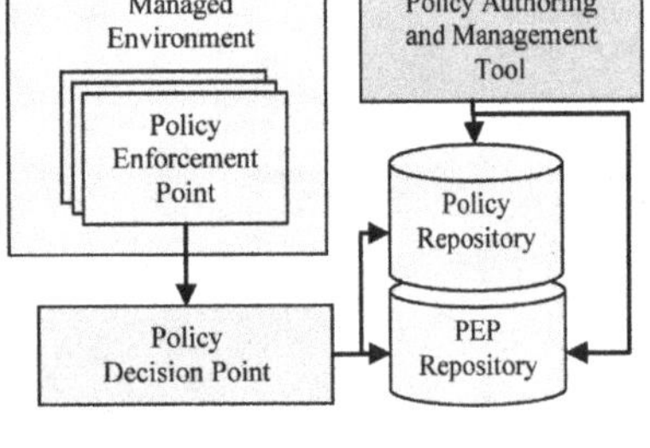

The reminder of this section describes software technologies that primarily assist the mission planning phase: authoring, analysis, negotiation and management of policies. Section 4 describes technologies that focus on policy-enablement of the tactical operations stage.

Policy Management Architecture and Library

A number of different components have been identified as needed to support the overall policy management lifecycle that includes policy authoring, analysis, transformation, and deployment. Such capabilities can be provided for each specific application instance, or as generic components that can be customized to meet the requirements of many particular domains of application. The latter approach allows for reusability of the components and flexibility in their deployment, and, building on general policy management architecture, the design and implementation of a policy management library consisting of the major elements that are needed for policy based management is presented in this section.

The policy management architecture builds on a fairly widely accepted model for policy runtimes. This model has the following key components described here and depicted in Figure 2. The principal concepts and system elements that are addressed in the architecture are described below:

- *Managed Environment*: a system (or network of systems) that requires the dynamic configuration and deployment of decision making. This might include the ability to change how a network firewall is configured, to decide who has access to a file system, or to downgrade fidelity of information when it is sent to a specific user.

- *Policy*: a dynamically deployable software component for execution within a managed environment. Policies are typically used for authorization, data filtering, and control operations within the managed environment. Evaluation is done over a set of instance data provided by the managed environment. The Simple Policy Language (SPL) (DMTF: CIM Simplified Policy Language, 2007) defines a specific form of policy that uses a condition specification and an action to execute when the condition evaluates to true.
- *Policy Enforcement Point (PEP)*: a point in the managed environment that needs to have a decision made, for example, to enforce access to a secured resource. It requests the answer to this decision from the Policy Decision Point. One implements such points in the managed environment as PEPs instead of a hard-coded decision so that the decisions can be configured flexibly and dynamically with policy. The PEP is responsible for providing instance data over which the policies are evaluated. There may be any number of PEPs in the managed environment.
- *Policy Decision Point (PDP)*: evaluates policies stored in a repository that are relevant to the requesting PEP (e.g., an access rights decision). The results of evaluation are returned to the PEP for enforcing the action (e.g., allowing or denying access). PDPs also decide which policies in the policy repository apply to the decision request.
- *Policy Repository*: holds the policies that are available for execution by the PDP. Policies may be activated or deactivated within the repository.
- *Policy Authoring and Management Tool (PAMT)*: an application that allows one to specify and edit policies and to manage policies available to the PDP by controlling the contents of the policy repository and the activation state of the policies contained therein. Policy analysis (conflict, coverage, etc.) may be provided by this tool to assure consistent policy deployments.

The policies and metadata are stored in a policy repository. A policy stored in a repository may be marked as active or inactive. The Policy Management Library (PML) implements a number of policy repositories including non-persistent in-memory, and a persistent relational database. Before requesting policy decisions, a Policy Enforcement Point (PEP) registers with a PDP and provides information including its type (i.e. authorization or obligation), a description of the instance data that will be provided on decision requests, and an optional set of attributes. When a PDP receives an evaluation request from a PEP, it performs a matching of the PEP metadata with the metadata (i.e., type, instance, attributes) and policy activation state of policies in the policy repository. This matching process must be done on every evaluation request and so needs to be as efficient as possible. PML caches the results of the matching process to speed evaluation.

Policy Authoring and Management

Buildings on the foundations of the policy library, a set of tools automate and support the management of policies during their lifecycle. Following, the policy authoring, analysis and management features are described in greater detail.

Policy Authoring

The policy authoring tool (PAT) is used to author policies in a structured high level format. The policies are defined using policy templates that capture the structure of the policy rules in terms of boolean combinations of sensor values and attributes, constituting the condition clause, and the actions that should be taken when the condition clause evaluates to "true". The policy author sees a natural language expression of the policy with key attributes indicated and values of these key attributes can then be chosen to determine a policy instance. The templates are usually defined by an administrator prior to authoring of the policies. For example, a template might be used to define *sensor information access control* which allows the policy author to specify who (person, data fusion element, etc), with what authentication, under what conditions, and in what form (i.e., raw, processed, fused) a particular sensor information can be accessed. For this template, the structured natural language policy statement looks like the following:

A US correlator asset may receive SOUND messages from a UK AcousticArray sensor between the hours of 8 and 17.

This statement is translated to the following policy phrase:

If {affiliation=="US" AND subscriber_type=="Correlator" AND info_feed_type=="Sound" AND affiliation=="UK" AND publisher_type=="AcousticArray" AND time==[8, 17]},

Then 'Allow access'

The underlined arguments are the so called sensor values and the values in double quotes are their values in the policy. Once the templates are created by an administrator, they will be stored in a repository and made available for use by other administrators and policy authors.

A policy author (or administrator) may analyze the stored policies for correctness using the PAT. The analysis operations check whether there are any conflicts, uncovered regions or dominated policies among the set of authored policies. The PAT uses the analysis operations from the Policy Management Library (PML) to identify any dominance, coverage and conflict issues within the authored policy set. Figure 3 illustrates the policy management system and the interaction among its subsystems.

Policy Analysis

- *Conflict check:* The conflict check analysis uses a set of rules which specify which policy actions conflict with one another (Chomicki, Lobo & Naqvi, 2003).
- *Coverage check:* The coverage check analysis uses a set of "coverage rules" which define the boundaries of the user's intended condition regions. These rules define a set of hyperspaces that must be covered by the policies. If the authored policies' condition regions do not fully cover the intended regions within these boundaries, the coverage check outputs the uncovered regions.
- *Dominance check:* The dominance check operation outputs all the dominated policies as well as the policies that dominate them and the condition region(s) which are dominated by other policies.

Policy Management

Policy management involves the administration of the policies deployed into the management environments and typically takes place after authoring and/or negotiation, but may also be performed during mission operations. A policy administrator is authorized to make changes to the policy environment including the addition and removal of policies. Policies may be left in the system but have their activation state changed. Deactivated policies are not considered for evaluation by the PDP.

Policy analysis may be performed any time the state of the policies in the managed environment changes. For example in deploying new policies

Figure 3. Policy management system diagram

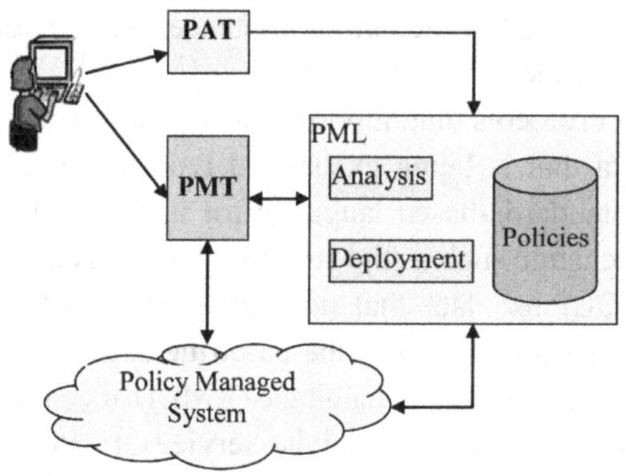

or activating already deployed policies. Any such change may result in the creation of conflicts, overlaps and redundancies.

Policy Negotiation

In coalition environments, a set of policies govern the operation of each of the partners. While planning for coalition operations, the coalition members need to suggest policies for the joint mission and determine the constraints on operations that exist due to the presence of their own policies and the interaction of these with the policies of other partners.

In order to expedite planning, the coalition members need the capability to analyze the policies that will be in effect, and understand the impact of those policies on the joint mission. Each partner must analyze the policies suggested by their coalition partners together with their own organizational policies to see whether they permit a desired set of operations to be performed, where they conflict, and whether those conflicts are acceptable. The other partners are also performing similar analyses of their own. Following such analysis, each partner will try to negotiate a modification of policies with the other coalition partners. The negotiation will result in either a set of modified operational policies that all partners would agree to use for the joint mission or a disagreement and a voided mission.

Negotiation in general is the process of making proposals, trading options, offering concessions and obtaining mutual agreement (Jennings, 1993), (Lomuscio, Wooldridge, & Jennings, 2001), (Jennings, Faratin, Lomuscio, Parsons, Sierra & Wooldridge, 2001). It takes place between two or more entities with an ultimate goal of satisfying all entities with mutual agreement. Based on the concepts of policy ratification (i.e., determining policy conflicts, coverage gaps and dominance (Agrawal, Giles, Lee & Lobo, 2005)) we can give a precise definition of what constitutes an offer, compare offers, test that a solution has been found, and based on that, define the protocols.

Before negotiation begins, policies authored by each party are checked for syntax errors and then analyzed for their interactions with each other for conflicts, coverage and dominance. Policies are ready for negotiation once they have passed these checks, and have been normalized to ensure that the two policy sets use consistent units, as the policies from the various partners can only be compared if they use the same units. Figure 4 shows the process of policy analysis and normalization when two entities engage in negotiation.

To ensure convergence and to eliminate infinite loops, the system needs to guarantee that the modified policies are not repeated in each iteration, by keeping a history of the modified policies in a database. A faster convergence method may check to make sure *every* conflicting region gets smaller, during each iteration of policy modification.

Two main forms of interaction exist among entities engaged in negotiations: the first approach is peer-to-peer/direct communication between the two entities; the second architecture is centralized, which uses a mediating third party agent. When the entities involved in negotiation do not have analysis capabilities, the centralized architecture will be used.

Figure 4. Analysis process when two entities engage in negotiation

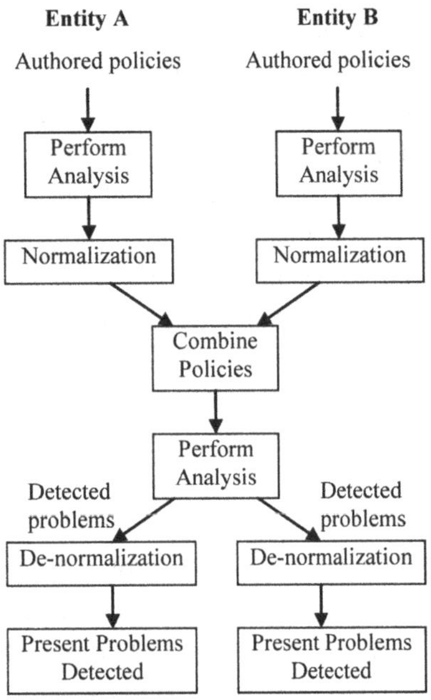

POLICY TECHNOLOGIES FOR COALITION OPERATIONS

Building upon the foundation of the aforementioned software components that provide policy lifecycle management, the approach for policy-enabling two popular protocols used for information federation and interoperation is presented in this section.

Various application-level protocols are used for information interchange among coalition networks. To control this information flow using policies, a model of these protocols is required, so that policies can be authored, deployed and executed against the elements of the environment that communicate over these protocols and are also being managed. This model constitutes the managed runtime (also referred to as managed environment with its respective object model) in the policy architecture (Figure 2). The granularity and expressiveness of the policies that can be applied against the managed environment critically depends on the functionality of the model as exposed by the runtime instance data provided to policies.

Protocols that impose a clear structure on the data that is being exchanged typically employ a standards-based language for information interchange such as the structured query language (SQL) for data that adheres to the relational model, publish/subscribe-based messaging, or the extensible markup language (XML) for coalition operations built around the Service Oriented Architecture (SOA) paradigm. Consequently, placing filters driven by policies on the application-layer protocol that encodes and transports data described in the aforementioned language formats (JDBC, JMS and SOAP respectively) enables controlled federation of these networks.

While the implementation of a coalition environment may benefit from the use of a Service Oriented Architecture (SOA), this chapter does not focus on such paradigm. A coalition environment can be considered as an interconnection of nodes providing services that are managed by different entities. and, as such, it could be described in terms of service requesters, service providers, and message exchange patterns. It is within the message exchange infrastructure that the security mechanisms would be implemented (e.g., authentication, authorization, and audit). Policies can be employed to make these mechanisms more flexible and adaptable. Policies can also be employed within the IT infrastructures of each node for domains other than security (e.g., configuration, monitoring, and performance). The applicable policy technologies would be much the same in SOA-based systems as in other types of systems. The difference is that in an SOA environment such concepts as service, service composition, and service orchestration are recognized as first class entities, making it easier to apply policies to such higher level architectural elements.

Two key component technologies are central in the effort to enable controlled federation and interoperation of coalition networks: the software library for policy management (PML), and a data federation system that implements at least one of the protocols described above. In the rest of this section, two representative systems that embody both of these technologies are presented: the first implements an application-layer firewall that filters information exchanged using the SQL language over JDBC (Fisher, Ellis & Bruce, 2003), the de facto protocol used for database-independent connectivity. The second system controls information flows over a publish/subscribe system developed for sensor networks.

Policy-Enabled Federated Database

Besides establishing connectivity to databases and other tabular data sources, the JDBC protocol is used to send SQL queries to the sources and process the returned results. Based on this mode of operation, there are two main approaches for performing filtering and controlling of the information flow: a) re-writing the SQL query (Wang, Yu, Li, Lobo, Bertino, Irwin, et al, 2008) that is submitted according to filtering policies, so as to limit at the source the results that are retrieved from the databases, and b) filtering the returned results on a row-by-row basis, before returning them to the requesting entity. The former approach has several advantages such as that it minimizes the results that are transmitted over the network. Also, it exhibits performance benefits, as the query optimizer at the data source identifies the most efficient execution plan for the filter-enhanced query. However, the latter method, namely filtering the returned result-set, provides the finest granularity of control as obligation and authorization policies are directly applied to the raw data at the individual cell level. Moreover, it allows for easy incorporation of contextual information that filtering policies can take into account, without requiring any extensions in the information sources (Wang, Yu, Li, Lobo, Bertino, Irwin, et al, 2008) . In what follows, details of the latter approach are presented through the description of a prototype system based on a federated database named GaianDB, which is particularly suited for network-centric operations.

GaianDB is a Dynamic Distributed Federated Database that combines the principles of large distributed databases (Becla & Lim, 2007), database federation (Bent, Dantressangle, Vyvyan, Mowshowitz & Mitsou, 2008), network topology (Newman, 2003) and the semantics of data (Shadbolt, Hall & Berners-Lee, 2006). The resulting database architecture is ideally suited to a range of applications that are required in support of networked operations. To implement a network of federated gateways, each node acts as a policy controlled gateway to the locally federated heterogeneous data sources. These sources can include databases, sensor message streams, etc. which are exposed as virtual tables to the rest of the network. Policy-controlled access to data is then performed from any gateway node using SQL-like queries and distributed stored procedure-like processing. Using the 'Store-Locally-Query-Anywhere' (SLQA) paradigm, access to a distributed policy repository is also maintained such that policies can be stored anywhere in the network and accessed from any gateway node.

To logically connect the database nodes into a network of databases, the GaianDB adopts a strategy that uses "preferential attachment" (Bent, Dantressangle, Vyvyan, Mowshowitz & Mitsou, 2008), such that, as new nodes join the network, all nodes remain within a limited number of logical network hops from each other and no single node becomes over-connected, thus avoiding bottlenecks. The average number of 'hops' between any two nodes in the logical network is a measure of network diameter and typically this rises logarithmically with the number of nodes(n) (Bent, Dantressangle, Vyvyan, Mowshowitz & Mitsou, 2008). Using controlled flood queries, the query execution time across the network is minimized, through parallel processing, to a linear function of the diameter.

In support of a distributed policy solution, the GaianDB allows the definition of contextual meta-data for data on a database node using its constant columns feature. For example, a database node may declare constant columns to describe its location and its processing capability that can be updated any time to reflect varying contextual conditions. The current implementation of the GaianDB is based on the Derby database engine, which implements a Virtual Table Interface (VTI). The VTI facilitates connections to data sources other than the Derby tables and has been extended to enable database nodes to self configure into a network of databases.

Enforcing global data access policies across this type of distributed federated database can be achieved using a distributed version of the type of policy architecture described Section 3. In this case the PEP and PDP components are deployed to all database nodes where policy is to be enforced. Each PDP then has access to a distributed policy repository which uses a local database connection for writes (Store Locally) and a Gaian database connection for reads (Query Anywhere). A Policy Management Tool can be used at any node in the distributed database to create, analyze and store policies locally, which will then be enforced globally at all PEPs.

The GaianDB incorporates a 'plug-in' mechanism that facilitates the implementation of various types of filters on the JDBC data that flows through the system, including the SQL query that is submitted and the ResultSet structure that is returned as the result to that query (Fisher, Ellis & Bruce, 2003). To realize the distributed policy enforcement architecture, a policy management module is incorporated into the GaianDB as a 'plug in' using the Watson Policy Management software library, as shown in Figure 2. The PDP and PEP components of the policy module are instantiated in the policy plug-in for each new query that arrives in the node. Then, for each row of the ResultSet that is returned for that query, a policy evaluation is requested based on the contents of that row and the context in which the query has been requested. The result of the policy evaluation might be either to return/drop the row as a partial result to the requestor (in case of an authorization policy), or even modify the contents of some (or all) of the columns of the row (in case of an obligation policy).

To support evaluations and decisions based on the requested data, an object model that describes the elements of the query and its results needs to be developed as the managed runtime. The model provides anchor classes for the managed runtime, for representing tables, rows and columns, which are fundamental elements of the relational schema. Columns can be referenced in the policy by their name in the query, and the values of their data can be changed on the fly before being returned, according to policy decisions. A generic anchor class represents the context in which the query is submitted, which can include arbitrary information such as the name of the user that requested it, her country affiliation and security clearance, or any other type of information that might be considered by the policies.

The policy repository is also implemented as tables in the GaianDB and queries requesting policies are propagated to all other nodes in the network. Local policies can be accessed by limiting the query propagation to the locally-attached database only (i.e. query depth of zero) whilst global policies can be accessed from any (or all) other policy repositories by performing a query of an unlimited query depth. The response time for the global query is then a linear function of the inter-network graph diameter. Policies are searched using the controlled flooded queries of GaianDB and, once retrieved, are stored locally in the policy repository of each gateway. Thus, for each query, policies are fetched only once over the network and policy evaluations are performed from the local cache to avoid further network latencies (Figure 5).

Figure 5. Policy endabled federated gateway node with PEP/PDP

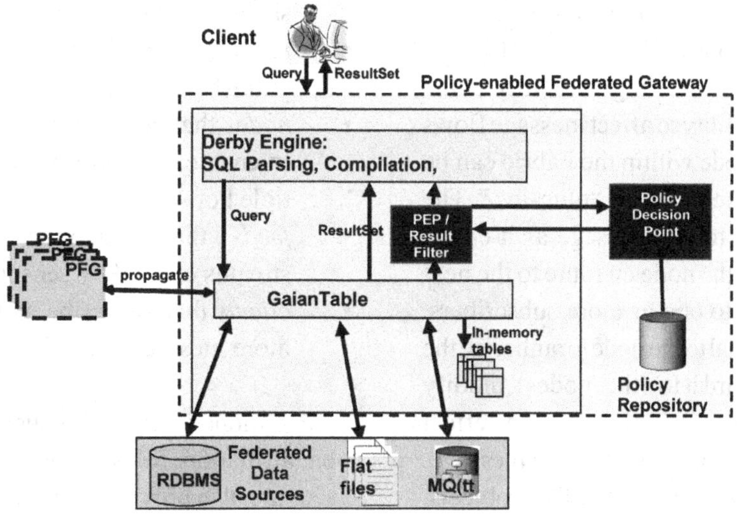

Policy-Enabled Publish/ Subscribe Middleware

Recently, policy-enabled publish/subscribe platforms has been an active area of research, with much of the work focusing on authorization of access to published data. (Opyrchal, Prakash, Agrawal, 2007) provides a policy-enabled system for controlling access to location-based information in order to meet privacy requirements and (Belokosztolszki, Eyers, Pietzuch, Bacon & Moody, 2003) details a generalized RBAC model for publish/subscribe middleware. However, they limit authorization policies to subscription time, which does not easily allow for dynamic changes that might later allow access to a particular topic. (Wun & Jacobsen, 2007) describes a novel approach to matching policies with topics by utilizing the matching capabilities of the underlying publish/subscribe system. While powerful, this approach is not sufficient when policies need to be matched against arbitrary message payloads about which the pub/sub system is not aware, for e.g. JPEG images.

Following the same approach as with the policy-enabled federated database, the second exemplary system for policy-driven data federation is based on the IBM Sensor Fabric (Bergamaschi, Conway-Jones, Gibson, Stanford-Clark, Verma, Calo, et al, 2008) middleware (Fabric for short), which enables message routing in a network of nodes following the publish/subscribe paradigm implemented through the MQTT protocol (MQ Telemetry Transport, n.d.). The Fabric presents a generalized publish/subscribe model in which publishers send messages into the Fabric and if there are subscribers for those particular messages, the messages are routed through the network to each of the subscribers. A message is published on a "feed" and subscribers must specify the feed they wish to subscribe to, but may also specify the node, and publisher to narrow their subscription if desired. The Fabric manages the publishers and subscribers that are connected to the Fabric providing facilities for tracking of node availability, discovery of node location and routing information and establishment of communication channels among the nodes. It also manages the inter-node connectivity to route the messages to the subscribers through one or more nodes (multi-hop) of the Fabric.

Under various circumstances it may be desirable to dynamically install configuration, authorization and operational code within the Fabric to affect message routing, authorized subscriptions and publications, and/or modify message contents. The Watson Policy Management Library (WPML) together with the plug-in facility of the Fabric provides the ability to affect message flows using policy. Each node within the Fabric can be configured to run one or more "plug-ins." The plug-ins gain access to the message as it enters and before it departs the node en route to the next node and ultimately to one or more subscribers. The plug-ins may signal to the node to authorize the message (i.e. to forward it to other nodes), modify the message contents, or trigger some side effect based on the information carried in the message. For messages containing Java serialized objects, JPEG byte code as well as comma-separated values both authorization and obligation policies can be enforced, while for messages of arbitrary type, only authorization policies are allowed.

Similar to the policy-enabled federated gateway system, an object model (managed runtime environment) for the Sensor Fabric must be defined (or be made available) to the policy subsystem, so as to allow evaluation of policy conditions and execution of policy decisions. The object model for the Sensor Fabric middleware includes its network elements, which are:

- *sensors* (or publishers) that publish messages into the network,
- *platforms* that host one or more sensors/publishers,
- *nodes* that connect with one another so as to enable transport of messages over multiple hops,
- *feeds* that constitute named message streams created by sensors/publishers, and
- *clients* that subscribe and consume one or more message feeds.

Additionally, attributes such as name, description, affiliation, roles and level of security clearances are also provided for the network elements of thc object model. Some of these elements and their interconnection are illustrated in Figure 6. Within a policy, the message runtime is the starting point for gaining access to the above information that is active during the evaluation. Routing information including the source of the message (sensor/publisher), feed type, the current, previous and next nodes in the route, and the destination node and client are also available. The SPL policy below combines the fabric resource model with WPML authorization mechanisms to demonstrate the authorization of messages between a publishers

Figure 6. Pub/Sub sensor fabric network with policy plug-ins

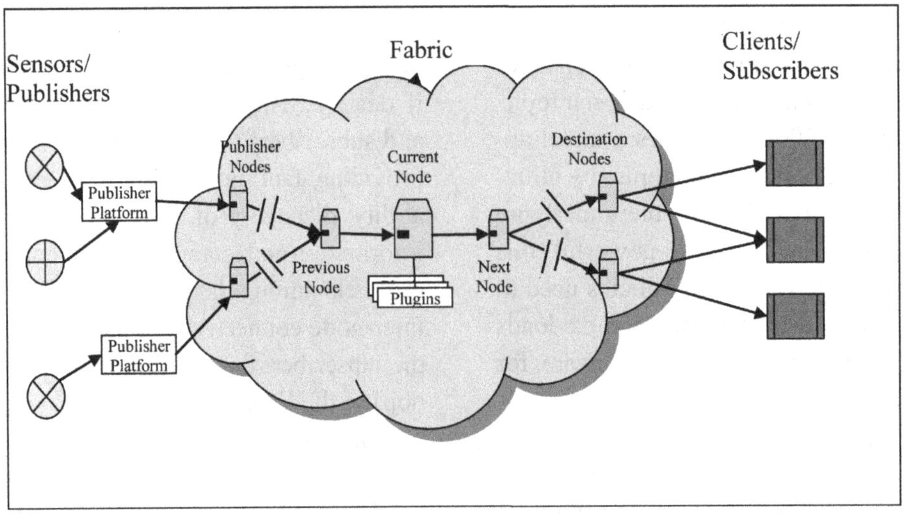

and subscribers of the same affiliation, regardless of message content.

```
Import Class fabric.bus.messages.facade.
IMessageFacade:msgRuntime;
Import Class com.ibm.watson.pml.policy.
types.IAuthorizer:authorizer;
Strategy Execute_All_Applicable;
Policy
{
Condition
{
(msgRuntime.getClient().getAffiliation()
==
msgRuntime.getPublisher().getAffilia-
tion())
}
Decision
{
authorizer.allow()
}
}:1;
```

The policy plug-ins are written to evaluate policies over the fabric runtime and any Java objects that may be represented in, or created from, the message. Java objects can be extracted or created from the message in a plug-in-specific manner. Plug-ins to extract serialized Java objects, JPEG image data as a byte array, and a string containing comma-separated values. An example of a policy to downgrade imaging transmitted from a US publisher to a UK subscriber is as follows:

```
Import Class fabric.bus.messages.facade.
IMessageFacade:msgRuntime;
Import Class com.ibm.watson.pml.policy.
types.Bool:msgRewrite;
Import Class com.ibm.watson.pml.fabric.
messaging.messages.IImageMessage:message;
Strategy Execute_All_Applicable;
Policy
{
Condition
{
msgRuntime.getClient().getAffiliation()
== "UK" &&
msgRuntime.getPublisher().getAffilia-
tion() == "US"
}
Decision
{
message.setGrayscale() &&
message.setResolution(0.2) -> msgRewrite.
setTrue()
}
}:1;
```

In the above example, note that because the policy imports an IImageMessage class object, the policy will only be applied to fabric messages containing an image. The msgRewrite flag when set to *true* causes the plug-in to notify the fabric of the change in message payload.

Additional plug-ins can be added to support other ways of extracting Java objects from the message. As mentioned, a policy plug-in can be one of two types: authorization or obligation. The basic flow of execution is as follows and depends, to some degree, on the type of plug-in:

1. Check if the message has already been discarded and if so, perform no action.
2. Extract any objects from the message.
3. If any objects are found, provide them along with the base runtime to the policy evaluator.
4. If the policy evaluation request returns false and the plug-in is an Authorizer, then the message is discarded, otherwise it is passed on through the Fabric.

In the case when both authorization and obligation plug-ins are enabled, their order of execution is prioritized in the Fabric so that authorization plug-ins are executed before the obligation plug-ins. This way, further unnecessary computations are avoided when messages fail authorization.

Performance Implications of Policy Enablement

Evaluation and enforcement of policies will impose some overhead on the federation systems, both the policy-enabled federated database, as well as the policy-enabled publish/subscribe middleware. However, from a preliminary evaluation of the effect of policy control on the performance of the federated database (Calo, Wood, Zerfos, Vyvyan, Dantressangle & Bent, 2009), there is a low, fixed cost per policy evaluation. Careful implementation with lightweight policy infrastructures employing mechanisms like proactive caching significantly ameliorates the computational burden.

Indeed, in the design of our system, caching at various levels has been aggressively employed. For example, in the design of the policy management library, when a PDP receives an evaluation request from PEP, it performs a matching of the PEP metadata with the policy metadata, using type, instance, attributes and policy activation state. This matching process must be done on every evaluation request and so needs to be as efficient as possible. WPML caches the results of the matching process, which requires that the repository provide a timestamp of the last policy modification. Since the policy repository is updated infrequently, this generally reduces a full read of the policy repository together with the matching process down to a single read of the timestamp from the repository and an in-memory cache look up. Our relational database implementation for the policy repository uses triggers to maintain the repository timestamps.

In a similar way, the GaianDB federated database provides the capability for a node to cache the rows of its federated data sources in-memory, and additionally to index these rows. This can dramatically improve performance for data retrieval and is best suited to reasonably static data (with infrequent data updates) to minimize the need for synchronization with the underlying data sources.

As many security mechanisms in particular are required in such environments, it is not clear that the more flexible policy based approach would be significantly more burdensome than static approaches.

CONCLUSION

Policies have the potential to play a key role in network-centric, coalition operations. They can serve to codify the "rules of engagement" for individual coalition members as well as the coalition as a cooperating force, and define levels of security clearance, protection and access to information, as well as authorization and control of assets.

The policy lifecycle in a coalition environment is a multi-staged process that starts with each member of the coalition authoring its national and coalition policies for a given mission, followed by analysis for conflicts and redundancies; then, policy negotiation takes place among the coalition members to reach agreement on the set of policies required for the mission, including an iterative refinement and analysis phase; finally, deployment and management of policies within the managed environment occurs over the course of the mission. The Policy Management Library from IBM Research provides the policy components (policy models, repositories, decision points, etc.) and tools (for policy authoring, negotiation and management) that are necessary to realize and implement the policy lifecycle.

Two exemplary technologies for information federation and interoperability in a dynamic, mobile ad hoc network coalition environment have been policy-enabled, based on the foundation of the policy management library: a policy-enabled, federated database system provides controlled dissemination, filtering and transformation of information that is exchanged among the assets of coalition members through SQL queries; and a policy-enabled distributed, publish/subscribe middleware that controls information published to various topics of interest, flowing both within and across coalition member networks.

The design and development of the policy library, tools and exemplary policy-enabled technologies for federation and interoperation that were built upon them provided us with several insights: first, distributed policy evaluation and enforcement is particularly appropriate for the dynamic, mobile ad hoc networks that are envisioned for use in network-centric operations. Indeed, frequent network partitioning, as well as limited and unreliable communications make the deployment of typical centralized solutions impractical and further introduce single points-of-failure. However, distributed policy evaluation and enforcement ensures robustness to network partitioning and a more efficient use of the limited network resources.

Second, the policy analysis step becomes a particularly important aspect of the policy lifecycle for balancing the conflicting needs for policy expressiveness and consistency. Not only is it useful for checking conflicts among multiple policies that are written by several authors, but it also provides invaluable support during the negotiation phase, in which policies from coalition members with possibly conflicting objectives are combined onto a common set that will be used for the mission.

Third, authoring of policies using templates offers a common basis for parties in a coalition setting to conduct negotiations and agree on a final set of policies for a mission. They provide the common policy structure within which various objectives can be expressed in a constraint manner that in turn facilitates the process of negotiation.

In conclusion, policies offer a natural way for members of a coalition environment to express their constraints and requirements with respect to information federation and network interoperation, over the course of joint, network-centric operations. A policy management library and its assorted tools can greatly facilitate the policy-enablement of technologies that are used for federation and interoperation, and support the lifecycle management of these policies.

REFERENCES

Agrawal, D., Giles, J., Lee, K.-W., & Lobo, J. (2005). *Policy Ratification*. Policy.

Becla, J., & Lim, K. T. (2007). *Report from 1st Workshop on Extremely Large Databases*. Held at Stanford Linear Accelerator Centre, Menlo Park, CA.

Belokosztolszki, A., Eyers, D. M., Pietzuch, P., Bacon, J., & Moody, K. (2003). *Role-Based Access Control for Publish/Subscribe Middleware Architectures*. Distributed Event Based Systems.

Bent, G., Dantressangle, P., Vyvyan, D., Mowshowitz, A., & Mitsou, V. (2008). A dynamic distributed federated database. In *Proceedings of ACITA'08*, London.

Bergamaschi, F., Conway-Jones, D., Gibson, C., Stanford-Clark, A., Verma, D., Calo, S., et al. (2008). Policy Enabled ITA Sensor Fabric, A Distributed Framework for the Validation of Experimental Algorithms Using Real and Simulated Sensors. In Proc. POLICY.

Calo, S., Wood, D., Zerfos, P., Vyvyan, D., Dantressangle, P., & Bent, G. (2009). Technologies for Federation and Interoperation of Coalition Networks. In *proceedings of The 12th International Conference on InformationFusion,* Seattle, WA.

Chiang, C.-Y., Chadha, R., Newman, S., Lo, R., & Bauer, R. (2007). Integrated Network Operations For Future Army Tactical Networks. In MILCOM'07, Orlando, FL.

Chomicki, J., Lobo, J., & Naqvi, S. A. (2003). Conflict Resolution Using Logic Programming. *IEEE Transactions on Knowledge and Data Engineering*, *15*(1), 244–249. doi:10.1109/TKDE.2003.1161596

DMTF. (2007). *CIM Simplified Policy Language*. CIM-SPL.

Fisher, M., Ellis, J., & Bruce, J. (2003). *JDBC API Tutorial and Reference* (3rd ed.). Upper Saddle River, NJ: Prentice Hall PTR.

Future Combat Systems. (n.d.). Retrieved from https://www.fcs.army.mil

Jennings, N. R. (1993). Commitments and Conventions: The Foundation of Coordination in Multi-Agent Systems. *The Knowledge Engineering Review*, *8*(3), 223–250. doi:10.1017/S0269888900000205

Jennings, N. R., Faratin, P., Lomuscio, A. R., Parsons, S., Sierra, C., & Wooldridge, M. (2001). Automated negotiation: prospects, methods and challenges. *Int. J. of Group Decision and Negotiation*, *10*(2), 199–215. doi:10.1023/A:1008746126376

Lomuscio, A., Wooldridge, M., & Jennings, N. R. (2001). A Classification Scheme for Negotiation in Electronic Commerce. In Agent-Mediated Electronic Commerce: A European Perspective.

MQ Telemetry Transport. Retrieved from http://mqtt.org/

Newman, M. (2003). The structure and function of complex networks. *SIAM Review*, *45*, 167–256. doi:10.1137/S003614450342480

Opyrchal, L., Prakash, A., & Agrawal, A. (2007). Supporting Privacy Policies in a Publish-Subscribe Substrate for Pervasive Environments. *Journal Of Networks*, *2*, 17–26. doi:10.4304/jnw.2.1.17-26

Shadbolt, N., Hall, W., & Berners-Lee, T. (2006). *The Semantic Web Revisited*. IEEE Intelligent Systems.

Wang, Q., Yu, T., Li, N., Lobo, J., Bertino, E., Irwin, K., & Byun, J.-W. (2008). On the Correctness Criteria of Fine-Grained Access Control in Relational Databases. In VLDB'07, Vienna, Austria.

Wun, A., & Jacobsen, H. (2007). *A Policy Management Framework for Content-based Publish/Subscribe Middleware* (pp. 368–388). IFIP Middleware.

Chapter 7
Policies to Enable Secure Dynamic Community Establishment

Alberto Schaeffer-Filho
Imperial College London, UK

Emil Lupu
Imperial College London, UK

Morris Sloman
Imperial College London, UK

ABSTRACT

Many coalition operations require the establishment of secure communities across the different networks that make up a coalition network. These communities are formed dynamically in order to achieve the goals of a specific mission, and frequently consist of mobile entities interconnected into a mobile ad-hoc network. Technologies are needed to create these communities, and manage their operations. In this chapter, the authors show how a framework for self-managed cells can be extended to provide this capability for coalition operations.

INTRODUCTION

Successful deployment of coalition operations depends on establishing secure interactions between autonomous entities in the context of mobile ad-hoc networks (MANETs). Dynamic coalitions may be established for a short period to achieve common goals and will need to interact and provide services to each other in order to accomplish these goals. These coalitions are dynamic in that entities may join or leave over short timescales. The inter-operation between coalition partners with different security policies is typically required, where security policy decisions are made on-line in real-time. These dynamic coalitions require policy-based algorithms for establishing communities of mobile entities that interact via wireless communications, e.g. foot soldiers carrying portable devices, manned and unmanned vehicles, or service providers from different coalition forces, for specific missions. Algorithms are needed for policy deployment in response to changing conditions in the operational environ-

DOI: 10.4018/978-1-61520-855-5.ch007

ment. The specific entities forming the coalition may not be known in advance, so they have to be discovered, authenticated and assigned to roles within the scope of this collaboration. Policies define what services and resources can be accessed by other entities in the coalition, what information can be exchanged, and the protocols to be used for communication. The system needs to be highly dynamic and support entities joining and leaving the interaction due to mobility, failures or being compromised if captured by hostile forces. Thus the coalition needs to be self-managing, self-healing and protect itself from attacks. Such communities can be built upon the concept of *Self-Managed Cells* (SMCs) (Lupu et. al. 2008), which provides the underlying policy-based framework for the construction of such complex systems.

This chapter describes a *community model* based on a set of task-oriented roles to which participants can be assigned, policies defining the obligations of the participants and their authorizations to access services and resources provided by other members, as well as constraints for ensuring the integrity of the community. Community roles include both application-specific roles as well as roles required for the management of the infrastructure itself. Entities are dynamically assigned to roles when they join a community and may be re-assigned during community operation, based on changes in context or operational requirements. Constraints are used in order to enforce separation of duty principles upon role assignments, and to guarantee that minimum requirements for community operation are satisfied.

A community may consist of large numbers of entities, with complex management relationships and complex interactions between autonomous sub-communities. This requires the use of software principles and *architectural abstractions* that facilitate systematic composition and federation of autonomous policy-based systems. There is also the need to check for correctness of the specification to detect flaws or possible unsuitability of the resources available for accomplishing a mission. *Model-checking* techniques are used to automatically analyze policy-based interactions before deployment to the devices that will execute them.

SELF-MANAGED CELL FRAMEWORK

The *Self-Managed Cell* (SMC) (Lupu et. al. 2008) framework provides the underlying infrastructure for building dynamic communities. An SMC consists of a set of hardware and software components that is able to work autonomously, based on a policy-driven feedback control-loop that determines which management or reconfiguration actions must be performed in response to changes in the context. Examples of SMCs are *unmanned autonomous vehicles* (UAVs) available in a mission, as well as other resources and equipments in the field environment, such as *Gumstix*[1] and PDAs carried by foot soldiers.

An SMC comprises a dynamic set of management services (Figure 1). A publish/subscribe *event-bus* is used to provide the underlying communication infrastructure within an SMC. This has the advantage of de-coupling the services and resources that are part of an SMC, as an event publisher does not need to have prior knowledge of the recipients when sending a message. This also permits adding new services to the SMC without disrupting existing ones. The SMC relies on a *policy service*, based on the *Ponder2*[2] system, and caters for two types of policies: *obligations* are adaptation rules which specify what actions must be performed in response to events of interest; and *authorizations* specify the conditions under which a subject entity is permitted or forbidden to perform an action on a target entity. Policies can be dynamically loaded, enabled, disabled and unloaded to change the behavior of the SMC without interrupting its functioning. Finally, a *discovery service* is used to detect new devices in the vicinity that are capable of joining the SMC. The discovery service is capable of distinguishing transient failures, which are com-

Figure 1. The self-managed cell architecture

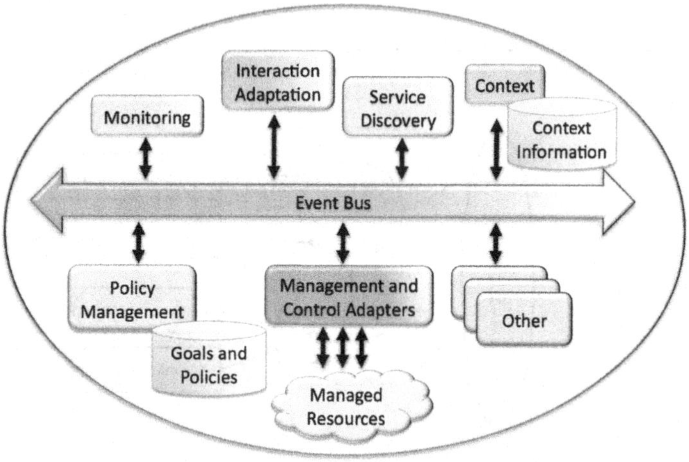

mon in wireless communications, from permanent departures (e.g. device out of range, switched off, or failure). Additional management services may be added to the core SMC architecture as required (Lupu et. al. 2008).

The SMC serves as an architectural pattern for building policy-based autonomous systems, where individual SMCs can be assembled into larger and more complex structures (Schaeffer-Filho et. al. 2007). This pattern allows the SMC to scale-down to a single autonomous vehicle or body-area network and scale-up to large applications of many thousands of autonomous systems. A basic SMC such as an unmanned vehicle or the equipment carried by a foot soldier, may be composed to form a more complex SMC such as a particular mission with many vehicles and foot soldiers. The next section describes how the SMC framework is used to build an infrastructure for the management of policy-based dynamic communities of autonomous nodes.

ROLE-BASED INFRASTRUCTURE FOR COMMUNITY MANAGEMENT

The specification and deployment of secure dynamic communities build upon a role-based infrastructure for community management. The following sections describe the basic framework and the main security functions it supports. This role-based model is described in greater detail in (Schaeffer-Filho et. al. 2008).

Community Model

The abstract model for policy-based collaboration relies on a set of task-oriented roles. A community is defined by a collection of *roles*, to which nodes represented as *Self-Managed Cells* can be assigned dynamically. Each role is associated with two classes of policies: *obligations*, which define adaptive actions that a role must perform in response to events of interest (e.g. device failure or changes in the context), and *authorizations*, which define what actions a role is allowed or forbidden to invoke on another role. The core functionality of the framework supports the same specification mechanism for both application-specific and management roles.

Assignment of nodes to roles is subject to constraints to guarantee integrity during operation. *Cardinality* constraints define the minimum and maximum number of nodes that can be assigned to a specific role at any time; and *separation of duty* constraints define conflicting roles to which

Figure 2. Role-based community model: solid lines represent the community specification, and dashed lines represent run-time behavior

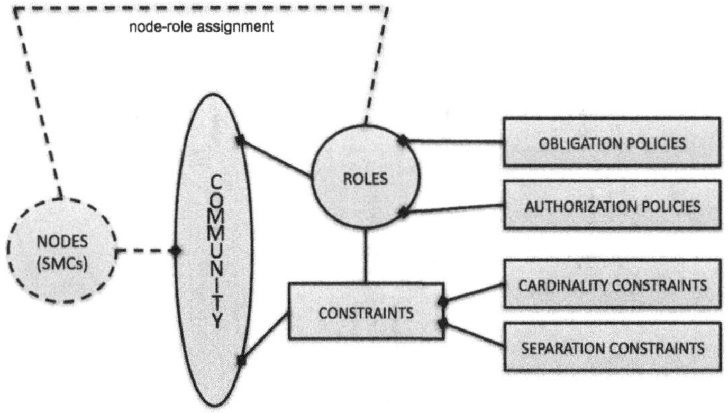

nodes cannot be assigned simultaneously. Figure 2 illustrates the abstract community model.

This role model facilitates explicit distribution of management responsibilities to different participants of the community by defining their role assignments during community operation. This increases the scalability of the model, as management tasks are partitioned and allocated to different nodes, and also its robustness, as the management duties are partitioned into different nodes consequently avoiding a single-point of failure in the community.

Secure Management

The mechanisms considered essential for the construction of secure dynamic communities include *authentication*, *membership management* and *access control*. They ensure that new members are authenticated before allowed to join the community, failures of current participants can be promptly detected, and that access control is applied to the services provided by the members of the community. Additionally, a set of *coordination* procedures is required in order to enforce essential community maintenance. The community model partitions these security mechanisms into roles, which can then be enforced by different nodes in the community, as explained below.

Coordination Role

Coordination is related to tasks for community bootstrapping, assignment of discovered nodes to roles and validation of community constraints. The node performing the coordination role periodically broadcasts messages to advertise the community to nearby nodes, and enforces policies that assign discovered nodes to roles available in the community, based on the capabilities each node possesses. When a node is assigned to a role, the policies associated with the role are automatically transferred into the node. The coordinator also checks whether the community constraints are met, and if not it may re-assign nodes to different roles. At community initialization time, the only node available is the one performing the coordinator role, which will possibly be assigned to other critical roles as well; however, as new members are discovered the coordinator may delegate one or more management roles to the new members.

Authentication Role

A common authentication mechanism based on public-key certificates is used to verify the identity of the nodes that want to join the community. The node assigned to the authenticator role is loaded with the public keys of the certification authorities (CAs) that are relevant to the community, and only nodes presenting certificates signed by these authorities are able to join. This avoids the need to contact a centralized certification authority and is necessary where access to a network infrastructure may not exist. However, this simple mechanism does not cater for key revocations, and non-PKI based authentication mechanisms are currently being investigated.

Membership Management Role

Membership management monitors the current members of the community and the roles they are playing. This is required to detect nodes that fail, move out of communication range or run out of battery power. Members periodically renew their membership with the node performing the membership manager role. If the membership manager detects that a node failed to renew its membership, the node is considered to have left and this information is propagated to other members. This may cause the community constraints to be reevaluated by the coordinator.

Access Control Role

Access control enforcement is distributed amongst all roles, as typically each node is interested in protecting its own resources. If a node is not able to enforce its access control policies, it may delegate them to another specific role in the community.

The role-based framework is extensible, and the community may be extended with additional management roles according to specific security requirements. However, policies and roles are not sufficient for engineering large systems, or collaborations that may involve cross-community interactions, which require structuring and reuse of complex SMC interactions. For example, a *rescue team* community may be a hierarchical composition of multiple search teams, a protection team and a medical team with each team being a community. The member communities would encapsulate their management and the rescue team would not be concerned with the details of the management in the internal communities. The next section will introduce the use of software patterns for engineering policy-based communities, which can be constructed by reusing and composing building block abstractions that exhibit well-defined properties.

PATTERNS FOR BUILDING LARGE-SCALE COMMUNITIES

This section describes the systematic building of policy-based SMC interactions to form large-scale collaborations, compositions and federations of autonomous systems as an infrastructure for dynamic communities in larger settings. The approach is based on software architectures which typically separate computation (*components*) from interactions (*connectors*) – the benefits of this have been widely recognized in the software community as means of structuring software development (Allen and Garlan 1994), (Garlan and Shaw 1993), (Shaw et. Al. 1995), (Mehta et. Al. 2000). Although components and connectors do not cater for the adaptive behavior of SMCs as expressed in roles and policies, similar principles can be applied for structuring and reusing SMC interactions to form larger collaborations.

Collaborations between SMCs rely on the exchange of *policies*, *events* and *interfaces*. Firstly, SMCs exchange policies, which allow them to dynamically change the behavior of interacting SMCs. An SMC may also generate events to notify remote SMCs of changes of context, possibly triggering management actions (policies) in response to these events. Finally, by exchanging interfaces, an SMC allows remote SMCs to invoke operations on it, or mediates access to its local internal resources, as interfaces are required for executing the actions prescribed by a policy.

Architectural styles (Schaeffer-Filho, Lupu and Sloman 2009) based on exchanges of policies, events and interfaces, can be used for defining SMC interactions. A catalogue of architectural styles provides several useful abstractions for defining management relationships across *Self-Managed Cells*. Architectural styles are similar to *software design patterns* (Gamma et. Al. 1995) in the sense that they provide a set of standard solutions for recurring problems. Three main aspects of SMC interactions are of particular interest:

- ***Task-Allocation:*** defines how policies are exchanged between SMCs and under which conditions these exchanges happen.
- ***Communication:*** defines event-forwarding patterns between SMCs, as events are required for triggering policies.
- ***Structural:*** specifies how SMCs are organized and structured and addresses issues such as interface mediation, filtering, encapsulation and SMC visibility, as interfaces are required for invoking the actions specified in a policy.

These categories can be seen as complementary *views* for defining management relationships between autonomous SMCs. They provide a better understanding of these relationships and promote the reuse of common abstractions for systematically building *large-scale communities*. The basic architectural styles discussed in this chapter deal with individual abstractions however complex collaborations can be built from the combination of styles as design elements. Table 1 presents a brief overview of these architectural styles.

In order to characterize the uses of each architectural style (i.e. clarify the problems they aim to solve and how they do it) and its rationale, each style is presented in the following format:

- ***Purpose:*** states a recurring problem that the style aims to solve and where it applies.
- ***Description:*** describes the style itself and details of how it is implemented in terms of policies, interfaces and events.

Table 1. Categorization of architectural styles

Category	Architectural Style	Description
Structure	Peer-to-Peer	Ordinary, symmetric mode of interaction between SMCs that exchange interfaces
	Composition	One SMC encapsulates another's interface and determines its visibility through mediation
	Aggregation	The inner SMC becomes resource of outer but without imposing encapsulation (allows sharing)
	Fusion	Combines the interfaces, policies, and managed objects of two constituent SMCs into a new SMC
Task-Allocation	Hierarchical Control	One top-level SMC controls the execution of a set of leaf SMCs
	Cooperative Control	One leaf SMC is controlled by a set of cooperating manager top-level SMCs
	Auction	Task allocation employing a negotiation approach (issuers and bidders)
	Mutual Control	Fully decentralized interaction where SMCs can both load and receive tasks from their partners
Communication	Diffusion	Provides a way of directly forwarding events to interacting SMCs
	Shared Bus	Provides a blackboard for decoupled event-based communication among SMCs
	Correlation	Individual events may be combined for generating a higher-level event
	Store-and-Forward	Useful in ad-hoc settings where SMCs may not have a permanent connection to their partners

- ***Examples:*** examples illustrating situations where the architectural style may be applied.
- ***Related styles:*** specific similarities or differences with respect to other styles.

A catalogue of architectural styles is presented below and the following section describes how these styles can be combined for composing and federating Self-Managed Cells.

Structure

Structural styles define how SMCs are organized and structured with respect to the access of their interfaces. These may involve abstractions such as encapsulation, or filtering and mapping of interfaces.

Peer-to-Peer (P2P)

Purpose: to allow interactions between SMCs that may exchange interfaces, where no specific semantics of ownership or encapsulation is applicable.

Description: the peer-to-peer architectural style (Figure 3) defines a relationship between peer SMCs, which may provide or request services to or from each other, while each peer retains its autonomy and is free to establish additional P2P interactions with other SMCs. Each SMC exposes an interface, which specifies actions that can be invoked and events it is able to raise. Invocations are subject to the existence of authorization policies allowing the partner to perform the actions but no predefined management or control relationships are implied.

Examples: UAVs in the same mission may collaborate and offer services to each other (e.g. one UAV may request updates about hazardous chemicals detected by a surveyor UAV, while the latter may also use the storage service provided by a third UAV to back-up this information during field operation).

Related styles: in contrast to a composition or an aggregation, the P2P style does not impose any hierarchical organization to the SMCs.

Composition

Purpose: to hide the complexity of a collection of SMCs, where components are encapsulated within an SMC and treated as owned, hidden resources.

Description: a composition (Figure 4) specifies an interaction between SMCs in which the outer SMC encapsulates the inner SMC, which then becomes a managed resource of the outer SMC. The inner's visibility is mediated by the outer SMC (i.e. the outer SMC defines the degree of encapsulation of the inner SMC and which aspects of the inner SMC's interface are exposed). An outer SMC may contain multiple SMCs, but a contained SMC can only be contained within a single outer SMC.

Examples: foot soldiers in the field may carry a personal computing device, e.g. *Gumstix*, providing general management functions, as well as sensors for health and environmental monitoring. In this case, the sensors may become inner resources

Figure 3. Peer-to-peer

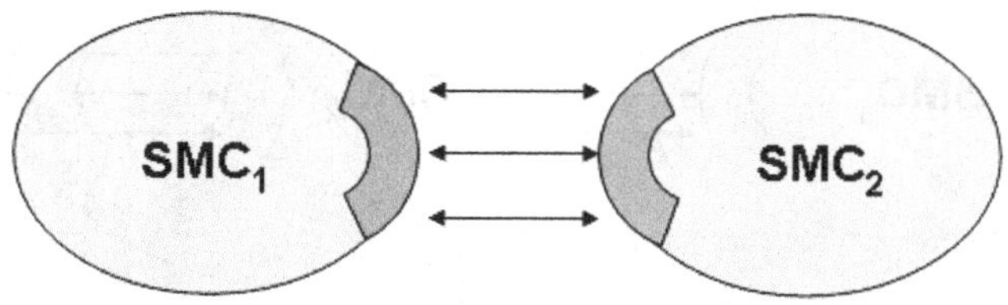

of the *Gumstix*, which also mediates interactions with them and effectively encapsulates them, as they are not externally visible.

Related styles: in contrast to the P2P architectural style, the composition defines ownership among SMCs.

Aggregation

Purpose: to allow SMCs to aggregate in a single, hierarchical structure, without preventing its component SMCs from interacting directly with SMCs outside the aggregation.

Description: an aggregation (Figure 5) specifies a relationship in which one of the SMCs (inner) becomes an owned resource of another (outer), and may be mediated by the outer's interface. However the aggregation does not impose strict encapsulation among SMCs, and the inner SMC may also interact directly with other SMCs outside the aggregation if necessary (as a shared resource).

Figure 4. Composition

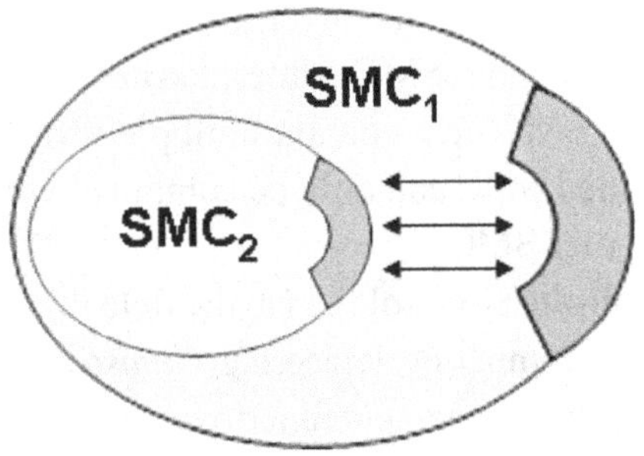

Figure 5. Aggregation

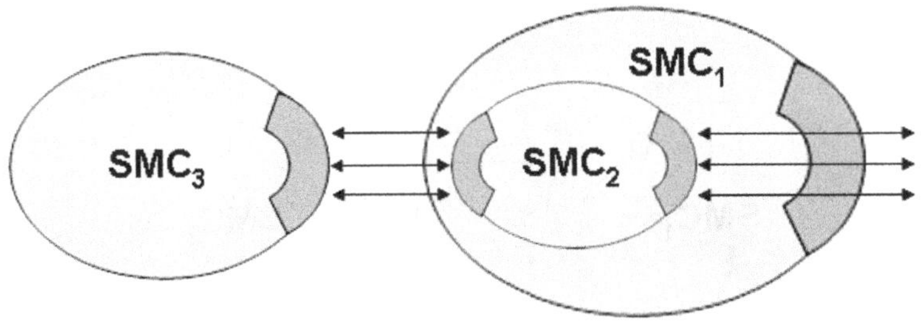

Examples: a UK team that owns and manages a particular UAV may allow this resource to be shared with cooperating partners. In this case, although the UK team encapsulates and manages the UAV most of the time, it may allow it to be accessed directly by an ally US team.

Related styles: aggregation is similar to composition in that one of the SMCs becomes an owned resource of the other and may be accessed through the outer's interface. However, the aggregation does not necessarily impose encapsulation among SMCs, which may be involved in other aggregations or P2P relationships with external SMCs.

Fusion

Purpose: to combine the components of two SMCs into a single new SMC, where the forming SMCs cease to exist.

Description: the fusion architectural style (Figure 6) combines two SMCs into one. The resulting SMC is effectively a union of the components of its constituents, so it must provide an interface combining the functionality previously provided by each constituent interface, and may need to enforce the sets of policies previously enforced by each constituent SMC. Finally, the resulting SMC must also take-over the collection of managed resources possessed by each constituent SMC.

Examples: two small platoons of soldiers on related missions may combine to form a single larger platoon with a single commander for all the soldiers.

Related styles: the fusion style implies that the constituent SMCs will vanish when the resulting SMC is created, which differs from the other styles, in which the SMC's identities were conserved.

Task-Allocation

Task-allocation styles define ways of loading tasks into interacting SMCs. The tasks exchanged between SMCs are specified in the form of *missions* (Lupu et. al. 2008). Missions are groups of policies that can be dynamically loaded to change the behavior of interacting SMCs at runtime. While missions define *what* tasks are being exchanged between SMCs, task-allocation styles specify *how* these tasks are exchanged. These abstractions are orthogonal to those provided by the structural styles, and styles of both categories may be combined.

Hierarchical Control

Purpose: to address management issues when one SMC delegates tasks and policies to multiple subordinate SMCs and can control the behavior of the subordinates.

Description: in the hierarchical control style (Figure 7) one top-level SMC controls the execution of a set of leaf SMCs by delegating tasks and policies to them. This implies that the top-level SMC has guaranteed rights of programmability over the leaf SMCs (i.e. the top-level SMC is authorized to load tasks and policies into them). It is a unidirectional interaction in that only the top-level SMC loads tasks into the leaf SMCs, and not the opposite. Also, this style does not imply that access to the leaf-SMCs is mediated by the top-level SMC. Finally, notice that multilevel hierarchies can be achieved by combining this style multiple times.

Examples: a commander and a set of lieutenant SMCs may form the top-level section of a military squad. In this case, lieutenants are subordinate to the commander SMC, which has guaranteed rights to manage the lieutenants.

Related styles: in contrast to the structural styles presented so far (P2P, composition, aggregation) the hierarchical control style concerns control and task-allocation relationships between the SMCs (those presented earlier concern how the interactions are structured and organized). Notice that these are independent concerns when defining collaborations between SMCs.

Cooperative Control

Purpose: to enable various SMCs to cooperatively delegate tasks and policies to one subordinate SMC.

Description: in the cooperative control style (Figure 8) a set of cooperating top-level SMCs control the execution of a leaf SMC by delegating tasks and policies to it and have the relevant authorizations to achieve this. Similar to the hierarchical control style, this is a unidirectional interaction. Note that these multiple top-level SMCs may cause conflicting tasks and policies to be loaded into the leaf SMC, and this architectural style must address synchronization of actions among

Figure 6. Fusion

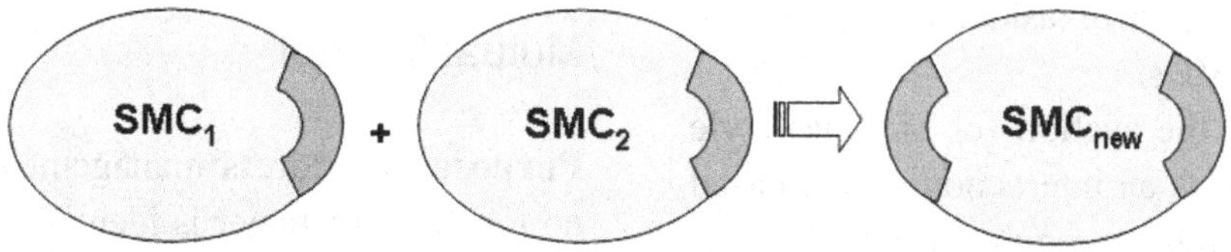

Figure 7. Hierarchical control

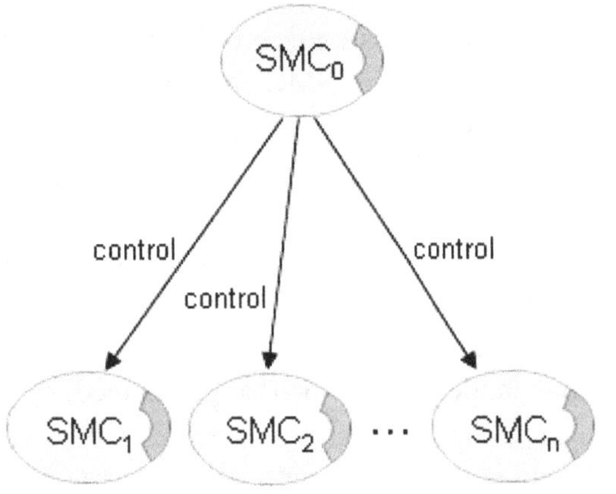

Figure 8. Cooperative control

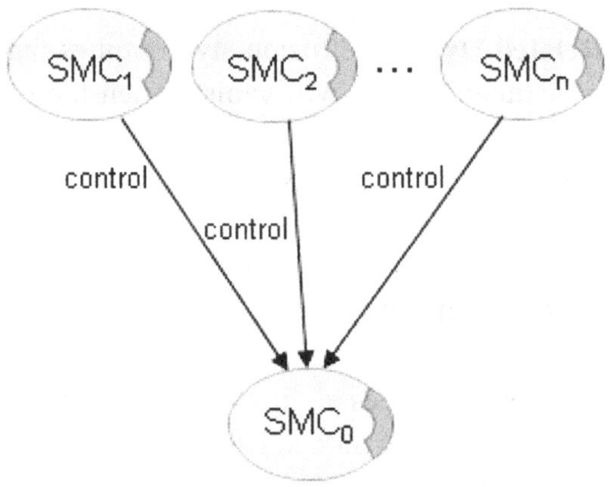

the top-level SMCs and mechanisms to solve the conflicts that may occur (such as prioritization of policies or use of application-specific meta-policies (Lupu and Sloman 1999)).

Examples: this style can be applied for the management of a shared UAV that is subject to policies loaded from two cooperating coalitions (e.g. a US and a UK coalition, which may be structured as aggregations themselves). In this case, the cooperative control style must address the conflict issues that may arise.

Related styles: similar to the hierarchical control, the cooperative control implies an asymmetrical authority among the SMCs. Additionally, in contrast to the aggregation and composition styles, there is no interface mediation, encapsulation or structural restrictions among the SMCs: the interactions may have a structural organization (P2P, composition or aggregation) that is defined independently of the task-allocation aspects.

Auction

Purpose: to enable an issuer SMC to trade and negotiate the execution of tasks and policies with a set of bidder SMCs.

Description: the auction (or bidding) style (Figure 9) provides an interaction model based on the *contract net* approach (Smith 1988). It facilitates task distribution employing negotiation, where tasks (or roles) are announced and, nodes decide whether to bid for the execution of these tasks. The top-level node (*issuer*) evaluates the bids and assigns the task to the most appropriate leaf nodes (*bidders*). The issuer SMC decides what tasks to offer and bidders decide which tasks to bid for, possibly after some policy defined negotiation. The decision of which bids to accept is made by the issuer, who cannot impose tasks on a bidder.

Examples: utility functions based on the capabilities available in autonomous robots were used in (Chaimowicz et. al. 2004) to bid for roles, although using a slightly different type of negotiation. Communities of UAVs may use a similar role assignment approach, based on a negotiation scheme that considers the capabilities of different UAVs.

Related styles: although it has the same topology of the hierarchical control style, its semantics is different in that the issuer does not have predefined rights to impose tasks on the leaf nodes but task assignment is negotiated (as represented by the double-ended arrows in Figure 9).

Mutual Control

Purpose: to address management issues when no top-level manager is identified and the man-

Figure 9. Auction or bidding

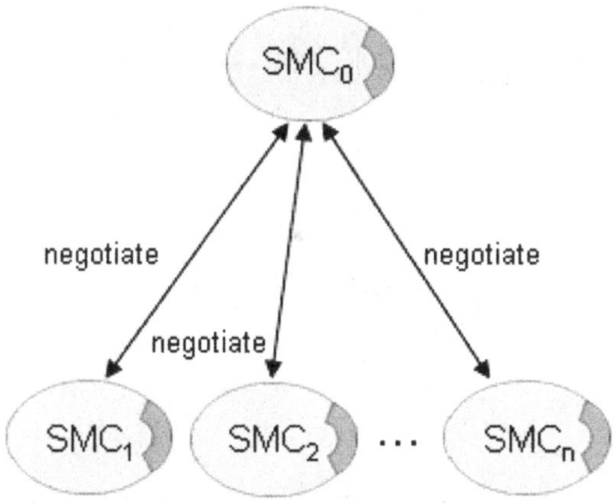

Figure 10. Mutual control

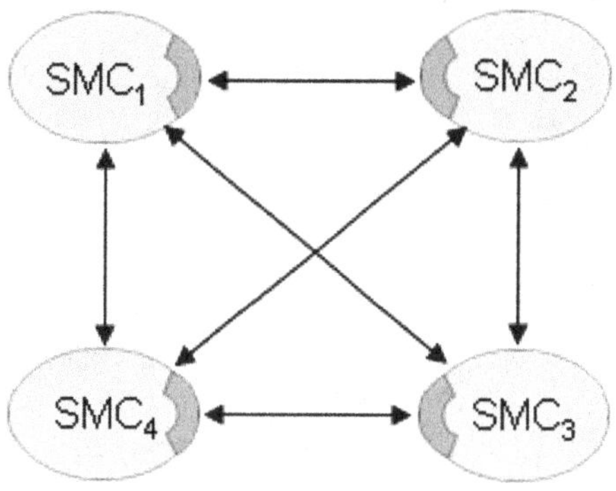

agement relies on a completely decentralized cooperation among SMCs.

Description: the mutual control architectural style (Figure 10) represents a fully decentralized interaction among a set of SMCs. The nature of such interaction is application-dependent, where both SMCs act as *managers* and *managed* resources simultaneously. Typically, multiple SMCs will be interested in mutually loading tasks into each other. Policies associated with this style address general authorizations amongst the SMCs, and general conflict resolution rules similar to those provided by the cooperative management style.

Examples: in an interaction between a soldier and a medic in the field, the medic may be interested in loading health-monitoring tasks into the soldier SMC, however the soldier may be also interested in loading sensor re-calibration tasks into the medic SMC.

Related styles: this is the only task-allocation style that involves bi-directional task and policy loading among SMCs, and it is also the most general in that it imposes less restrictions and semantics on the interaction.

Communication

Communication styles define patterns that specify how events are exchanged and how the event buses of various SMCs are interconnected.

Diffusion

Purpose: to support broadcasting of messages and dissemination of events in a collaboration of SMCs.

Description: the diffusion architectural style (Figure 11) defines the forwarding of events to all (or a subset of) neighbor nodes. Figure 11 shows an example of the general case of branching diffusion, but other topologies could be defined, e.g. a linear diffusion resembles a pipeline structure. This architectural style specifies rules regarding the general forwarding strategy and routing protocols.

Examples: ad-hoc sensors monitoring environmental data in the field are likely to use the diffusion style to propagate the monitored data until it reaches processing units.

Related styles: the diffusion differs from the styles presented so far in that it is only concerned with event relaying, and does not address task-allocation or structural issues. The diffusion style is not tied to a specific topology such as a shared bus.

Figure 11. Diffusion: (a) branching diffusion and (b) pipeline diffusion

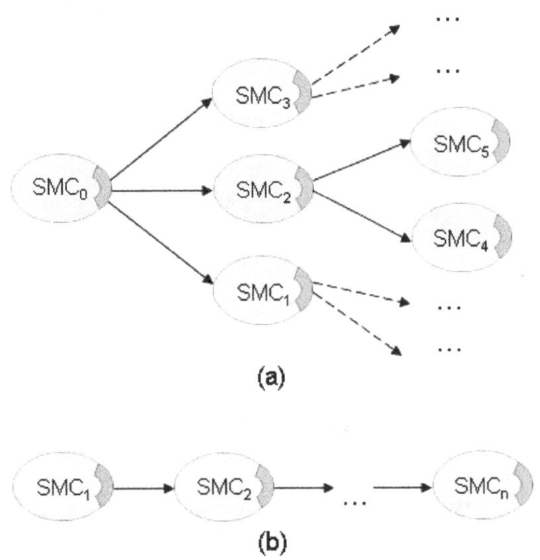

Shared Bus

Purpose: to support sharing of events among a set of SMCs using a blackboard-like event-sharing scheme.

Description: the shared bus style (Figure 12) allows SMCs to interconnect their event buses to a central SMC, which relays events published in its event bus to all SMCs connected to it. SMCs can use the shared bus as a blackboard for de-coupled interactions, because the publisher SMC does not require prior knowledge of the recipient SMCs when sending an event. Only selected types of events may be published in the shared bus, and SMCs have the choice of publishing specific events only to their local event bus.

Examples: this style can be used for communication among surveyor UAVs collecting layout information of an area. Data about any new obstacle found by a UAV is published in the shared bus, which relays the information to all the other surveyors. Thus, all surveyors share a blackboard of obstacles, keeping a consistent view of the environment.

Figure 12. Shared bus

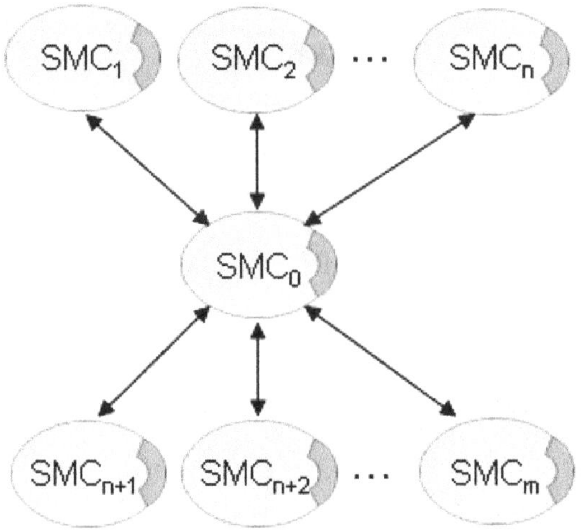

Related styles: the shared bus is the only style that relies on a centralized SMC to route events to a set of other SMCs.

Correlation

Purpose: to support correlation of events between SMCs, where individual events may be combined for generating higher-level ones.

Description: the correlation style (Figure 13) addresses event correlation between SMCs. For example, it can be used for collecting events from different sources and generating higher-level events or for collecting patterns of events over a period of time and generating some other event containing synthesized information.

Examples: the information collected from various physiological sensors in a body-area network may be correlated in a coarser-grained event, such as a periodical report on the health condition of the soldier.

Related styles: in contrast to the diffusion style, the correlation style defines abstractions for combining several source events into a new event, rather than just forwarding raw events.

Figure 13. Correlation

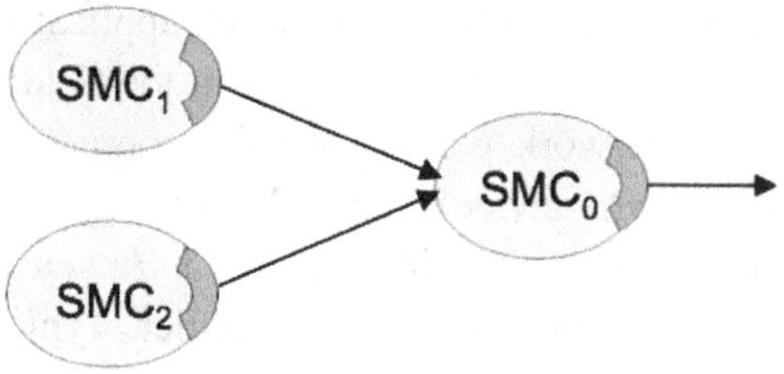

Store-and-Forward

Purpose: to support ad-hoc communication among SMCs that do not have a permanent and direct connection to each other.

Description: store-and-forward (Figure 14) addresses issues on event transfer between SMCs using *delay-tolerant networking*. It is particularly useful in ad-hoc settings where an SMC does not have permanent connection to all other SMCs it wishes to communicate with. The store-and-forward style enables SMCs to communicate via multi-hop interactions, where some SMCs act as physical carriers of events targeted to other SMCs. This style specifies how to store and transfer events to neighbor SMCs according to their availability, ensuring that undelivered events are retained in the SMC for later delivery.

Examples: this may be used if UAVs in a military community do not have a permanent connection to the coordinator, and then depend on other nodes to relay their communication.

Related styles: it is worth noting that these styles are not mutually exclusive, and the store-and-forward style could be combined with the diffusion style, for example, using more complex topologies and more sophisticated routing algorithms.

Figure 14. Store-and-forward

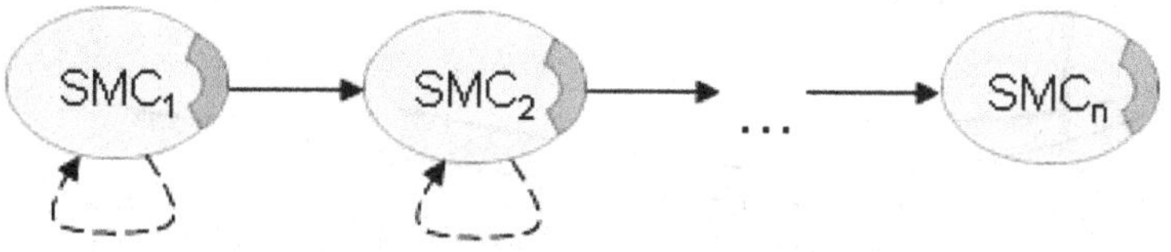

FEDERATION AND COMPOSITION

Based on the catalogue of architectural styles described in the last section, this section describes a methodology for engineering large-scale SMC interactions by combining architectural styles as design elements of these systems. Under this perspective, each architectural style is seen as a template interaction that can be instantiated between specific SMCs. Each style enforces a particular abstraction for either *(a)* the exchange of policies or *(b)* the exchange of events required for triggering policies or *(c)* the exchange of interfaces for performing actions prescribed by policies. The instantiation of a particular style in an interaction thus causes a specific abstraction with well-defined properties to be established between sub-sets of SMCs.

In the proposed interaction model, each SMC maintains *roles* in its local domain, which are *placeholders* for remote SMCs that will be discovered at runtime. Each architectural style also defines its own set of style-specific roles: for example, a *composition style* defines the roles *outer* and *inner*, while a *hierarchical control* style defines the roles *manager* and *managed* and a *diffusion style* defines the roles *source* and *target*. To specify an interaction domain roles can then be *bound* to style-specific roles, thus prescribing how SMCs that will eventually be assigned to those roles should behave with respect to policy, interface and event exchanges. Roles in the local domain will thus combine the behavior of the overall interaction, which will be defined by the collection of architectural styles bound to those roles. Figure 15 presents an overview of how these concepts are interrelated.

When remote SMCs are dynamically discovered and assigned to roles in the local domain, the semantics specified by each architectural style bound to those roles will be enforced. This may involve the exchange of interfaces, policies or events. For example, a structural style will enforce the exchange of interfaces between a set of SMCs (enforcing any associated abstraction defined by the style, e.g. encapsulation). The same is valid for the architectural styles in the other management layers. Notice that each architectural style is parameterized according to the abstractions it enforces: this typically involves the *methods* to be filtered or mapped (for a structural style), the specific *policies* to be loaded in which *conditions* (for a task-allocation) and the *events* to be shared with other SMCs (for a communication style).

This model permits specifying the management aspects of large-scale policy-based applications by reusing building block abstractions. Consider for example the requirements for an application for health monitoring of a soldier in the field: a body-area network would typically comprise a PDA or *Gumstix* device hosting SMC management services, interacting with various *body sensor nodes*[3] (BSNs) attached to the soldier. Typical sensors include heart-rate, temperature, oxygen saturation and acceleration. The body-area may include actuators as well, such as a defibrillator SMC, which can be activated if necessary, according to sensor measurements. Other devices available in the operational field may also be used to host application services, e.g. for performing diagnostics. A medic SMC would load monitoring tasks for collecting sensor information before soldiers are deployed in a mission, allowing the soldier to be self-monitored in his operational environment. These tasks continually run on the soldier's SMC during field operation. Monitored results may be pre-processed using diagnostic devices locally available, for synthesis and subsequent delivery to the field hospital, or they may be used to automatically request medical assistance

Figure 15. Composition model: (1) architectural styles are bound to roles in the local domain; (2) remote SMCs are assigned to these roles; (3) behavior associated with each style is enforced in these SMCs

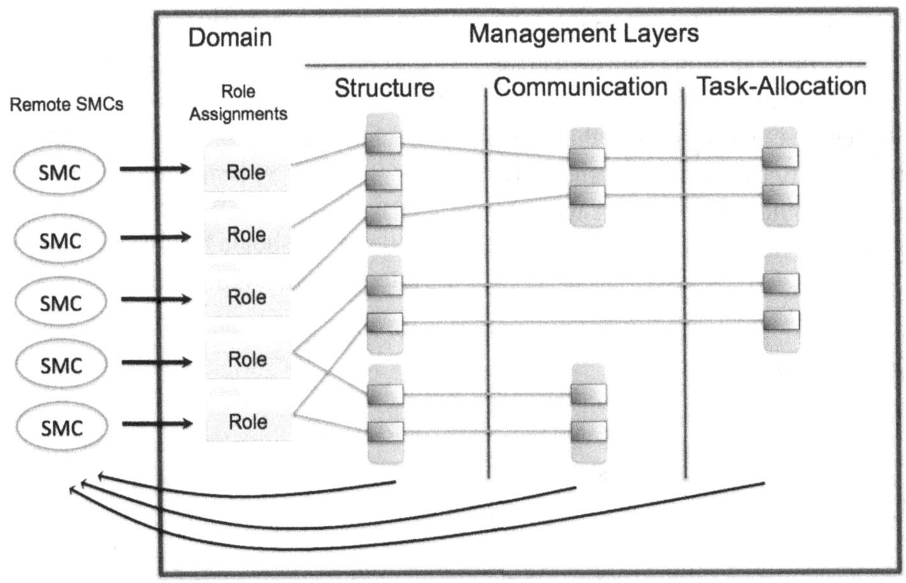

when the soldier is injured. An overview of the SMCs involved in this application is illustrated in Figure 16.

While *Self-Managed Cells* provide a suitable abstraction for representing policy-based autonomous components, architectural styles are used to systematically specify interactions between these components. Figure 17 presents an architectural representation of the scenario where each one of the five columns corresponds to one of the triangles in Figure 16. For example, the structural aspects of the body-area network between the soldier device (*outer*) and a heart-rate and accelerometer SMCs (*inners*) is specified as a *composition*. This was the natural choice, as a body-area network typically requires that sensor devices be encapsulated so that the sensors of one soldier do not establish interactions with those of other soldiers in the platoon. In this body-area network, typically sensors (*sources*) forward their measurements to the soldier SMC (*target*) using a simple *diffusion* of events. The structural requirements of an interaction between medic and soldier SMCs are different, as the notion of composition and encapsulation is not suitable for this interaction. Instead, a *peer-to-peer* structure with simple exchange of interfaces is more likely to occur. Possibly a medic will enforce task-allocation styles (e.g. for ECG monitoring and defibrillator activation) in the form of a *hierarchical control* style, where the style is parameterized with the specific mission to be loaded. The example can be elaborated by selecting the abstractions that are suitable for realizing the other interactions. A dynamic collaboration between groups of unmanned autonomous vehicles can be designed in a similar way, by using architectural styles to define how these devices exchange interfaces, policies and events.

Architectural styles proved a useful abstraction for designing and deploying large-scale SMC interactions. However, the successful operation of these collaborations depends both on their correct specification and on the suitability of the participating SMCs. The next section presents how model-checking techniques have been applied to verify the correctness of the community specification and deployment.

Figure 16. Soldier-monitoring scenario: each triangle represents an interaction that requires a different combination of management abstractions

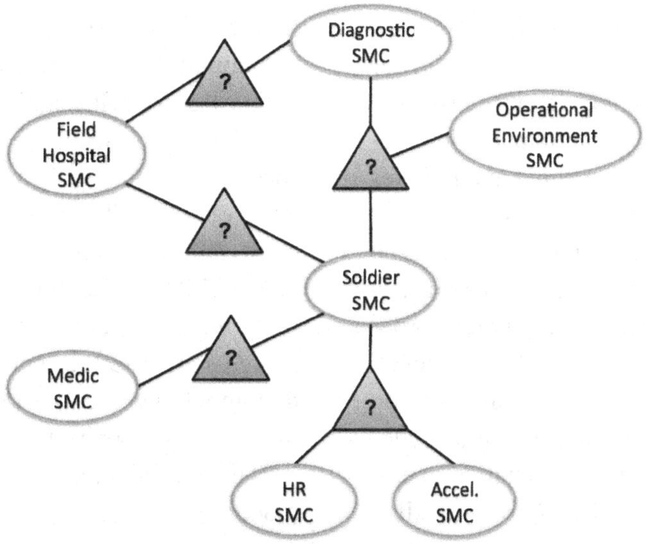

Figure 17. Architectural representation of the soldier-monitoring scenario: the illustration shows the overall configuration of the collaboration with the selected abstractions for achieving policy, event and interface exchanges

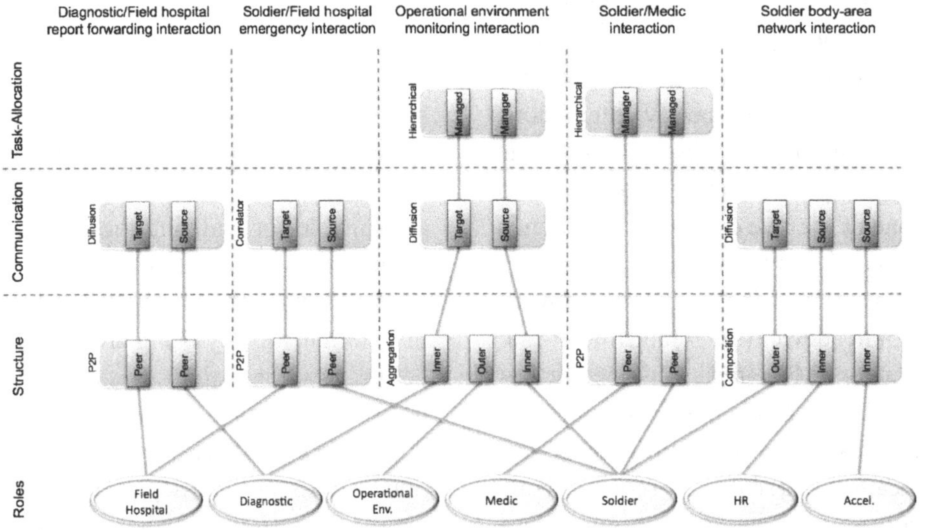

FORMAL SPECIFICATION AND MODEL-CHECKING

Dynamic collaborations typically involve SMCs belonging to different administrative authorities, for example using UAVs and equipment from both a US and a UK coalition. When SMCs are composed or federated, inconsistencies, conflicting policies or unsuitability of the resources available in the mission may prevent these SMCs from operating as expected. A formal model for specifying SMC interactions assists in their design and enables the analysis of anticipated interactions before these are implemented and policies are deployed in the physical devices.

The *Alloy Analyzer* (Jackson 1999) was chosen for the formal specification of SMC behavior. Alloy is a modelling language based on first-order logic and is different from *pi-calculus* (Milner et. al. 1992), *ambient calculus* (Cardelli and Gordon 1998), *channel ambient calculus* (Phillips 2006) and others which model the computation operationally. Instead, Alloy is a declarative modeller, and that is more natural and concise for describing SMC interactions and the integrity constraints related to SMC management. Its analyzer can automatically check models for correctness, generate instances of model invariants, execute guided simulations involving a number of operations defined in the model and check user-specified properties of a model. The toolset also provides a visualization tool, which helps to understand the solutions found by the analyzer. This tool was used to automatically generate the figures that appear in this section (with small hand edits to make the visualization easier). Defining a formal Alloy specification for the behavior of Self-Managed Cells enables: *(1)* formally capturing the static and dynamic aspects of the structure and behavior of the SMC interactions; *(2)* automatically verifying the consistency of SMC collaborations by using its analyzer; *(3)* simulating SMC behavior in complex interactions (Schaeffer-Filho, Lupu, Sloman and Eisenbach 2009).

The formal model is divided in three sub-models:

- A **Basic Self-Managed Cell Model** defines the elementary concepts such as SMC, Role, Interface, and the basic predicates for discovery of a new SMC, departure of an SMC, role assignment and role de-assignment.
- A **Policy Model** that imports the basic SMC definitions and adds to it the specification and analysis of obligation and authorization policies in collaborations of SMCs.
- An **Architectural Model**, which imports the basic model and adds to it the architectural aspects of an interaction, such as architectural styles, bindings, etc. This model focuses on the interaction specification (bindings between style-specific roles and domain roles) and interaction enforcement (causing a new behavior to be added in the form of forwarding of events, loading of policies or mapping of interfaces).

Each model written in Alloy consists of a set of *signatures* and a set of *predicates*. Signatures define the structure of the model, where each signature represents a concept in the model and its relation to other concepts. Predicates are used to define the behavior of a model when operations are executed. A common technique to represent dynamic behavior is to show how the *"state"* of the system before the operation differs from the *"state"* after the operation, i.e. what properties hold before and after an operation is executed. Typically an additional signature is used to represent the entire system being modelled, which corresponds to an interaction between a set of SMCs. A *predicate* defined for each operation takes as arguments an instance *S* and an instance *S'* of this signature, and shows how *S* differs from *S'* in this predicate. This is equivalent to showing what properties hold before and after the execution of an operation. This can be used to indicate the changes that happen when an SMC is discovered, when an SMC departs, when an SMC is assigned to a role, and so on.

Basic Self-Managed Cell Model

This first sub-model defines the basic concepts such as SMC, Role, Interface (containing Events, Operations and Notifications supported by that interface), and the basic operations for discovery of a new SMC, departure of an SMC, role assignment and role de-assignment. The main component in the model is the SMC, represented by the signature *SelfManagedCell* below.

```
abstract sig SelfManagedCell
{
provides: some Interface,
requires: some Role,
obligations: set Obligation,
authorizations: set Authorization
}
```

In Alloy, a signature body defines a number of *relations*, which are similar to the attributes of an object in the OO paradigm. The signature *SelfManagedCell* specifies four relations. The first two define respectively which interfaces an SMC *provides* to remote SMCs and which roles an SMC *requires* to be fulfilled (by remote SMCs). The other two relations define the policies an SMC is enforcing (discussed later). The *SelfManagedCell* is an **abstract** signature, so it can be extended to define specialized components in the model, such as *MedicSMC*, *SoldierSMC* or *SensorSMC*.

An SMC may provide one or more interfaces, which can then be assigned to a role in a remote SMC. The signature *Interface* defines the *operations* (methods that can be invoked), the *events* (which can be published externally) and the *notifications* (which are external events of which the SMC can be notified) supported by that interface.

```
Sig Interface
{
operations: set Operation,
events: set Event,
```

```
notifications: set Notification
}
```

Figure 18 shows an example of a valid configuration of SMCs automatically generated from the Alloy specification. In this example, the SoldierSMC requires the roles Sensor and Medic, and provides *Interface1* (which defines the operation *startECG*). Similarly, *MedicSMC* requires the role *Soldier,* and provides *Interface0* (which defines the operation *load*, the event *loaded* and the notification *stopped*). In the example, *Interface1*, provided by *SoldierSMC,* is assigned to the role *Soldier*, required by the *MedicSMC*.

The predicates defined in this model cater for the specification of discovery and departure of SMCs, and assignment and de-assignment of SMC interfaces to/from SMC roles. This basic model, however, merely serves as the foundation for building more complex configurations among SMCs, which rely on policy-based interactions and different management relationships that are encoded as architectural styles. These concepts are defined in the following sub-models.

Policy Model

This model covers the specification of obligation and authorization policies, and the predicates related to policy loading and unloading, activation and deactivation. In particular, the signature *ConcreteObligation* defines the subject and target roles for a policy, the event that triggers the policy and the action to be invoked in response.

```
sig ConcreteObligation extends Obli-
gation
{
subject: one Role,
event: one Event,
action: one Operation,
target: one Role
}
```

Similarly, the signature *ConcreteAuthorization* defines a subject role, a target role, an action and the modality of the policy (which can be either positive or negative).

```
sig ConcreteAuthorization extends Au-
thorization
```

Figure 18. Alloy graphical representation of a simple interaction between SMCs

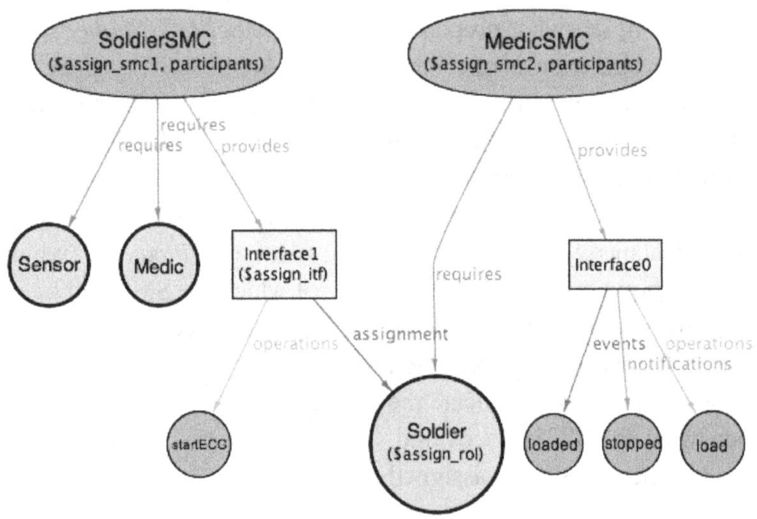

```
{
modality: one Modality,
subject: one Role,
action: one Operation,
target: one Role
}
```

The formal specification of policies in Alloy enables several types of analysis. Its focus, however, is not on the ability to detect policy conflicts, but instead on the verification of SMC collaborations and whether their participants are able to enforce the policies. For example, it is possible to verify whether the SMCs assigned to roles are capable of enforcing the policies associated with those roles, or whether all obligations enforced by collaborating SMCs have a corresponding authorization policy. For example, consider the interaction illustrated in Figure 19. The *MedicSMC* enforces the obligation *Obl*, which states that the subject role (*Medic*) must invoke on the target role (*Soldier*) the action *startECG* in response to the event *highHR*. *Interface2* provided by the *MedicSMC* is locally assigned to the subject role, and *Interface0* provided by the remote *SoldierSMC* is assigned to the target role in the *MedicSMC*. The *SoldierSMC* enforces the authorization *Aut*, which states that the subject role (*Medic*) is allowed to invoke the action *startECG* on the target role (*Soldier*). The remote *Interface2* provided by the *MedicSMC* is assigned to the *Medic* role in the *SoldierSMC*, whereas the local interface *Interface0* is assigned to the local *Soldier* role, which is also the target role of the policy being enforced by this SMC.

Verifications are specified as additional *predicates* (not shown here) in the Alloy model, which define properties to be checked in a specific collaboration of SMCs. For example, the role assignment can be type-checked to verify whether the interfaces assigned to roles satisfy the requirements for enforcing the policies associated with those roles. The assignments given in the example above satisfy this property, as *Interface2* (which is assigned to the *Medic* role in the *MedicSMC*) supports the event *highHR* (which is required for triggering the obligation policy in that SMC). Similarly, *Interface0* (which is assigned to the *Soldier* role in the *SoldierSMC*) supports the action *startECG* (which is the action to be allowed execution by the authorization policy in that SMC). Note that each obligation has a corresponding authorization policy. More traditional types of policy analysis, such as modality or application-specific conflicts can be defined in a similar manner.

Architectural Model

This model defines the architectural aspects of an interaction, in particular the architectural styles,

Figure 19. Alloy graphical representation of a valid policy configuration between SMCs

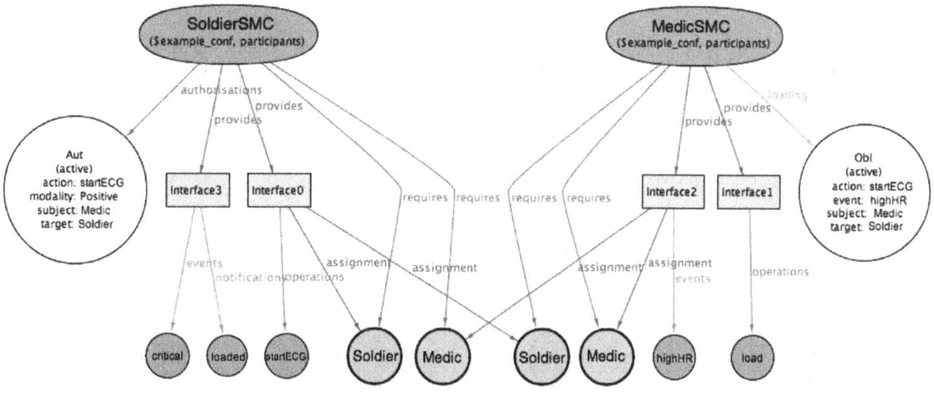

and bindings between style-specific and domain roles in an SMC, and includes some new concepts and predicates in addition to those defined in the base models.

The most noticeable difference is the addition of the management abstractions (i.e. architectural styles). Each architectural style defines a set of style-specific roles (e.g. the *Composition* style defines the roles *Outer* and *Inner*, the *Diffusion* style defines the roles *Source* and *Target*, the *HierarchicalControl* style define the roles *Managed* and *Manager*, etc.). A *pattern* of interaction is defined as a specific combination of architectural styles, arranged in a particular manner (e.g. a body-area network will often have the same structure, task-allocation and event forwarding aspects).

Two distinct steps related to SMC interactions are of particular importance: *(1)* interaction specification and *(2)* interaction enforcement. The *interaction specification* is defined by the bindings between style-specific roles and the roles required by an SMC – this specifies how SMCs assigned to these roles will be expected to behave. The *interaction enforcement* is represented by the deployment of styles previously bound to a set of roles, when actual SMCs are assigned to these roles. The deployment causes a new dynamic behavior to be added to the current interaction (when the system passes from state *S* to *S'*). This behavior can be forwarding of events, mapping of interfaces or exchanges of policy, depending on the architectural style being deployed.

Figure 20 shows an architectural configuration generated by the *Alloy Analyzer*. In this interaction the *SoldierSMC* requires two roles, *Soldier* and *Sensor*. The *Sensor* role is currently bound to *Source* (through a *Diffusion* style), *Managed* (through a *HierarchicalControl* style) and *Inner* (through a *Composition* style). Similarly, the *Soldier* role is bound to *Target* (through a *Diffusion*

Figure 20. Alloy graphical representation of an architectural configuration between SMCs

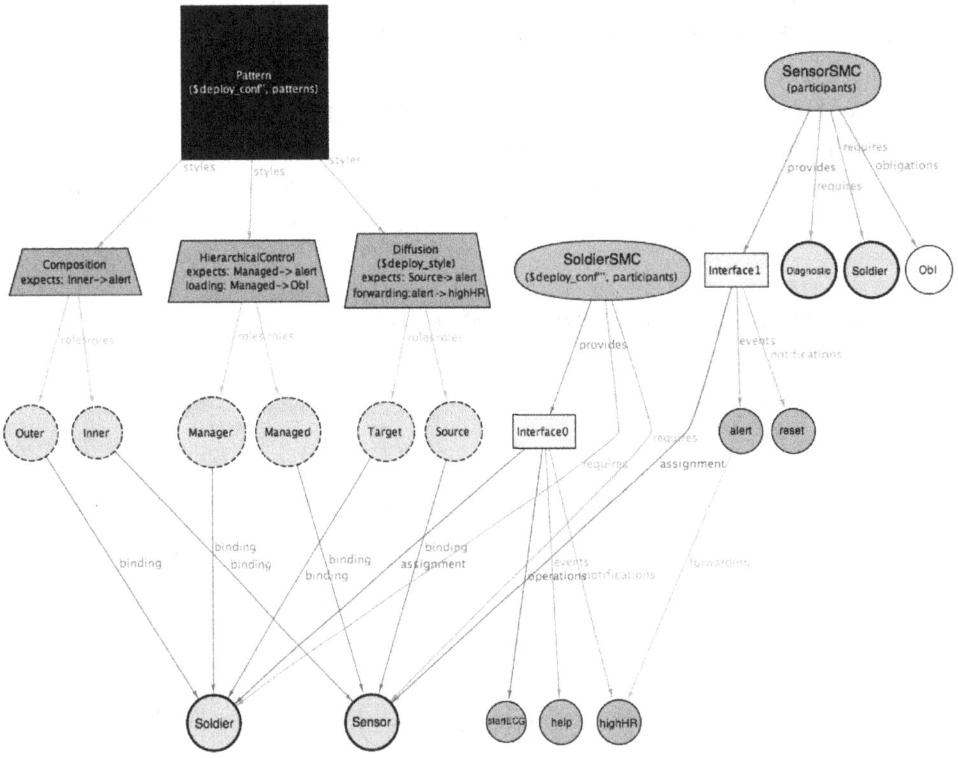

style), *Manager* (through a *HierarchicalControl* style) and *Outer* (through a *Composition* style). This architectural configuration means that whichever SMCs are assigned to the *Soldier* and *Sensor* roles required by *SoldierSMC*, these SMCs will behave accordingly to the behavior defined by each architectural style bound to their roles. Each style also defines what it *expects* from the SMCs that will be eventually assigned to the roles, as well as the *behavior* to be added to the interaction after the style is deployed (forwarding, loading or mapping). The example in particular illustrates the deployment of the *Diffusion* style (the style is labelled as *"$deploy_style"*). This style is marked as *"expects: Source → alert"*, which states that, for whichever SMC is playing the role *Source*, its interface must provide the event *alert*. By following the binding of the style-specific role *Source* to the domain role *Sensor* and noting that *Interface1* provided by *SensorSMC* is assigned to this role, one can see that this interface indeed provides the event *alert*, satisfying the expectations of the style. Similarly, this style is also labelled as *"forwarding: alert → highHR"* (this corresponds to the behavior that must be added to the interaction after the style is deployed). The deployment of this behavior can be seen through the arrow labelled *"forwarding"* between event *alert* (provided by *Interface1*, which is assigned to the role *Sensor*, itself bound to the style-specific role *Source*) and the notification *highHR* (provided by *Interface0*, which is assigned to the role *Soldier*, and bound to the style-specific role *Target*). The behavior added by the deployment of other types of architectural styles can be shown in a similar manner, in the form of mapping of interfaces, forwarding of events or loading of policies.

This model can be used to check whether all the SMCs enforcing policies have the required events forwarded to them (as these events are required for triggering the policies) as well as whether all SMCs have access to the interfaces required for enforcing the actions prescribed by policies. These verifications increase the confidence in the robustness of policy-based coalitions of SMCs, as the interactions can be rigorously verified prior to instantiation and deployment in actual devices.

EVALUATION

Evaluating the support for scaling-up SMC interactions includes aspects such as reuse of code and ease in rapidly instantiating different types of interactions. The experience gained in developing several policy-based SMC applications shows that the use of architectural styles satisfies these two criteria. For example, manually setting an interaction such as an event sharing scheme similar to the *SharedBus* style, using only primitive abstractions not only requires a considerable amount of code to be written, but it is also error-prone: this is because the programmer is responsible for using the primitive abstractions for correctly setting up policies for event forwarding and installing the required *event templates* in different locations to allow the sources to forward the events and the receivers to handle these events appropriately. The architectural style itself is responsible for enforcing the semantics for a given interaction instead of relying on the programmer to use the primitives appropriately.

The use of architectural styles also permits to reduce the complexity and size of the interactions specifications, by structuring and decreasing the number of necessary bindings between SMCs. For example, the size of an interaction for event sharing between SMCs may be drastically reduced by using an abstraction such as the *SharedBus* style, when compared to a non-structured solution. Table 2 shows a comparison between the numbers of necessary bindings to achieve the event exchanges in both cases.

A similar comparison can be made between abstractions for structuring interactions between peer-to-peer collaborations and compositions. Indeed, one of the motivations for compositions is to hide the complexity of large SMCs that

Table 2. Comparison between the number of bindings using a SharedBus style and a non-structured approach for event forwarding

#SMCs	#Bindings (Diffusion)	#Bindings (SharedBus)
2	1	1
3	3 = 2 + 1	2
4	6 = 3 + 2 + 1	3
5	10 = 4 + 3 + 2 + 1	4
6	15 = 5 + 4 + 3 + 2 + 1	5
7	21 = 6 + 5 + 4 + 3 + 2 + 1	6
...	...	...
n	(n - 1) + (n - 2) + (n - 3) + ... + (n - n)	n-1

may comprise a set of smaller, yet autonomous, components, e.g. a body-area SMC. The number of interface exchanges for a completely unstructured interaction, e.g. peer-to-peer, is given by the formula

$$2 \times ((n - 1) + (n - 2) + (n - 3) + ... + (n - n))$$

by comparison, partitioning an interaction between two compositions of one level only reduces the number of interface exchanges in the best case to

$$2 + 2 \times ((n - 2) + (n - 4) + (n - 6) + ... + (n - n)).$$

This is indicted in Figure 21, which illustrates the number of interface exchanges for interactions involving from 2 to 6 SMCs, arranged either as peer-to-peer collaborations or compositions.

These results indicate that the use of architectural styles mitigates the problems of scaling to larger systems, with respect to both programming complexity and the number of interactions that must be established among components. Engineering SMC interactions through the use of architectural styles and patterns thus provides a measurable gain over unstructured solutions.

CONCLUSION

A prototype based on the *Ponder2* framework was implemented in order to demonstrate the feasibility of the model and test its applicability. Because the framework and the algorithms are meant to be applicable to small devices as well as larger systems, it is important to assess the use of this model in constrained devices, with limited computational power and memory – typical examples are *Gumstix* and *Koala*[4] robots. A library of architectural styles has been implemented, in *Ponder2*, to compose these SMC interactions. This library is divided into structural, task-allocation and communication categories. These architectural styles are individually instantiated to create communities whose behavior is dictated by the abstractions defined by each style. The implementation of each style defines a specific algorithm or protocol to be executed by the participant SMCs. This ensures that the semantics specified by an architectural style is enforced by the interacting SMCs.

Engineering collaborations of autonomous system in order to establish dynamic communities is still a research challenge. The use of patterns of structure, task-allocation and communication to design and deploy large-scale policy-based systems is a promising approach. The catalogue of architectural styles, using the Self-Managed Cell as a building block abstraction, caters for a better

Figure 21. Interface exchanges in compositions and peer-to-peer interactions

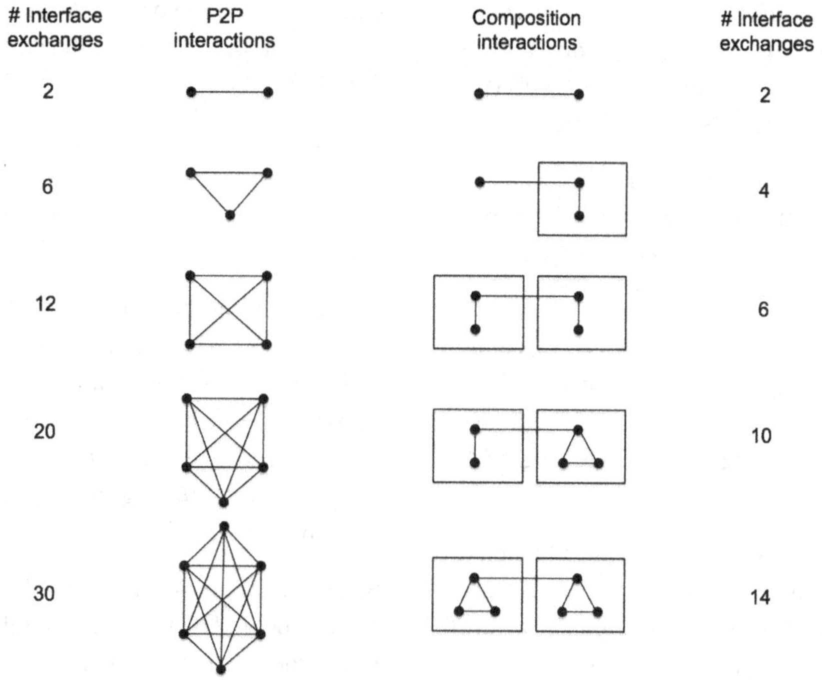

understanding of the relationship between SMCs and promotes the reuse of common abstractions. Collaborations of SMCs may interact with other collaborations, and re-apply the architectural styles recursively. This facilitates defining large-scale composable systems by reusing and combining common abstractions.

The different categories of architectural styles can be seen as orthogonal views for modeling the management aspects of SMC interactions. The catalogue of patterns identifies some common types of interaction that are useful for building communities of SMCs. This is not meant to be exhaustive but focused solely on the frequently occurring patterns identified in the scenarios encountered. The investigation of other application scenarios could result in many new patterns. Although the Self-Managed Cell framework was used as the infrastructure for building autonomous communities, the patterns proposed here have broader applicability for engineering pervasive and autonomous systems in general.

The end goal is to be able to form pre-defined collaborations of SMCs by reusing existing patterns of interaction, which are suitable for a particular application domain. This may involve patterns for deploying a surveillance mission of UAVs or for health monitoring of a soldier in the field, for example. These patterns will enable off-line checking and analysis of collaborations prior to their instantiation in physical devices, reducing the amount of run-time checks and increasing performance during execution.

REFERENCES

Allen, R., & Garlan, D. (1994). Beyond definition/use: architectural interconnection. In *Proceedings of the ACM Workshop on Interface Definition Languages*, New York, NY, (pp. 35–45).

Cardelli, L., & Gordon, A. D. (1998). Mobile ambients. In *Proceedings of the 1st International Conference on Foundations of Software Science and Computation Structure (FoSSaCS)*, (pp. 140–155). London, UK: Springer-Verlag.

Chaimowicz, L., Kumar, V., & Campos, M. F. M. (2004). A paradigm for dynamic coordination of multiple robots. *Autonomous Robots*, *17*(1), 7–21. doi:10.1023/B:AURO.0000032935.30271.a5

Gamma, E., Helm, R., Johnson, R., & Vlissides, J. M. (1995). Design Patterns: Elements of Reusable Object-Oriented Software, (1st ed., ser. Professional Computing Series). Reading, MA: Addison-Wesley.

Garlan, D., & Shaw, M. (1993). An introduction to software architecture . In Ambriola, V., & Tortora, G. (Eds.), *Advances in Software Engineering and Knowledge Engineering* (pp. 1–39). Singapore: World Scientific Publishing Company.

Jackson, D. (2002). Alloy: a lightweight object modelling notation. *ACM Transactions on Software Engineering and Methodology*, *11*(2), 256–290. doi:10.1145/505145.505149

Lupu, E., Dulay, N., Sloman, M., Sventek, J., Heeps, S., & Strowes, S. (2008). AMUSE: autonomic management of ubiquitous systems for e-health. *J. Concurrency and Computation*, *20*(3), 277–295. doi:10.1002/cpe.1194

Lupu, E., & Sloman, M. (1999). Conflicts in policy-based distributed systems management. *IEEE Transactions on Software Engineering*, *25*(6), 852–869. doi:10.1109/32.824414

Mehta, N. R., Medvidovic, N., & Phadke, S. (2000). Towards a taxonomy of software connectors. In *Proceedings of the 22nd ACM International Conference on Software engineering (ICSE)*, New York, (pp. 178–187).

Milner, R., Parrow, J., & Walker, D. (1992). A calculus of mobile processes, I. *Information and Computation*, *100*(1), 1–40. doi:10.1016/0890-5401(92)90008-4

Phillips, A. (2006). *Specifying and implementing secure mobile applications in the channel ambient system*. Ph.D. dissertation, Imperial College London.

Schaeffer-Filho, A. Lupu, E., Dulay, N., Keoh, S. L. Twidle, K., Sloman, M., Heeps, S., Strowes, S. & Sventek, J. (2007). Towards supporting interactions between self-managed cells. In *Proceedings of the 1st IEEE International Conference on Self-Adaptive and Self-Organizing Systems (SASO)*, Boston, (pp. 224–233).

Schaeffer-Filho, A. Lupu, E., Sloman, M., Keoh, S.-L. Lobo, J., & Calo, S. (2008, July). A role-based infrastructure for the management of dynamic communities. In *Proceedings of the 2nd International Conference on Autonomous Infrastructure, Management and Security (AIMS)*, Bremen, Germany (LNCS pp. 1–14). Berlin: Springer.

Schaeffer-Filho, A., Lupu, E., & Sloman, M. (2009). Realising management and composition of self-managed cells in pervasive healthcare. In *Proceedings of the 3rd International Conference on Pervasive Computing Technologies for Healthcare*, London.

Schaeffer-Filho, A., Lupu, E., Sloman, M., & Eisenbach, S. (2009, July). Verification of policy-based self-managed cell interactions using alloy. In *Proceedings of the 10th IEEE International Symposium on Policies for Distributed Systems and Networks*, London.

Shaw, M., DeLine, R., Klein, D. V., Ross, T. L., Young, D. M., & Zelesnik, G. (1995). Abstractions for software architecture and tools to support them. *IEEE Transactions on Software Engineering*, *21*(4), 314–335. doi:10.1109/32.385970

Smith, R. G. (1988). The contract net protocol: high-level communication and control in a distributed problem solver. In Distributed Artificial Intelligence, (pp. 357–366).

ENDNOTES

1 http://www.gumstix.com
2 http://www.ponder2.net
3 http://vip.doc.ic.ac.uk/bsn/
4 http://www.k-team.com

Chapter 8
Policy Technologies for Security Management in Coalition Networks

Seraphin B. Calo
IBM Research, USA

Clare-Marie Karat
IBM Research, USA

John Karat
IBM Research, USA

Jorge Lobo
IBM Research, USA

Robert Craven
Imperial College, UK

Emil Lupu
Imperial College, UK

Jiefei Ma
Imperial College, UK

Alessandra Russo
Imperial College, UK

Morris Sloman
Imperial College, UK

Arosha Bandara
The Open University, UK

ABSTRACT

The goal of policy-based security management is to enable military personnel to specify security requirements in terms of simple, intuitive goals. These goals are translated into the concrete system settings in a way that the system behaves in a consistent and desirable way. This technology minimizes the technical expertise required by military personnel and automates security management while allowing a high level control by the human in the loop. This chapter describes a framework for managing security policies, and an overview of two prototypes that simplify different aspects of policy management in the context of coalition operations.

DOI: 10.4018/978-1-61520-855-5.ch008

INTRODUCTION

Secure, reliable and adaptable communications is needed to support dynamic mission-based coalitions of partners from different military and non-military organizations. If sensitive information is communicated to the wrong person/device, it could cost the lives of the personnel involved in the mission. Likewise, if necessary information is not communicated and shared with the right people, it could also lead to loss of lives. Policy-based security management should enable military personnel to specify security requirements in terms of simple, intuitive goals that are translated into the concrete system settings in such a way that the system behaves in a consistent and desirable way. The objective is to minimize the technical expertise required by military personnel, and to automate policy management as far as possible. This is dependent on being able to specify and analyze policies to ensure that they prescribe correct and desirable behavior. For example, inconsistencies should not arise because the available communication devices cannot support the specified policies. We assume that military personnel specify goals using a structured natural language aimed at non-technical people. Goals are automatically translated into a formal, logic-based abstract language for refinement and analysis. Our past experience has indicated that logic languages, while good for reasoning, are not amenable to efficient implementation, particularly on small hand-held devices. Thus abstract policies must be translated into concrete implementable policies described in languages such as Ponder2 (Twidle et al, 2008), XACML (OASIS XACML TC, 2005), or CIM-SPL (Agrawal et al, 2007).

We start with the presentation of a policy-based security management framework for complex, dynamic, ad hoc systems. This provides the platform supporting mechanisms for adapting system behaviors to meet high-level user-specified security policies through the enforcement of low-level controls in coalition networks. To accomplish this, end-to-end policy mechanisms are described that capture the security requirements of the system, transform them into constraints on the system resources and executable policies, which are then disseminated to and executed upon the appropriate distributed entities within the coalition network. Yet without analysis much of the benefit of using policy-based techniques and declarative policy languages may be lost. Arguably, the lack of effective analysis tools accounts in part for the lack of wider adoption of policy-based techniques. In order to perform analysis, policies must be expressed unambiguously in a manner that captures their semantic meaning. We describe an approach based upon a logical construct for the specification and analysis of security policies. This construct is developed over a very expressive policy language that may be used to represent policies and systems at many different levels of abstraction and stages during the refinement process.

In order to bring together and demonstrate the various policy technologies for security management, two concept prototypes are discussed. These served to integrate some capabilities and provided an end-to-end view of the policy lifecycle for coalition systems. They were based on user scenarios. The purpose of a user scenario is to provide sufficient context about a problem-space being investigated so that researchers can identify the scientific, technical, and feasibility questions they must address in the research. A user scenario can be a valuable research and design tool for scientists to employ to understand the trade-offs about different options in the context of the targeted end user. Details of these prototype demonstrations are presented, as are a number of research questions that arose during their development.

FRAMEWORK FOR SECURITY POLICIES

Our framework for policy analysis and refinement captures policies at various levels of abstraction

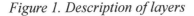

Figure 1. Description of layers

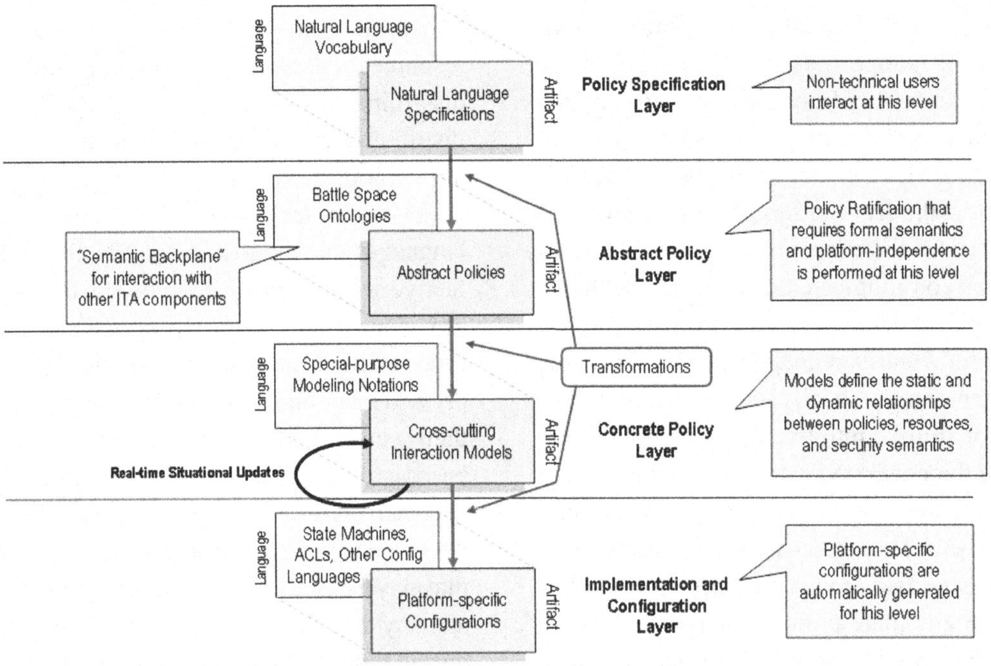

(or layers), which identify the key stages in the process of refining policies from goals to implementations. The layered architecture is useful in identifying the architectural elements, software components, and system models that are needed to support the refinement and analysis processes, and to facilitate the management of policies. In this section we describe and illustrate the framework, identifying the levels comprising it and the services that we expect to be able to perform at each level.

DESCRIPTION OF THE LAYERS

We can identify four key stages, or layers, in the process of refining policies: *Specification, Abstract, Concrete,* and *Executable*. Policies are defined with respect to a system model, which is a formal representation (or description) of the system within which the policies are enforced. Given that policies exist at multiple levels of abstraction, it is natural that each layer of policy specification should exist in the context of system models at that layer. These models are elaborated at the appropriate level of abstraction, so that they become more detailed at the lower layers (Figure 1).

Policy Specification Layer. This is the most abstract level of policy specification; it is at this layer that we would expect the military user to have the most interaction. Typically, the military user would define the mission security goals, in some intuitive format at this layer. We would not expect policies specified at this layer to make any reference to security mechanisms, or any other concrete system features. However, it should still be possible to assign formal semantics to the specified policies. The system model at this layer should only refer to high-level operations and services which are meaningful to a non-technical military user. Policies defined at this level can be referred to simply as *security goals*, or *end-user goals*.

These policies must be interpreted in context. Ontologies are being developed for capturing the semantics of the battle space. These can be used in establishing semantics for transforming the policies at the specification layer into the abstract policies described below. The sets of information relating the various aspects of the military environment, including policies, constitute a *semantic backplane*. This consists of the set of concepts and relationships that are used by the military in carrying out coalition missions.

Abstract Policies. To implement a policy we need candidates for the system entities and actions that will be used to enforce the policy. The abstract policy models are possible refinements of the end-user goals but enriched with suitable information about candidates for implementation. Descriptions of these candidates will, of course, reside in the system models. Note that there may be multiple abstract policies with respect to corresponding system models in support of a single security goal.

Concrete Policies. Inherent in the notion of refinement is the resolution of choices. The abstract policies capture the range of choices that may be available in implementing an end-user goal. These are refined to concrete policies when the candidates from the abstract layer have been restricted to singletons. The system model at this layer refers directly to objects and services which are to be found in the target system. The system model at this layer may also include classes of user, types of risk, and so forth.

Executable Policies. This layer contains the policies which will actually be deployed. It may further refine the policies from the concrete layer by setting low level system parameters, expressing policies in a format so that they can be used conveniently by mechanisms in the system.

As an example of the refinement process, we consider policies for Intelligence, Surveillance, and Reconnaissance (ISR) assets being used in a coalition operation. One high level goal defined at the specification layer can lead to a number of policies at the lower layers.

- *Policy Goal:*
  - High quality sensor information is reserved for US use only
- *Abstract Policies:*
  - For coalition members, video sensor output will be degraded to reduce image resolution.
  - For coalition members, location sensor output will be degraded so that coordinates only identify a proximity region.
- *Concrete Policies:*
  - If a UK control asset requests an image from a US camera then it will be set to a resolution of .8 times its native resolution
  - If a UK control asset requests the location of a US asset then the GPS data will be transformed into data identifying a region of 10 square meters

The executable policies would be in whatever form is required by the mechanisms in the system that would enforce them. They would identify specific assets with specific characteristics.

DESCRIPTION OF THE FUNCTIONALITY AT EACH LAYER

The layers are intended as a means of identifying key milestones in the refinement process; but, in addition, they also underpin effective solutions to the broader challenges of policy management. For instance, policy analysis is a crucial part of policy management, and, as a rule, it highly desirable to analyze policies at as high a level of abstraction as possible, because abstract policies are simpler and therefore more tractable. Furthermore, should a given type of analysis detect a violation, then it is easier to remedy that violation at a more abstract level, than at a more concrete level. For example, the conflict detection problem is more tractable for abstract policies, and should two policies be found to conflict, the remedy for the conflict is easier to identify and implement when the policies are more abstract.

There are various policy management services which are expected to be supported by the layers; and, in general, implementing these services will require interaction between the layers. The services, and their interactions, are as follows:

Policy Specification Layer

1. Policy authoring
2. Translating/mapping policies into formal semantics.
3. Basic analysis: tools such as model-checkers or logic programs can be adapted to analyze security goals in the goal specification language for detection of conflict and other policy properties. In addition it may be possible, using the semantics developed, to analyze sets of end-user goals for coverage – in other words does this set of end-user goals capture the complete range of security variables that need to be controlled.
4. Conflict resolution: if conflicts can be identified, it may also be possible to perform conflict resolution.

Abstract Policies

1. The information added at this layer should enable substantial conflict (and inconsistency) detection services to be provided. There are interesting problems arising from attempting to check a policy that is being refined, and so is rather abstract, against one that is executable and so rather concrete.
2. Conflict resolution is the complementary service to conflict detection, but it may be necessary to coordinate conflict resolution with the layer above. In other words, for some conflicts it may make more sense for the resolution to be provided by the specification layer.
3. It may be necessary to perform coverage analysis at this layer as well. In general it is desirable to perform as much coverage analysis at the specification layer as possible; however, the layers below this layer may yield negative results about validity, which may obligate modification of policies at this layer (or above), and that in turn may require coverage checking at this layer.
4. Incorporating user preferences or accommodating risk-based decisions from the user would happen at this level. While the ultimate objective is that polices are managed automatically (and specifically their refinement is automatic), in reality full automation is probably not possible. The compromise to is to provide users with an interface to the refinement (and management) services; this will allow users to insert their preferences (either preemptively or on-demand), and it will also be the means for enabling users to make risk-based policy decisions.
5. Mappings to and from the formal semantics are likely to be necessary here.

Concrete Policies

1. While we would hope that most conflict detection is done above this layer, so that the concrete policies derived here are known to be conflict free, it may be necessary for some types of conflict and inconsistency detection to be done here.
2. Validity checking is done most naturally at this layer, since we expect the system model at this layer to refer directly to objects in the system.
3. Compliance checking is done either at this layer, or the layer below.
4. Refinement checking. We would hope that the policies generated would, by construction,

be a correct refinement of the security goals. However, in practice there may be changes to policies, or violations of policies, at runtime (it is most likely that manual override must be permitted). In this case it makes sense to be able to check that modified policies, or indeed, system behavior, still constitutes a refinement of the original security goal.

5. Mapping to and from the formal semantics is a basic service, required here as well, especially to support refinement checking.

Executable Policies

1. Validity checking.
2. Compliance checking.
3. Refinement checking.
4. Mapping to and from the semantics; we would expect that at this layer the results of the refinement algorithms, which operate on the formal semantics, are translated into executable policies; however, it may also be necessary to be able to map from the policies back to the formal semantics in order to perform the indicated analyses.

As indicated in the above, various forms of analysis must be performed at each of the refinement layers. The ability to carry out the required analyses is thus a basic consideration in the management of policies within the system. An in depth discussion of analysis mechanisms and the underlying policy representations that are required to support them is undertaken in the next section.

ABSTRACT POLICIES AND POLICY ANALYSIS

The selection of an abstract policy language will determine the quality and scope possible during analysis. The goal is to be able to have an expressive abstract language that captures the intuitive understanding of security descriptions expected by the end user and at the same time a language that is amenable to efficient implementations. Hence, we need to balance expressiveness with efficiency of policy evaluation. We also need a powerful analysis component. In this section we describe a formal framework for representing and analyzing authorization and obligation policies.

Policies are represented in an expressive logical language, which can capture complex dependencies amongst policy rules. These policy rules are then joined to a specification of the behavior of the system which the policies regulate, and many different types of analysis and verification may then be performed on the joint system. We have designed an abstract language which we hope to be expressive enough that policies written in many different formalisms (such as Ponder2 (Twidle et al, 2008), XACML (OASIS XACML TC, 2005), or Cassandra (Becker & Sewell, 2004)) can be translated into it, where they can then be analyzed for the presence of modality conflicts, coverage gaps, and other properties.

A crucial component of our framework is the representation of dynamic system behavior. Policies have constraints on their applicability that often depend on the local system state. Therefore, a model of the system is required as a component of an analysis framework, both in order to determine whether policies conflict, and to be able to analyze a given set of policies for properties that the system would satisfy when they are enforced. An analysis framework should also represent policies and the systems they regulate in a separable way, so that the behavior of a policy on different systems, and the implementation of different policies in the same system, can easily be studied.

The properties for which we can perform analyses include both foundational, policy focused principles, and also application specific features, dependent on the particular system to which the policy applies and its structure. Specifically (though not exhaustively) these include:

- **Modality conflicts,** such as the joint authorization and denial to perform some action, or the presence of an obligation to act without the permissions necessary for its fulfillment.
- **Separation of duty clashes**, including static separation of duty, dynamic, and many other classes (see (Simon & Zurko, 1997) for terminology and instances).
- **Coverage gaps**, where no policy exists to dictate what the correct response to a request should be.
- **Policy comparison**, including the question of whether two policies are equivalent or one is contained in the other.
- **Behavioral simulation**, where specific sequences of requests and events in the policy regulated system are entered, to see the policy decisions that would arise during in the actual system.

LANGUAGE

Our operational model broadly follows the architecture and principles of operation of XACML (OASIS XACML TC, 2005), PolicyMaker, and KeyNote (Blaze et al, 1999). There is a policy component, consisting of policy decision and enforcement points (PDP/PEP), and the system to which policies refer and which they modify. The PDP has access to a policy repository. Authorization decisions are made in response to requests for a subject to perform an action on a target, using the policies, and these decisions are then enforced by the PEP. The PDP also monitors whether obligations of subjects to perform actions have been met or not.

We distinguish between regulatory predicates, used to describe the state of the PDP/PEP, and non-regulatory predicates that express the state of the policy-governed system. Regulatory predicates are subdivided into input regulatory, state regulatory, and output regulatory; similarly, for non-regulatory predicates. In general, a system moves between states depending on the occurrences of actions and events. Non-regulatory state predicates represent properties of states, and non-regulatory event predicates describe the occurrence of events. (The need for event predicates arises because, in general, not all occurrences which modify the state of a system are controllable by the policy mechanism.)

We use many-sorted first-order predicate logic as our base language, and clearly distinguish the policy representation language from the domain description language. This allows us to detach policy representations from system representations, and compare the implementation of a policy in different systems. The policy representation language, L^{π}, includes at least the sorts Subject, Target, Action, the sets of subjects, targets and actions, respectively, and the sort Time, given as R+ , the set of non-negative Real numbers, with constants including numerical constant symbols (0, 1, . . .) and variables such as T, with super- and subscripts as needed. Standard arithmetical functions (+, –, /, ∗) and relations (=, =, <, etc.) are presumed. Variables which range over Subject, Target, Action are usually, respectively Sub, Tar, Act, again possibly with subscripts or superscripts. We insist on the possibility of representing the revocation of obligations, so that for all Sub, Tar, Act = revoke(· · ·) and T_s, T_e, revoke(Sub, Tar, Act, T_s, T_e) ∈ Action. The predicates of L^{π} are as defined in Table 1. For a formal definition of the language the reader is referred to (Craven et al 2009).

The predicates permitted, and denied are self-explanatory. A particular instance req(sub, tar, act, t) means that a request for sub to perform act on tar is made at time t. Two instances obl(sub, tar, act, t_s, t_e, t) and fulfilled(sub, tar, act, t_s, t_e, t) (resp. violated(sub, tar, act, t_s, t_e, t)), denote that at time t, sub is placed under an obligation to perform act on tar between t_s and t_e, and that the obligation with these parameters has been fulfilled (resp. violated). Finally, an instance cease_obl(sub, tar, act, t_{init}, t_s,

Table 1. Policy analysis language L^{π}: the predicates

Input regulatory	State regulatory	Output regulatory
req(Sub, Tar, Act, T)	permitted(Sub, Tar, Act, T)	do(Sub, Tar, Act, T)
	denied(Sub, Tar, Act, T) obl(Sub, Tar, Act, Ts, Te, T)	deny(Sub, Tar, Act,T)
	fulfilled(Sub, Tar, Act, Ts, Te, T)	
	violated(Sub, Tar, Act, Ts, Te, T)	
	cease obl(Sub, Tar, Act, Tinit, Ts, Te, T)	

t_e, t) is true at time t, if an obligation initially contracted by sub at t_{init} to perform act on tar between t_s and t_e is no longer binding. The language L^{π} also includes an auxiliary predicate reqInBetween, whose instance reqInBetween(sub, tar, act, t_0, t) represents that there is a request for sub to perform act on tar, at some time between t_0 and t.

The domain description language, $L^D = L^D_{EC} \cup L^D_{stat}$, is used to represent both changing and unchanging properties of the system being regulated by the policy, the changes occurring both as a consequence of actions authorized and enforced by the PDP and PEP, and also as a result of events which are not under policy control. We use the Event Calculus (EC) (Kowalski & Sergot, 1986) to model this dynamicity in our domains. The language includes sorts Fluent (for representing dynamic features of states), Event (for system events not regulated by policies), Occurrence (for representing system events which are regulated by policies) and Time (as before).

The predicates of LDEC which model changing actions of the system and stem from the EC, are given in Table 2; the predicates of LDstat, representing unchanging properties of systems, are user-defined. For a formal definition of LD the reader is referred to (Craven et al, 2009).

The policy analysis language $L = L^{\pi} \cup L^D$, is the union of the policy representation and domain description languages, with the sort Occurrence given by the terms Sub:Tar:Act for every Sub $\in$ Subject, Tar $\in$ Target, and Act $\in$ Action.

AUTHORIZATIONS

We first define time constraints, and then the structure of authorization rules.

Definition 1. A time constraint C is an expression of the form $\tau_1\ \rho\ \tau_2$, where each τ_i is a constant or variable of type Time, or an

Table 2. Domain description language L^D : the predicates

Event Calculus predicates	
initially(F)	State property F holds at time 0.
holdsAt(F, T)	State property F holds at time T
happens(E, T)	Event E occurs at time T .
broken(F, T1, T2)	State property F ceased to hold at some time between T1 and T2 .
initiates(E, F, T)	At time T, the occurrence of an event E would cause F to begin to hold.
initiates(Occ, F, T)	At time T, the occurrence of an action Occ regulated by PDP would cause F to begin to hold.
terminates(E, F, T)	At any T, an event E would cause F cease. terminates(Occ, F, T) At any T, action Occ regulated by PDP would cause F to cease holding.

arithmetic linear expression built using +, –, /, *, Time constants and variables, and where ρ is one of =. ≠, ≤, <, ≥, >

Definition 2. An authorization rule is a formula:

[permitted/denied](Sub, Tar, Act, T) ← L_1, . . ., L_m, C_1, . . ., C_n.

1. L_i are atoms or atoms preceded by the negation-by-failure not taken from the policy language, the domain language or hodslAt, happens, broken; the C_i are time constraints;
2. any variable appearing in a time constraint must also appear somewhere other than in a time constraint;
3. Sub, Tar, Act, T are terms of type Subject, Target, Action and Time respectively;
4. for the time argument T_i of each L_i not in L^D_{stat}, we must have $C_1 \wedge \cdots \wedge C_n \models T_i \leq T$; if $C_1 \wedge \cdots \wedge C_n \models T_i = T$ then the L_i must not be a regulatory output predicate; if it is from the EC language it should be holdsAt or broken.

Where such a rule has permitted in the head, it is a positive authorization rule; otherwise, it is known as a negative authorization rule. (Additional constraints of local stratification will be imposed later.)

Condition 1, on the predicates which may appear in the body of our authorization policy rules, excludes the initiates and terminates predicates which are used to express the laws of system evolution. Condition 4, on the time arguments of our predicates, is imposed to ensure that authorization does not depend on 'future' properties.

Example 1. "Alice may delete classified data files from her device if she sends a notification to the supplier of the data 10 minutes in advance, and the supplier does not respond to the notification asking Alice to retain the file."

We represent this as follows:

permitted(alice, device, delete(F), T) ←
holdsAt(fileDesc(S, F, class), T_n),
do(alice, S, notify(delete(F)), T_n), T = T_n + 10,
not reqInBetween(S, F, retain(F), T_n, T).

The predicate reqInBetween is related to the operator Since of temporal logics (Goldblatt, 1992); the utility of having such a predicate has been demonstrated to us repeatedly in many formalizations of policy rules. To capture its semantics, the following rule is always included in our framework:

reqInBetween(Sub, Tar, Act, T', T) ← req(Sub, Tar, Act, T_r), T' ≤ T_r ≤ T .

An instance reqInBetween(Sub, Tar, Act, 0, T) means that a request (with the relevant parameters) was made at some time before T ; this is related to the modal temporal operator expressing that a property held at some previous time.

When gathering together authorization rules to form an authorization policy, it is normal to include a number of other, more general rules. Some of these are used to specify the behavior of the PEP in response to the PDP—such as whether a request to perform an action is accepted (and the action performed) by default if there is no explicit permission in the policy rules; or whether explicit permission is required; what the behavior is in the presence or absence of a policy rule stating that the request is to be denied, and so on. We see it as a virtue of our framework for policy analysis that many different rules which embody the action of the PEP can be represented, and that no one approach is fixed as part of the formalism. This flexibility is crucial if we need to cover the behavior of many policy systems in heterogeneous

Table 3. Default availability rules

do(Sub, Tar, Act, T) ← req(Sub, Tar, Act, T), permitted(Sub, Tar, Act, T).	Basic availability
do(Sub, Tar, Act, T) ← req(Sub, Tar, Act, T), not denied(Sub, Tar, Act, T).	Positive availability
deny(Sub, Tar, Act, T) ← req(Sub, Tar, Act, T), denied(Sub, Tar, Act, T).	Negative availability

environments. Consider the three example availability rules in Table 3.

The basic availability rule is more stringent: according to it, an action is enforced by the PEP only when it has been positively permitted by the PDP—similar to (Loscocco & Smalley, 2001). The positive availability rule is less strict: it enforces the performance of an action as long as that action has not been expressly denied by the policy rules. The negative availability rule describes one possible response to the denial of a request for action; we imagine that the main function of the regulatory output predicate deny will be for auditing purposes, for instance to record when an action which was denied had to be overridden and permitted in an emergency.

Definition 3. A policy regulation rule has one of the predicates do or deny in the head and a body as in Definition 2.

Many more policy regulation rules are possible than those given as examples in Table 3; all are optional inclusions in an authorization policy.

Definition 4. An authorization policy is a set Π of authorization rules, together with the req-in-between rule, and a set of policy regulation rules, such that Π is locally stratified.

Notice that it is possible to add general authorization rules to a policy, enabling a representation of very fine-grained defaults controlling responses to requests. For example, if a user belongs to the root system group, one may want to permit all the actions of that user by default, unless they are explicitly denied:

permitted(Sub, Tar, Act, T) ← group(Sub, root), not denied(Sub, Tar, Act, T).

A set of rules is locally stratified if in the set of all ground instances of the rules (i.e., where all variables are replaced by all their possible values) there is no head of a rule that depends directly or indirectly on the negation of itself. Testing for local stratification is computationally hard but large classes of rules can be identified easily based on the time index (Nomikos et al, 2005).

OBLIGATIONS

The obligations we represent are on a subject to perform an action on a target, a class which includes a large number of practical obligation policies (Irwin et al, 2006). As in most deontic logics, obligations may be fulfilled or not, allowing us to represent the behavior of systems of which humans are a part. Our approach to modeling obligations is similar to (Irwin et al, 2006).

Definition 5. An obligation policy rule is a formula

$\mathrm{obl}(Sub, Tar, Act, T_s, T_e, T) \leftarrow L_1, \ldots, L_m, C_1, \ldots, C_n$.

where the conditions 1–4 as for Definition 2 hold, with the addition that T_s and T_e should be variables of type Time. (That $t_s < t_e$ is not syntactically required, but sensible, obligation policy rules will always include constraints which make this true.)

Two domain-independent rules accompany the obligation rules, defining the fulfillment and the violation of an obligation:

fulfilled(Sub, Tar, Act, T_s, T_e, T) ←
obl(Sub, Tar, Act, T_s, T_e, T_{init}), do(Sub, Tar, Act, T'),
not cease_obl(Sub, Tar, Act, T_{init}, T_s, T_e, T'),
$T_{init} \leq T_s \leq T' < T_e, T' < T$.
violated(Sub, Tar, Act, T_s, T_e, T) ←
obl(Sub, Tar, Act, T_s, T_e, T_{init}),
not cease_obl(Sub, Tar, Act, T_{init}, T_s, T_e, T_e),
$T_{init} \leq T_s < T_e \leq T$.

An obligation is fulfilled when the action a subject has been obliged to perform is executed (notice that the do in the body of the rule here means that the execution of such an action must first be authorized by the system). An obligation is violated when no such action occurs. The rules for fulfilled and violated use cease obl as a subsidiary predicate, defined by the following rules:

cease_obl(Sub, Tar, Act, T_{init}, T_s, T_e, T) ←
do(Sub, Tar, Act, T'), $T_s \leq T' < T \leq T_e$.
cease_obl(Sub, Tar, Act, T_{init}, T_s, T_e, T) ←
do(Sub^0, Sub, revoke(Sub, Tar, Act, T_s, T_e), T'),
$T_{init} \leq T' < T \leq T_e$.

cease_obl is a state regulatory predicate used to mark the fact that something has occurred which would cause an obligation to cease. This can happen if the obligation is fulfilled or revoked; there are therefore two clauses defining cease obl. The cease_obl rule for revocation makes use of the revoke members of the sort Action, introduced in Section 3; revocation occurs when the PDP has authorized the request for a revocation action. The subject requesting a revocation might be the one bound by the obligation, a central administrator in the system, or an entirely different agent. The parameters of the revoke argument identify the obligation to be revoked.

Example 2. "A connecting node must re-indentify itself within five minutes of establishing a connection to the server; otherwise the server must drop the connection within one second."

There are two obligations in this example, one for the node making the connection and the second for the system dropping the node making the connection if the node does not fulfill its obligation. A formalization of the obligations looks as follows:

obl(U, serv, sub2ID(U, serv), T +0.1, T +300, T +0.1) ←
holdsAt(node(U), T), do(U, serv, connect(U, serv), T).
obl(serv, serv, disconnect(U, serv), T_e, T_e +1, T_e) ←
violated(U, serv, sub2ID(U, serv), T_s, T_e, T_e).

The EC predicate holdsAt is used to represent dynamic properties of the system: in this case, which nodes are registered. The obligation begins just after the server connects to the node—in the rule, we have assumed there is a delay of 0.1 seconds, but in practice this interval can be made as small as possible without being equal to zero, reflecting the limitations of the system clock.

Definition 6. An obligation policy Π is a set of obligation rules, with the 'fulfillment', 'violation' and 'cease_obl' rules, such that Π is locally stratified.

Definition 7. A security policy $\Pi = \Pi_a \cup \Pi_o$ is any union of an authorization policy Π_a and an obligation policy Π_o.

Table 4. Event calculus core axioms

holdsAt(F, T) ← initially(F), not broken(F, 0, T). (1)
holdsAt(F, T) ← initiates(Sub:Tar:Act, F, T_s), $T_s < T$, do(Sub, Tar, Act, T_s), not broken(F, T_s, T). (2)
holdsAt(F, T) ← initiates(Event, F, T_s), $T_s < T$, happens(Event, T_s), not broken(F, T_s, T). (3)
broken(F, T_s, T) ← terminates(Sub:Tar:Act, F, T^0), do(Sub, Tar, Act, T^0), $T_s < T^0 < T$. (4)
broken(F, T_s, T) ← terminates(Event, F, T^0), happens(Event, T^0), $T_s < T^0 < T$. (5)

SYSTEM DESCRIPTIONS

We use the Event Calculus (EC) to model the systems regulated by the policies. In the EC, effects of events are defined by two predicates, *initiates* and *terminates*. *Initiates* describes which state properties become true by the occurrence of an event, and *terminates* which properties become false. These predicates are defined with rules similar to policies:

initiates(X, F, T) ← $L_1, \ldots, L_m, C_1, \ldots, C_n$.

terminates(X, F, T) ← $L_1, \ldots, L_m, C_1, \ldots, C_n$.

with the limitation that the Ls in these rules are holdsAt or static predicates. These rules, combined with the axioms in Table 4 show how initiates and terminates affect the holdsAt predicate.

Consider the following set of axioms partially describing the RBAC model. We can represent user-to-role assignments with fluent hasRole(Subject, Role), and the permission-to-role assignments with the fluent hasPerm(Role, Resource, Action). Adding and removing users to and from roles can be modeled in EC with the axioms:

initiates(S:R:assignUser(U), hasUser(R,U), T).
terminates(S:R:unassignUser(U), hasUser(R,U), T).

Similarly we need axioms for adding or removing permissions from or to roles:

initiates(S:R:assignPerm(Res,A), hasPerm(R,Res,A), T).
terminates(S:R:unassignPerm(Res,A), hasPerm(R,Res,A), T).

The access control can be express by the following policy:

permitted(Sub, Resource, Act, T) ←
holdsAt(hasRole(Sub,Role), T),
holdsAt(hasPerm(Role, Resource, Act), T).

Simple static separation of duty of two roles, inspector and support staff can be modeled with rules:

denied(Admin, inspector,assignUser(U), T) ←
holdsAt(hasRole(U,support_staff), T).
denied(Admin,support_staff,assignUser(U), T) ←
holdsAt(hasRole(U,inspector), T).

ANALYSIS

The task of analyzing a domain-constrained policy $P = \Pi \cup D$ to see, for instance, whether there are no modality conflicts (e.g. permits and denials over the same resource, or obligations over resources for which a subject has no authorizations), can be converted into the task of seeing whether (stable) models of the domain- constrained policy verify a number of properties. For instance, we may wish to prove:

∀T (¬(permitted(sub, tar, act, T) ∧ denied(sub, tar, act, T))) for ground terms sub, tar, act. If this property is provable, then all well and good. If not, then we wish to have diagnostic information about the circumstances in which it fails to be true. Checking whether the system verifies this property converts into the task of checking whether there are inputs Δ^D and Δ^π (as described in Section 4.5) such that the property is not true, i.e., whether

$$\text{model}(P \cup \Delta^D \cup \Delta^\pi) \models \exists T\ (\text{permitted}(sub, tar, act, T) \wedge \text{denied}(sub, tar, act, T))$$

This is equivalent to showing that the previous formula is false, and can be solved using Abductive Logic Programming (with constraints—ACLP), which computes the sets Δ^D and Δ^π . The output to the algorithm will be these sets together with a number of constraints (expressed as equalities and inequalities) on the possible values of the time-arguments appearing in the answers. We currently use an abductive constraint logic programming proof procedure based on that found in (Kakas et al, 2000) to implement our system. (For the details of the algorithm, see (Craven et al, 2009).)

Following the same schema, we can define formulas to check for other properties. For example, violations of dynamic SoD can be checked with:

$$\text{model}(P \cup \Delta^D \cup \Delta^\pi) \models \exists T\ (\text{permitted}(sub, roles, activate(role\ a), T) \wedge \text{permitted}(sub, roles, activate(role\ b), T)$$

Also, coverage analysis can be performed using:

$$\text{model}(P \cup \Delta^D \cup \Delta^\pi) \models \exists Sub, Tar, Act, T\ (\text{req}(Sub, Tar, Act, T) \wedge \neg\text{permitted}(Sub, Tar, Act, T) \wedge \neg\text{denied}(Sub, Tar, Act, T))$$

This query will find instances of requests for which there is no associated permission or denial according to the authorization rules of the policy. Other properties can be handled in similar ways.

The above treatment of policy representation and analysis has been somewhat detailed because we need mathematical precision in order to be able to analyze and refine policies while maintaining their semantic meanings. The manners in which analysis will be incorporated in the overall flow of a policy management system was investigated in two prototypes that will be described in the next section.

THE END-USER VIEW OF POLICY MANAGEMENT

As a means of demonstrating technologies for adapting system behaviors to meet high-level user-specified security and networking goals in a coalition environment, two scenario-driven, conceptual prototypes are described which were developed to investigate different areas of the policy management space. Such scenario-based prototypes are seen as critical artifacts in fundamental human-computer interaction (HCI) research (Carroll, 1995). The goal of the policy research is an easy to use, end-to-end policy management environment that enables security requirements of the system to be authored as a set of policies. The authoring capability would be supported by methods for analysis and visualization of the policies to help identify and resolve potential issues, transformation of these policies into constraints on the system resources, and execution of appropriate actions that enforce these policies when deployed over wireless networking systems. There are many research challenges that must be addressed to realize such a vision, and the implementation of specific instances of a policy management system allows us to explore the range of such issues.

The scenario-based, proof-of-concept prototypes described served as a means of exploring both the kinds of capabilities necessary and the value different capabilities contribute towards achievement of the research goal. Scenario-based HCI research is conducted to identify the key **users**, **tasks**, and **contexts of use**, and the associated key functionality required in a system or tool for users to achieve defined task goals (Carroll, 1995; Karat, 1991). To design a usable system, it must first be understood who the target users of the system are and their user roles, as well as key characteristics of these users that pertain to the tasks to be completed in the context of use (Jacko and Sears, 2008). User characteristics might include education, skills, and experience; language and culture; and physical abilities (ability to see and hear, use arms and hands or voice to interact with the tool).

The context of use is a critical component to being able to design a usable system. Understanding whether a technological capability must be rugged and extremely simple and reliable for users to work with silently in extreme temperatures outdoors both in sunlight and in complete darkness, as opposed to a typical office situation, identifies key requirements for the design of a usable system. Focus on the context of use improves the probability of success in designing usable technology for the demands of the situation. These three factors together – the target users, their core tasks, and the context in which the tasks will be completed – provide the requirements that must be addressed in the creation of usable new technology.

INITIAL SECURITY POLICY PROOF-OF-CONCEPT PROTOTYPE

This scenario (Brodie et al., 2008) framed both user and system requirements, and was an extension of a high-level hostage rescue scenario used across the ITA project to guide the fundamental research efforts (Roberts, Locke, and Verma, 2007). Target user/resource attributes, critical policy-related tasks, and some contextual information about fictitious security level variables were employed to highlight some of the concepts in the prototype.

Four different user roles were identified as necessary for the scenario:

- Mission administrator
- U.S. policy administrator
- U.K. policy administrator
- Security administrator

The scenario highlights the following core user tasks employing proposed policy management capability:

- Authoring and updating of policies for U.S. personnel and equipment, U.K. personnel and equipment, and the rescue mission
- Viewing the policy set
- Analysis of the policies and resolution of the issues identified
- Transformation of policies as appropriate
- Deployment of the policies

A set of fictitious security levels and three fictitious encryption algorithms for coalition equipment were created and used in the policies in the prototype. The following section provides a condensed view of the scenario and discusses how the prototype explored the user and system requirements.

BACKGROUND ON THE SCENARIO

In the scenario, coalition groups from different countries prepare to rescue humanitarian aid workers who have been kidnapped in a fictitious country. Media reports have been received that indicate 4 aid workers have been kidnapped by insurgents in the capital city of Malekabad in Holistan. Early reports state that the four hostages were taken from the offices of the International Commission of the Red Cross (ICRC) and from Medicines Sans Frontieres. During the kidnapping, one Holistani national was killed and two wounded. It is believed that the four hostages are comprised of three men and one woman and their nationalities are Holistani, Italian, French and American.

Unconfirmed reports have said that a statement has been received from an insurgent group with demands for the release of the hostages. It is not known if there is a timescale linked with those demands. The Joint Coalition/Holistani Commanders, with the support of the coalition intelligence community, direct that a hostage release operation be mounted immediately.

The Commanders intent for the mission is "To mount a hostage release operation in order to rescue the hostages, with minimum loss of life." The coalition forces assign the task of preparing the security policy aspects of the rescue mission to a mission administrator who works for the mission commander. This mission administrator will coordinate with the U.S. policy administrator, the U.K. policy administrator, and a security administrator as necessary to complete the security tasks related to the rescue mission.

This team will work with the three existing security levels of the coalition equipment. Currently, the top secret security level is handled with 256 bit encryption, the secret security level works with 128 bit encryption, and the public–no confidentiality level of security has no encryption. All laptops support all levels of encryption. There are three classes of PDA's: Class 1 supports all levels of encryption, Class 2 supports levels 2 and 3, and Class 3 has only password authentication. Also, three high-level abstractions (Alpha, Bravo, Charlie) are employed for the three encryption algorithms (Blowfish, RC4, and DES3) respectively to be used in the mission. A meta-policy based transformation prior to policy deployment aligns them as follows: Alpha is Blowfish, Bravo is RC4, and Charlie is DES3.

DESIGN OF THE SECURITY POLICY PROTOTYPE

The overall view of the capability illustrated by the security policy prototype is as follows. The mission policies are written in natural language using the SPARCLE Policy Workbench technology (Karat, Karat, Brodie, and Feng, 2005; Karat, Karat, Brodie, and Feng, 2006). The policies that are relevant to the mission are transformed into machine readable code by SPARCLE and then analyzed using policy analytics for possible issues of conflict, dominance, and coverage (Agrawal, Giles, Lee, Lobo, 2005).

Once issues have been resolved, the policies arc transformed to add lower level details and are deployed on the rescue mission equipment. Figure 2 provides an overview of the prototype architecture. The different aspects of the prototype are discussed in more detail below.

The prototype was conceived as a web-based portal. The portal provided users access to a server-centric functionality described in the remaining sections via customizable, coordinated, robust window-like dialogs. Policy authoring, analysis, viewing, transformation and deployment were all

Figure 2. Components of the security policy proof-of-concept prototype

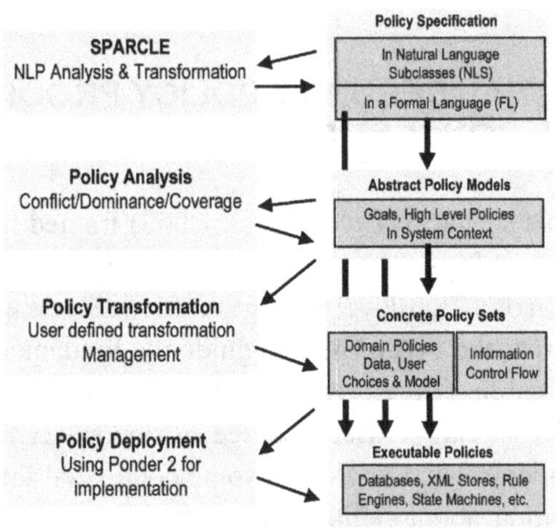

available as windowed panes within the portal that were presented to users as a set of logical steps to complete tasks. Any web-capable client could be used for the portal.

In the scenario, the assumption is that both the U.S. and the U.K. have their own policies, and a set of policies must be developed for the coalition mission. The US mission administrator needs to log in and then use the policy authoring tool to view the existing US and UK policy sets prior to updating the rescue mission policy set. He coordinates with the U.S. and U.K. policy administrators in updating the rescue mission policy set. He does this using the SPARCLE Policy Workbench which provided the authoring capabilities within the portal.

SPARCLE enables users to author policies in natural language. Figure 3 below provides a screen shot of policy authoring in natural language using a guide. The main text window on the screen is basically a word processing application. Each policy statement is created as a sentence following the guide above the text entry area. Each carriage return identifies a new policy statement. Once the author has completed policy authoring, he clicks on a button and the tool then parses the policy statements using natural language parsing technologies, and the reconstructed policy statements are provided in another window for the author to review.

A second window, besides allowing the author to review and update policies already created through guided natural language authoring, also provides a second method for authoring policies called structured format. This method is illustrated by a screen shot in Figure 4. Each of the elements in the policy format has a structured list. The author can select one or more items in each list, and the tool composes the policy sentence based on these selections. An author can compose a group of policy statements in either method, and the tool keeps the policies created using the two methods in synch at all times.

The author can then review the policies to check that they have been parsed correctly. The policies can be modified, and the sets of policies can be visualized in various ways. Additional policies may be authored in natural language or they may

Figure 3. Policy authoring in natural language using a guide

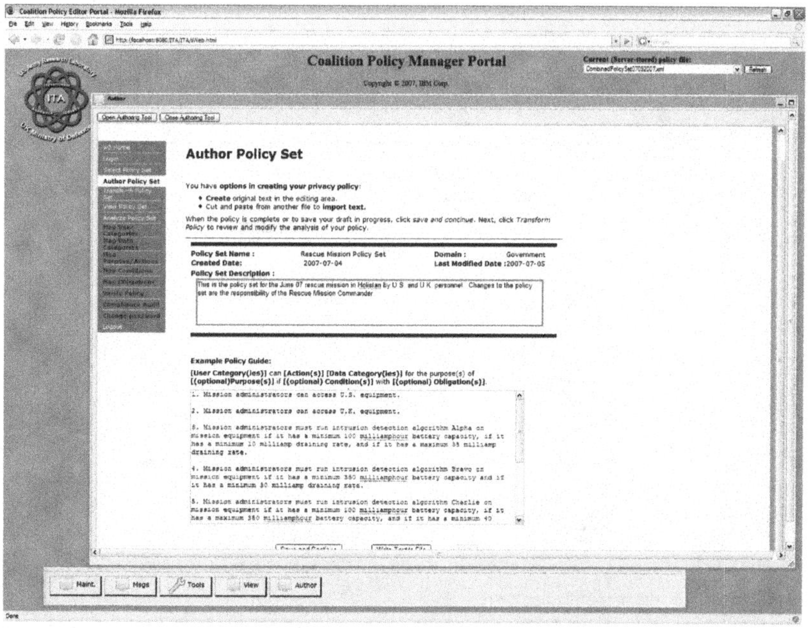

be authored directly in the structured format by selecting appropriate elements from each of the designated semantic categories.

The policies can then be viewed in a two-dimensional table which provides flexibility in setting the axes to two of the policy elements in the policy statements (see Figure 5). For example, one axis might be users and another the resources involved in the policies. One or more policy sets can be viewed together, and this capability enables users to discuss the overall scope and content of the polices. After authoring and review of the three policy sets was completed, they were transformed into a an XML based machine-readable policy language (we used ACPL, Agrawal et., 2005). For further information on the SPARCLE Policy Workbench, please see Karat et al., 2005 and Karat et al., 2006.

Continuing with the flow of the scenario now, the mission administrator views the policy sets that the U.S. policy administrator and the U.K. policy administrator have created, and authors the mission policy set. Examples of the three sets of policies are listed below:

- U.S. Policy Set:
  1. U.S. personnel can access U.S. equipment.
  2. U.S. personnel must set security level top secret for mission communication.
- U.K. Policy Set:
  1. U.K. personnel can access U.K. equipment.
  2. U.K. personnel must set security level secret for mission communication.
- Rescue Mission Policy Set:
  1. Mission administrators can access U.S. equipment.
  2. Mission administrators can access U.K. equipment.
  3. Mission administrators must run encryption algorithm Alpha on mission equipment if it has a minimum 100 milliamphour battery capacity, if it has a minimum 10 milliamp draining rate, and if it has a maximum 35 milliamp draining rate.
  4. Mission administrators must run encryption algorithm Bravo on mission

Figure 4. Policy authoring using the structured list method

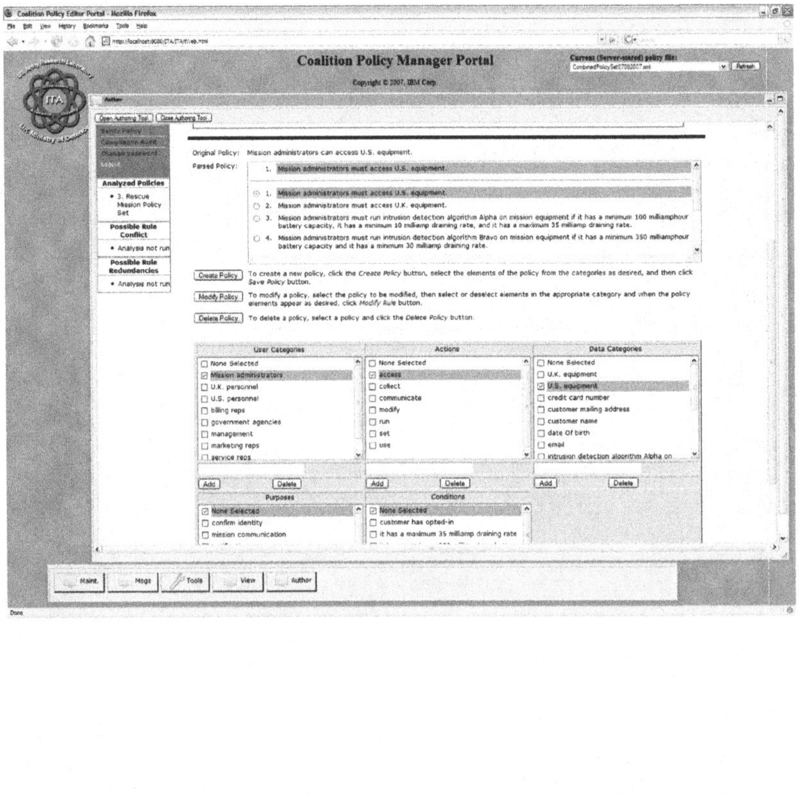

Figure 5. Viewing the policy set as a whole

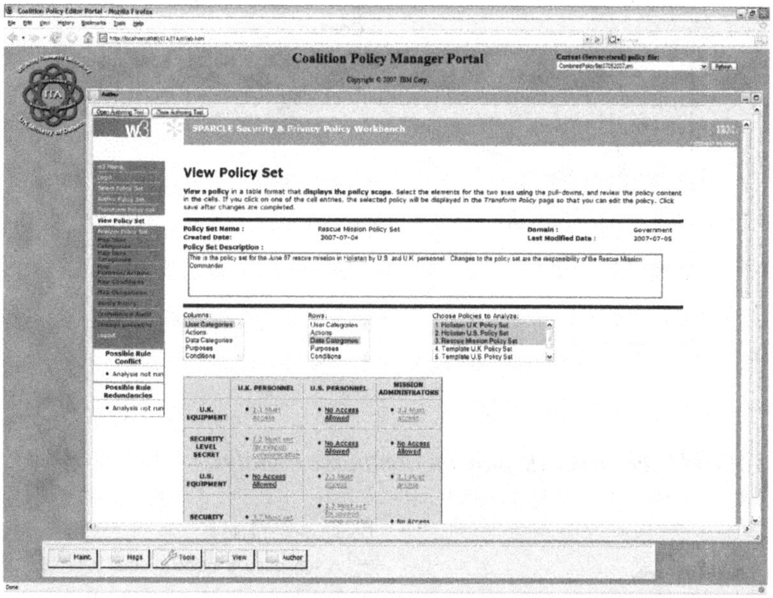

equipment if it has a minimum 350 milliamphour battery capacity and if it has a minimum 30 milliamp draining rate.

5. Mission administrators must run encryption algorithm Charlie on mission equipment if it has a minimum 100 milliamphour battery capacity, if it has a maximum 350 milliamphour battery capacity, and if it has a minimum 40 milliamp draining rate.
6. Mission administrators must run encryption algorithm Bravo on mission equipment if it has a minimum 400 milliamphour battery capacity, if it has a minimum 75 milliamp draining rate, if it has a maximum 95 milliamp draining rate.
7. U.S. personnel and U.K. personnel must set security level top secret for mission communication.
8. U.S. personnel and U.K. personnel can communicate mission data if they are in sector xray.

After authoring and viewing the policies, the mission administrator would now like to analyze the policies and resolve any issues identified in them, prior to deploying the policy sets. He selects all three policy sets to be analyzed, and they are combined into one group for this purpose. The combined policy set can then be analyzed by various tools. The prototype included examples of Conflict Detection, Dominance Analysis, Coverage Analysis, and rule-based Transformation. The conflict analysis routine looks at all of the policies in the set and determines which are simultaneously applicable and incompatible. Figure 6, 7, and 8 show examples of the visualizations that the portal provided to users to illustrate policies in conflict, policies being dominated by other policies, and coverage issues in the policy space. Explanatory text identified the particular policies involved so that the mission administrator can coordinate with the US and UK policy administrators to reach agreement, update the policies to resolve the policy issues, and rerun the analyses to confirm that no conflicts remained in the policy set.

Dominance Analysis checks to see if any of the policies are redundant. A policy rule is dominated

Figure 6. Example of end user visual display on conflict analysis

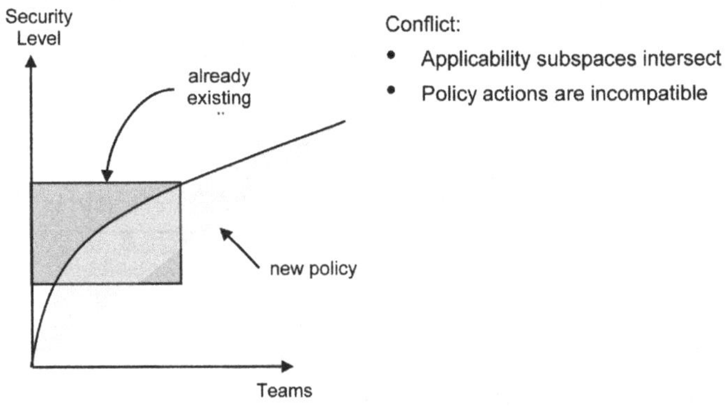

Figure 7. Example of end user visual display on dominance analysis

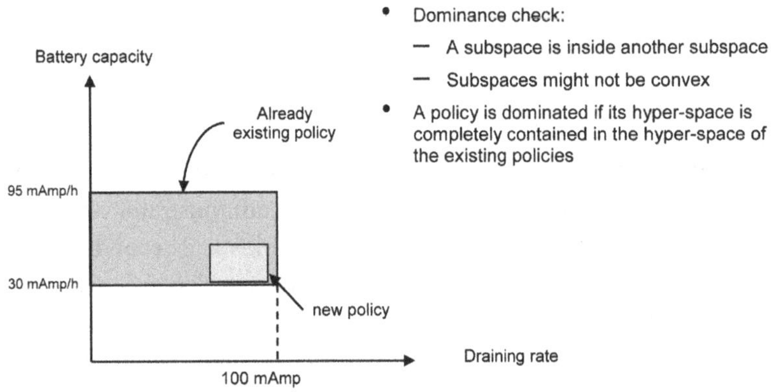

by another set of policy rules if adding the new rule to the set will not change the behavior of the system in any way. Figure 7 shows an example of the prototype's illustration of dominance issues in a policy set.

Coverage Analysis provides information on whether all ranges of the variables in the policies selected are represented. The current version of the portal prototype incorporates a feedback channel from the analysis component to the authoring component that suggests new policies to the user to replace sets of policies with incomplete coverage (see Figure 8).

After the analyses are completed and issues resolved, the mission administrator then uses the portal to transform the policies so they are ready for deployment onto mission equipment. The portal provides a rule-based transformation routine. The rules are called meta-policies, and in the scenario the security administrator has previously defined these for the policy space (Beigi, Calo and Verma, 2004). These meta-policies allow higher level abstractions to be mapped to the particular security mechanisms that will implement them as described in the scenario background regarding confidentiality levels of the coalition equipment and the encryption algorithms employed in the policies. For example, the high level abstraction "Alpha" is transformed using the meta-policies to the "Blowfish" encryption mechanism. As stated previously, fictitious security levels and encryption algorithms were created for the prototype for

Figure 8. Example of end user visual display on coverage analysis

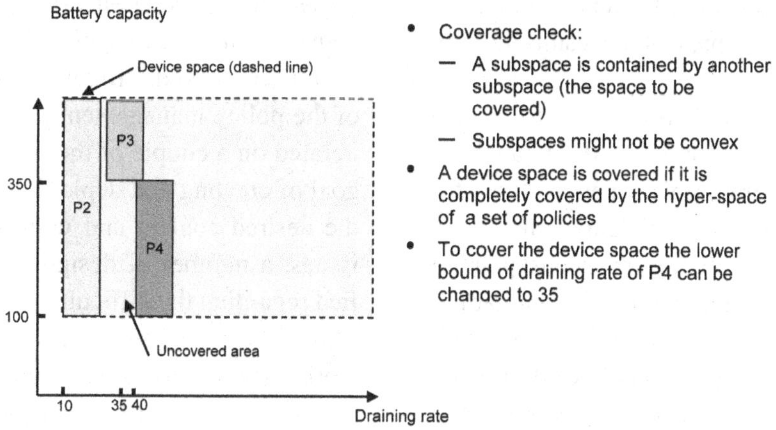

illustrative purposes. The mission administrator can view the result of the transformation in the portal.

Finally, the portal is used to deploy the set of policies onto the coalition equipment for the mission. The portal prototype worked with Ponder2 technology to demonstrate deployment (Twidle et al., 2008). This was the last task to be illustrated in the initial proof of concept prototype.

LESSONS LEARNED

The user scenario was developed to identify user requirements regarding the capability and the interaction methods needed for the proof-of-concept prototype. During the course of the design and creation of the prototype, lessons were learned in two main areas. The first concerned the display of information to the users regarding the results of the different types of policy analysis. The second concerned the user experience of the flow of the policy management process from authoring through transformation, analysis and deployment.

In the prototype elements from a number of research efforts were brought together. These included a policy authoring component along with several different policy analysis components. We had envisioned that the policy management capability formed by integrating the components would be flexible for the users to work with if we created a central control point for it within a web application. This design decision seemed to be substantiated. Different user groups may want to work with a tool such as this on different types of equipment based on the demands of a particular situation. The authoring capability was provided by the SPARCLE prototype which had been completed over several years with target customer feedback, thus, it also was well suited to the task. In the SPARCLE prototype there is a limited consideration of policy analysis which operates only at the structured language level. Policies are compared to identify potential conflicts (where there are two policies about the same matter, and one policy is more restrictive than another) and redundancies (one policy is the same as another or is a redundant child of a more comprehensive parent policy) between pairs of policies. The SPARCLE prototype presents the pairs of policies with conflicts and redundancies to the user to review with the elements in question highlighted. The user could click on the policy to go to the authoring page to adjust the policies, and then rerun the analyses to confirm that issues were resolved. Since the analysis is carried out directly on the natural language representation of the policy, it is easy to present the results directly in this format.

In this security policy proof-of-concept, we wanted to go beyond the SPARCLE analysis and employ the more sophisticated analyses for conflict, dominance, and coverage that were computed by running algorithms against pairs of policies that had been transformed from natural language into an abstract policy model. The challenge then became how to provide the information back to the user about potential policy issues and how to resolve them. We conducted several iterations of design regarding this research issue, and determined that it would be very valuable to the user to see a visual representation of the analysis space and be able to read a textual description of the pairs of policies with potential issues. We added figures similar to those in Figures 6, 7 and 8 above to the analysis feedback in the prototype, and represented problems by referring to terms from the natural language representation rather than the abstract model representation of the policy. For conflict and dominance, the analyses are somewhat simpler than for coverage, and our analysis suggested that a target user could reasonably determine how to evaluate the pairs of policies and resolve the issues as needed. For the coverage analysis, the analysis algorithm is more complicated, and in the case of a potential coverage issue, there are multiple methods to resolve it. For this problem area, an innovative method of presenting the user with feedback about how to resolve identified issues was created, by presenting a number of alternative solutions. In the design of the proof-of-concept prototype, one of the core guidelines was that the user is always in control. Therefore, even when a solution to a potential problem is provided, the user may select to address a coverage issue in another manner or not at all, if that is the right decision in the particular context.

The second major area of lessons learned concerned the user experience of the flow of the policy management process within the web application. As a user worked through the steps in the flow of the policy management process, and perhaps iterated on a couple of them while pursuing the goal of creating and deploying a policy set with the desired content and with no known policy issues, a number of design issues were identified regarding the difficulty of completing steps. This difficulty might arise due to complexity in working between different transformations of the policies or different functions in the process. We identified steps that required simplification for the user to complete, for example, how to seamlessly switch between functions within the prototype (e.g., from analysis back to authoring). We also identified steps such as the use of meta-policies in coverage and in the deployment transformation, that needed to be automated for the user due to the time or dependency on another user's knowledge (e.g., in the creation of meta-policies).

As this was a proof-of-concept rather than a fully functional prototype, we created interaction methods to simplify the user interactions required between the different components in the system and we finessed the issue of the meta-policies at this point, by providing static meta-policies that fit the user scenario. We learned that in the future, if there was interest in this becoming a functional tool, that the issue of how meta-policy authoring could seamlessly integrate with policy authoring would need to be addressed. We thought that providing a dynamic process for the creation of default meta-policies, with the option of user tailoring, might be the best solution in the future.

We also discussed and iterated on the design of the deployment function. In this early proof-of-concept prototype, the goal was to illustrate the idea of transforming the policies into a format with the necessary detail and then demonstrate that they could be deployed onto mission equipment. At this point, only basic functionality was provided for querying the deployed policies to indicate that they had been received by a policy engine (based on PONDER2). In the future, the deployment functionality would be enhanced with additional capabilities to view and work with deployed policies through a policy management tool for this purpose.

SENSOR FABRIC PROOF-OF-CONCEPT PROTOTYPE

The second conceptual prototype was based upon an extension of the Intelligence, Surveillance, and Reconnaissance (ISR) scenario (Roberts et al., 2007). To this basic scenario we added specific policy considerations and user interface features in an exploration of new research concepts.

The user roles identified for this scenario included:

- U.S. and U.K. headquarters policy administrators
- Local U.S. and U.K. coalition members in camps and on patrol

The user tasks illustrated through the scenario-based prototype included:

- U.S. and UK. headquarters and local authoring and updating of policies for U.S. personnel and equipment, U.K. personnel and equipment, and sensors involved in local coalition missions to patrol the Camp perimeter and secure the Camp site.
- Viewing the policy set
- Analysis of the policies and resolution of the issues identified
- Deployment of the policies on the mission equipment and sensors. These policies include automated changes in operational policies as context of mission changes.
- Accessing data from the sensors, interpreting it, and taking follow-on action.

BACKGROUND ON THE SCENARIO

The scenario involves an ad hoc team consisting of US and UK forces that form to conduct joint ISR operations. These operations include activities at several forward coalition camps (one coalition, one US, and one UK), and coalition patrols in

Figure 9. ISR network scenario example

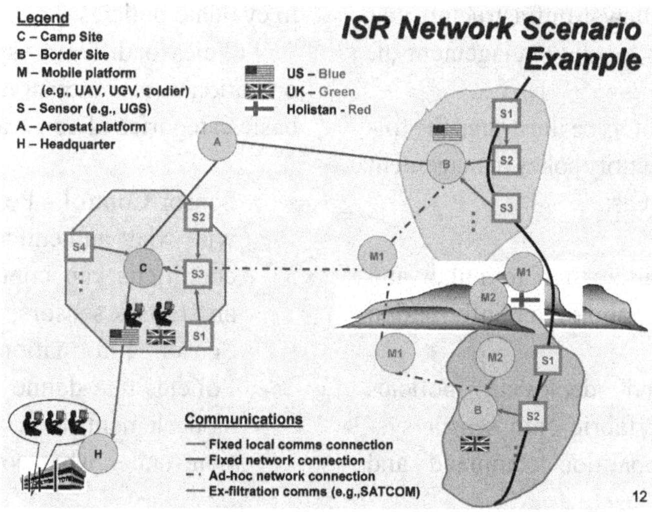

rugged areas between the camps (see Figure 9). The commander's intent is "The Coalition Team is to conduct border ISR to identify infiltrations and complete surveillance to protect the camps from hostile actions. The Team will respond to infiltrations and threats as needed."

In preparation for a wide range of coalition missions, coalition headquarters staff in the U.S, and U.K. have developed sensor policies to control their own country's sensors. The U.S. headquarters policy administrator will coordinate with the U.K. policy administrator to enable the necessary security policies related to headquarters access to, and local control of, all sensors in the field of operations.

There are two types of sensors that the coalition mission team has access to: acoustic sensors and video cameras. Each sensor has different characteristics. These two types of sensors can support the Commander's Intent and the coalition forces in the field by identifying the origin of attacks on camps and patrols. In our scenario, the U.S. and U.K. each own acoustic sensors and video sensors in the region.

The scenario presents the policies that must be specified, analyzed, deployed and enforced in order to meet the needs of the coalition mission in patrolling the camp perimeters and securing the camp sites. This demonstration goes beyond the previous one in being able to deploy the policies to a policy-enabled middleware infrastructure that provides unified access to, and management of, sensor networks.

A scenario-based prototype illustrated the following concepts in the security policy management and the sensor fabric areas:

- Enabling headquarters staff to author and analyze security policies on information access;
- Transforming and deploying policies across the sensor fabric onto sensor systems used by coalition command and control, camps and patrol teams in the theatre of operations;
- Using policies to trigger certain sensors to change orientation based on dynamic events; and,
- Allowing a coalition commander to override access to sensor data in an emergency situation.

DESIGN OF THE SECURITY POLICY – SENSOR FABRIC PROTOTYPE

The components involved in the security policy – sensor fabric prototype are illustrated in Figure 10. Although changes were made to several components of the tool, the main addition was the deployment and execution of policies in the sensor fabric (as opposed to a simulated PONDER2 agent). The Sensor Fabric is a middleware infrastructure for wireless and wired sensor networks providing a common environment for ISR applications. As seen in the Theatre of Operations figure for the demo, the Sensor Fabric spans the network with nodes representing a Joint Command and Control (CoCon) camp, three forward camps (US, UK and Coalition), a Patrol Team, and Devices and Sensors. The Sensor Fabric enforces policies at defined Policy Enforcement Points (PEPs). PEPs rely on a PONDER2 engine to evaluate policies.

Policies for different aspects of sensor network operation were demonstrated. They covered three basic categories of constraints, namely:

- Sensor Control - Policies that define who, with what authentication, and under what conditions can control, configure, move, and re-task sensors;
- Sensor Information Access Control - Policies that define who (person, data fusion element, etc), with what authentication, under what conditions, and in what

Figure 10. Components of the security policy – sensor fabric prototype

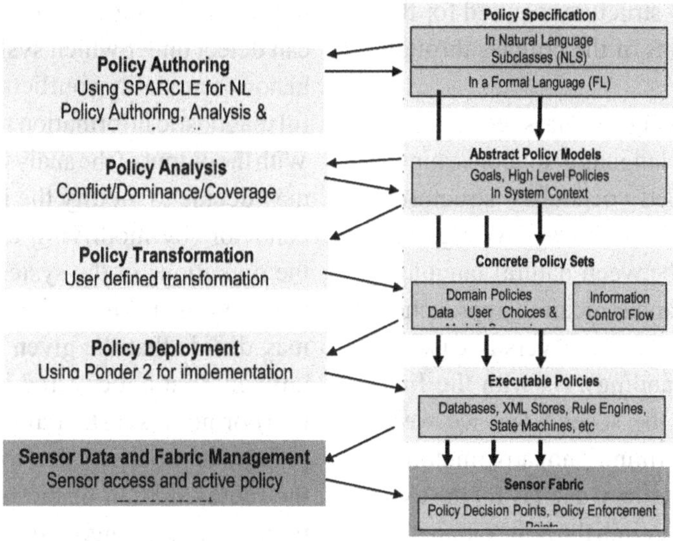

form (i.e., raw, processed, fused) can access sensor information; and,

- Information Flow Protection - Policies that define how information flows are to be protected (confidentiality, integrity, etc);

In this scenario-based prototype, similar initial steps and tasks are completed to author and update policy sets, and analysis is then done. The policies are deployed on the sensor network, and when events change, local users are able to work within the policy set to change possible actions as needed. The users involved during the planning of a coalition mission need to take into account the differences in the policies of the different coalition members and determine any conflicts among them. The conflicts in policies need to be resolved and a set of policies conformant to the requirements of all coalition members needs to be developed.

Subsequently, during the planning of an ISR operation, users need to determine which coalition member is capable of performing each type of required operation under the existing set of policies, identify any conflicts of operational needs with mission policies, and determine the right configuration of ISR assets to support the needs of the missions. Finally, as the operation is underway, the different ISR assets need to be configured to support the desired policies. When the operational situation changes (e.g., a sensor failure), policies need to be changed to adapt to the new mission environment.

LESSONS LEARNED

Much of the HCI effort related to the second proof-of-concept prototype concerned the investigation of options and tradeoffs in adapting the authoring task to the policy content specific to the sensor domain as well as investigating options and tradeoffs in making the complexity of the transformations occurring between the levels of policy transparent to the target user. Regarding the first issue, a variety of formats for policy authoring were explored to cover the range of sensor policies that might be required in different contexts. The format needed to meet the target users composition model for written language, be comprehensible to other users who would need to understand policies written in the format, and be sufficiently expressive to include the necessary information and policy structure required for the series of transformations of the policies through to executable code.

The policy authoring formats considered ranged from natural language to constrained templates and controlled vocabulary structured entry. A focus of our research was investigating the various tradeoffs between natural language authoring approaches and structured approaches (e.g., parsing accuracy for NL versus ease of authoring and understanding). As with the first concept prototype, in the second one we were again challenged to determine innovative methods to provide a usable, functional display for the user to employ in working through the policy management process from authoring to enforcement of policies in the deployed environment. Through iterations of design that identified the requirements to enable centralized and local control of the sensor capability, we were able to create usable content for the policy process flow and the target users with different roles and attributes. We identified many possibilities for the different user views and displays of information, gained insight, and are actively investigating these ideas in current research.

CONCLUSION

In this chapter we have presented a view of policy management that extends from authoring through enforcement. We defined a layered architecture that is useful in identifying the architectural elements, software components, and system models that are needed to support automated policy refinement and analysis, and to facilitate the management of policies.

We presented a formal framework for the analysis of properties of an interesting class of authorization and obligation policies. The purpose of this framework is to enable us to prove properties of our policies and the way they interact with the systems which they regulate. We see it as a significant benefit of our approach that it can detect under which system states and system histories modality conflicts arise, providing useful diagnostic information to the system designer. With the output of the analysis, the system designer may decide to modify the policies or the system behavior (by modifying the implementation of the operations in the system so that the conflict will not arise). Alternatively the system designer may decide that the given sequence of inputs is unlikely (but not covered in the system description) or not a substantial problem and leave the policies unmodified. Our formalization separates the representation of the laws of system evolution, and constraints on the system state, from the authorizations and obligations which define policy decisions. This gives us the ability to switch systems descriptions easily and study the behavior of policies on different systems. The use of temporal constraints and an explicit representation of time has enabled us to express complex dependencies of policy decisions on changing system states, as well as on other policies.

There are several extensions that will be useful to have in the language. One is the possibility of having obligations that, instead of incorporating explicit temporal arguments marking the beginning and end of the period during which an action ought to be performed, this period is dependent on the occurrence of events. The two approaches should be able to be mixed in our policies. This will simplify the encoding of languages such as XACML in which obligations are explicitly associated with the access authorization event (an obligation is triggered when a subject is authorized to do an operation). A second extension is the inclusion of aggregation so that policies that depend on counting such as voting protocols can be implemented. Apart from obligations, the language only has a single predicate involving two time points: reqInBetween. Other predicates that require more than one time variable are predicates to represent time operators like some time in the past, always in the past and the since operator. These are typical operators of linear time temporal logic. This is a third extension that can be accomplished with rules similar to the rule defining reqInBetween. Of course, all these extensions should be made while ensuring that analysis is still feasible. We will also look in more detail at technical questions related to the underlying abductive constraint logic-programming technology that is used, and consider whether alternative approaches to abduction may yield increases in efficiency.

Two proof-of-concept prototypes were completed to illustrate the technological capabilities that might be created to support users in policy authoring, analysis, transformation, and deployment. The prototypes were grounded in realistic scenarios involving dynamic coalitions charged with carrying out missions in the field. The scenarios helped to identify necessary policy capabilities for users of the tools, and these capabilities were designed based on knowledge of the system model as illustrated in the policy framework and on an analysis of the user skills and abilities, core policy tasks, and context of use. The human-computer interaction research on the concept prototypes focused primarily on how to enable simple and rugged policy authoring in very different content domains, how to provide meaningful feedback on policy analysis results to enable users to take appropriate action as needed, and how to design the overall flow of the iterative policy lifecycle process. The proof-of-concept prototypes demonstrate innovative design ideas for valuable policy management technology that is efficient and effective. Critical research challenges lay ahead. The scope of the policy research must expand to include topics such as policy refinement from high-level mission statements into executable policy, the incorporation of wide-ranging policy analyses into the policy lifecycle flow, and inter-organizational management capabilities such as policy negotiation for policies in conflict among coalition partners. Research in these topics will increase the value of the policy domain knowledge, and will likely motivate changes in policy lifecycle models, as the iterative research from policy model to specific concept demonstrations and back again provides the opportunity for scientific and technical breakthroughs in the mechanisms and methods for policy lifecycle management.

REFERENCES

Agrawal, D., Calo, S. B., Lee, K.-W., & Lobo, J. (2007). Issues in Designing a Policy Language for Distributed Management of IT Infrastructures. In Integrated Network Management, (pp. 30-39).

Agrawal, D., Giles, J., Lee, K., & Lobo, J. (2005). *Policy Ratification* (pp. 223–232). POLICY.

Becker, M. Y., & Sewell, P. (2004). Cassandra: Flexible trust management, applied to electronic health records. In CSFW, (pp. 139–154). Washington, DC: IEEE Computer Society.

Beigi, M., Calo, S., & Verma, D. (2004, June). Policy Transformation Techniques in Policy-based Systems Management. In Policy Workshop 2004, Yorktown, New York.

Bias, R., & Mayhew, D. (Eds.). (2005). *Cost-Justifying Usability: An Update for the Internet Age* (2nd ed.). San Francisco: Morgan Kaufman.

Blaze, M., Feigenbaum, J., Ioannidis, J., & Keromytis, A. D. (1999). The role of trust management in distributed systems security. In Secure Internet Programming, (pp. 185–210).

Brodie, C., George, D., Karat, C.-M., Karat, J., Lobo, J., Beigi, M., et al. (2008).The Coalition Policy Management Portal for Policy Authoring, Verification, and Deployment. In *IEEE Workshop on Policies for Distributed Systems and Networks, POLICY 2008*, (pp. 247 – 249).

Carroll, J. (Ed.). (1995). *Scenario-Based Design: Envisioning Work and Technology in System Development*. New York: Wiley and Sons.

Casey, S. (1998). *Set Phasers on Stun: True Tales of Design, Technology, and Human Error*. Santa Barbara, CA: Aegean Publishing Co.

Craven, R., Lupu, E., Ma, J., Russo, A., Sloman, M., & Bandara, A. (2009). *An expressive policy analysis framework with enhanced system dynamicity* (pp. 239–250). ASIACCS.

Goldblatt, R. (1992). *Logics of time and computation* (2nd ed.). Stanford, CA: Center for the Study of Language and Information.

Irwin, K., Yu, T., & Winsborough, W. H. (2006). On the modeling and analysis of obligations. In *Proc. of the 13th ACM Conf. on Computer and communications security*, (pp. 134–143).

Jacko, J., & Sears, A. (Eds.). (2008). *The Human-Computer Interaction Handbook: Fundamentals, Evolving Technologies and Emerging Applications* (2nd ed.). New York: Erlbaum Associates.

Kakas, A. C., Michael, A., & Mourlas, C. (2000). ACLP: Abductive constraint logic programming. *The Journal of Logic Programming*, *44*(1-3), 129–177. doi:10.1016/S0743-1066(99)00075-8

Karat, C., Karat, J., Brodie, C., & Feng, J. (2006). Evaluating Interfaces for Privacy Policy Rule Authoring. In *the Proceedings of the Conference on Human Factors in Computing Systems,* (pp. 83-92). New York: ACM Press.

Karat, J. (Ed.). (1991). *Taking Software Design Serously: Practical Techniques for Human-Computer Interaction Design*. New York: Academic Press.

Karat, J., Karat, C., Brodie, C., & Feng, J. (2005). Privacy in information technology: Designing to enable privacy policy management in organizations. *International Journal of Human-Computer Studies*, *63*(1-2), 153–174. doi:10.1016/j.ijhcs.2005.04.011

Kowalski, R. A., & Sergot, M. J. (1986). A logic-based calculus of events. *New Generation Computing*, *4*, 67–95. doi:10.1007/BF03037383

Loscocco, P., & Smalley, S. (2001). Integrating flexible support for security policies into the linux operating system. In *Proceedings of the FREENIX Track: 2001 USENIX Annual Technical Conference*, Berkeley, CA, (pp. 29–42).

Lupu, E., Dulay, N., Sloman, M., Sventek, J., Heeps, S., & Strowes, S. (2008). AMUSE: autonomic management of ubiquitous e-Health systems. *Concurrency and Computation*, *20*(3), 277–295. doi:10.1002/cpe.1194

Nomikos, C., Rondogiannis, P., & Manolis Gergatsoulis, M. (2005). Temporal stratification tests for linear and branching-time deductive databases. *Theoretical Computer Science*, *342*(2-3), 382–415. doi:10.1016/j.tcs.2005.05.014

OASIS XACML TC. (2005). *extensible access control markup language (XACML) v2.0.*

Simon, R., & Zurko, M. E. (1997). Separation of duty in role-based environments. In CSFW, (pp. 183–194). Washington, DC: IEEE Computer Society.

Twidle, K., Lupu, E., Dulay, N., & Sloman, M. (2008 June). Ponder2 - A Policy Environment for Autonomous Pervasive Systems. In Policy, (pp. 245 – 246).

Section 3
Human and Cognitive Issues in Coalitions

Chapter 9
A Model for Culturally Adaptive Policy Management in Ad Hoc Collaborative Contexts

John Karat
IBM TJ Watson Research, USA

Winston Sieck
Applied Research Associates, USA

Timothy J. Norman
University of Aberdeen, UK

Clare-Marie Karat
IBM TJ Watson Research, USA

Carolyn Brodie
IBM TJ Watson Research, USA

Louise Rasmussen
Applied Research Associates, USA

Katia Sycara
Carnegie Mellon University, USA

ABSTRACT

In this chapter, the authors consider an approach to advancing the value of context-sensitive policy management technology for collaborative mission planning and execution through integration of algorithms based on cultural models and collaborative decision making. Three research teams collaborated to leverage their research frameworks and results in policy lifecycle management, cultural analysis, and decision support in this effort. The chapter describes the three technical areas, and the results of a theoretical analysis of the potential value of their integration in a new perspective, combined with a set

DOI: 10.4018/978-1-61520-855-5.ch009

of research questions that might be addressed in further inquiry in this new area. The theoretical work identifies opportunities for addressing challenging issues in policy, culture, and collaborative decision making. The authors conclude with a view of future research that might provide a breakthrough in this intersection of disciplines and lead to the creation of a culturally aware policy management system for collaborative activities.

INTRODUCTION

Collaborative teams are increasingly called on for military and non-military operations with minimal lead time. Such coalitions are increasingly composed of different partners that need to form ad hoc teams to fulfill the mission (e.g. in Iraq, for one mission US soldiers are working with British, Swedish and Iraqi personnel, in another mission British soldiers are working with Italian and Australian coalition forces). These diverse mission teams operate with different operational policies in many domains of coalition activities. Currently, for each mission, each coalition partner "inherits" the overall policies of his/her military that are not mission-specific. This may result in having policies that are more restrictive than they need to be. The restrictive policies, in turn, can lead to the possibility of impeding effective collaborative planning and execution. Since coalition teams operate in an ad hoc manner and may not have co-trained, there is a need for rapid adaptation, including arriving at ways to work effectively despite the differences in policies. This highlights a need for a policy authoring and management infrastructure that enables (a) rapid (and possibly automated) translation of high level policies into operational ones, (b) rapid policy customization, given a particular mission and cultural background of the participants, and (c) detection of policy conflicts and suggested policy reconciliation along with support for policy negotiation activities. Additionally, during mission planning and execution, there is a need for automated support of the coalition team members to enable them to collaborate effectively despite the cultural and policy differences.

In this ad hoc collaborative context, our policy management model begins with an assumption that military policy analysts develop and refine policies that begin with high-level specifications and are transformed into operational policies that govern the way coalition systems behave in the specific mission context. Such policies might include constraints or guidance on the appropriate use and sharing of physical and information resources involved in a mission. In this research we are interested in how context-sensitive support for policy authoring, analysis, and execution might be provided. For example, in a collaborative context involving members with known cultural differences, policy management technology might include components which review and advise on policy content with respect to the known cultural differences. With the variety of partners involved in coalition operations, cultural issues can be expected to influence how policies are developed and analyzed, as well as the relative acceptability of policy specifications among coalition partners. As an example, a culturally aware policy management system might offer guidance to policy authors which suggested how information sharing policy typically relates to organizational structure in different groups impacted by a policy. This information might be included in data structures describing the organizations (e.g., rank or mission role descriptions). Such guidance might be provided at the time policies are authored, or might be incorporated into mechanisms which facilitate policy negotiation or conflict resolution. Similarly, the nature of appropriate support for decision making in the field can be influenced by many elements of the mission context. Situational context (e.g., various emergency situations), might call for additional policies to address circumstances calling for exceptional collaboration. A model of such collaboration could be used in guiding authoring of appropriate policy and enforcement of it in the field. For example, access to certain sensor information might be granted to a coalition team under enemy fire by another team member. Describing the context in which such exception policies might be active would be a way to incorporate collaboration and policy management.

To enable this vision of culturally adaptive policy planning to be realized, there is a need for the integration of a number of research areas. In this chapter, we report on a theoretical analysis conducted by the group of researchers to examine the research frameworks and results to date in three related arenas and consider the potential to integrate the areas in a unifying multidisciplinary perspective. As a part of the analysis, the team created a set of core research questions from which hypotheses could be defined to test the validity of the theorized integrated approach to policy management systems that take culture and collaborative decision making into account. Finally, we discuss high level ideas for future research in this multifaceted domain.

The three research themes include end-to-end policy management research, culture and coalition planning research and collaborative decision making research. Specifically, we see coalition policy making as a rich context for understanding cultural issues in policy management and in developing tools which are usable in dynamic coalition contexts. In this view, we see an important research challenge in the effort to attempt to extend culture and collaboration models so that they can be incorporated into policy management models. We see the act of specifically asking what one model (e.g., a model of cultural planning differences) has to say about another (e.g., a model of policy management) as important to making progress in broad arenas.

Within the ITA there have been explorations of both planning (Chapter 11) and policy management (most specifically security policy management (Chapter 8). We see coalition planning activities and policy management activities as being very similar. This does not mean that we equate plans and policies, but we do see both as involving artifacts which are authored, analyzed, and deployed. In general, the terms "plan" and "policy" and the activities "planning" and "policy management" might correspond to very specific artifacts and processes in an organization and the meanings of the terms might vary significantly across organizations. However, we believe that research from one area (planning or policy management) can be highly relevant to the other. Rather than try to provide a single definition of each of these terms here, we will say that both plans and policies can be seen as to including a variety of information (from high-level mission goals to specific protocols or rules), and that both involve complex human problem solving. Following this, we see both planning and policy management as proceeding from high-level goals through operational rules following a goal oriented refinement process. Certainly there are some differences, but for the purpose of advancing a model of how culture or collaboration models can be incorporated in either a planning or policy management process, we can view both as activities which create operational guidance to meet higher level goals through a goal refinement process.

From a policy management perspective, our discussion focuses on supporting the development of policy-based security management that provides mechanisms for adapting system behaviors to meet high-level user-specified security goals. From the perspective of cultural models of planning activities, the research challenge involves applying cultural network analysis to the domain of policy authoring in order to capture and represent cultural differences and commonalities in policy conceptions. From a collaborative decision making perspective, the research involves understanding the kind of situations which might arise in missions that could benefit from support through intelligent agents. These research perspectives and the understanding derived from integrating them will provide knowledge needed for further research on context-aware tools for coalition policy management.

A working assumption of the fit within coalition operations is that of a basic two-tier system:

- Policies (e.g. access control, network security) are developed within a coalition headquarters. High level policies are negotiated and developed in some form of collaborative decision making process among multinational partners. The degree of actual integration in the process is unclear: A less integrative system could be a system where each partner nation sets their own policy, informs the others, and only adjusts if there is any major conflict with other nations. A more integrative system would be where shared policy is decided by consensus among nations, and then adopted by all.
- Once established, the high-level policies are translated into low-level policy rule sets that then drive what can/cannot be done with technologies that are used in the coalition headquarters, and those used out in the field by soldiers working at tactical levels. In the ITA context, we envision the policy-affected technologies at this level to be part of MANETS.

AN APPROACH TO MISSION PLANNING AND OPERATION SUPPORT

In our conceptual analysis of the problem space, we consider the integration of successful methods and results from various research projects that are applicable to this context. These include Cultural Network Analysis (CNA) methods employed by researchers studying mission planning in coalition contexts and Contextual Inquiry (CI) methods employed by researchers in policy lifecycle research that have been used in the design of policy management systems in non-military organizations. Initial analysis of the methods indicates that these methods are likely to be quite compatible. For example, studies with mission planners from the US and UK suggested differences in plan specificity as a cultural difference (Chapter 11). These results relate to findings in research previously conducted which identified policy differences in information sharing within the different norms in the financial and healthcare industries (Karat et al. 2005). We believe that the cultural models can be adapted to inform the design of tools for policy management in specific domains as a part of supporting transformation from high-level plans to operational policies. The analysis performed as a part of this research should provide Domain Models for authoring policies specific to particular aspects of the overall plan – in this case security policies (e.g., access control). This work should involve comparing coalition policies, identifying conflicts, and formulating common policies. Specifically, we believe that analysis of natural language policy statements can yield formal representations which can be used to manage policies relevant to dynamic Communities of Interest and MANETS. Figure 1 summarizes our conceptual view of the potential integration of policy authoring, refinement, analysis and deployment with cultural adaptation and decision support.

Formulating and refining such a model of culturally-sensitive policy management involved a three step approach. First, theoretical work comparing data collected in the military planning studies was compared with research in commercial policy management studies in order to develop specific hypotheses about policy management issues that might arise in military coalition environments.

Figure 1. A model of culturally-adaptive and decision-support based transformations from high level policy specification through policy deployment

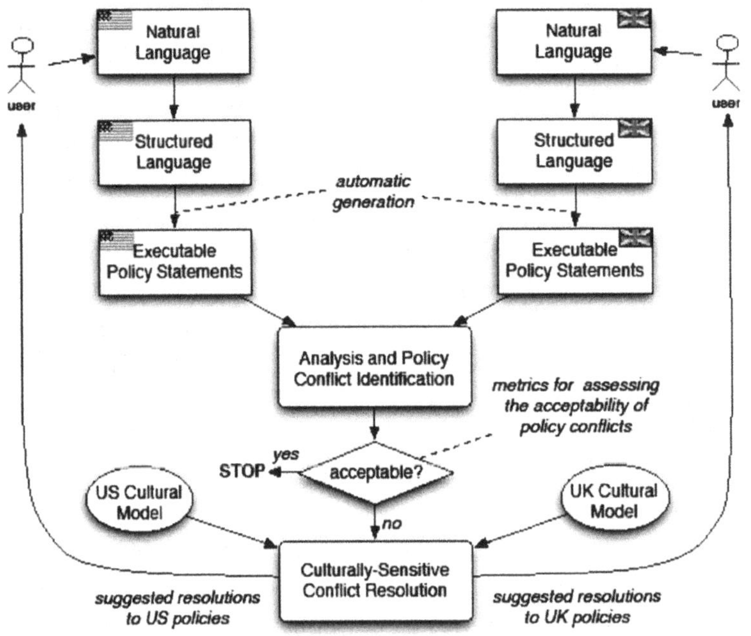

Second, this conceptual effort was translated into ideas for adapting the models from the planning and policy management domains to the target domain of culturally informed policy management. Third, we ultimately want to transition this basic research effort to a project with the goal of designing and developing a policy management system for use in coalition headquarters.

In this chapter, we present the first two steps: examining research results to identify common themes for policy management. We see this research as aimed at addressing three fundamental questions:

1. Can **cultural analysis** and **collaborative decision making** models inform the design of systems to support **policy management** in coalition operations?
2. Can policy management systems be designed that are **usable** and **acceptable** among collaborators in coalition environments?
3. Does support software need to have culturally distinguished **modes of interaction** (e.g., based on willingness to accept agent support)?

In the remainder of this chapter, we describe the research perspectives integrated in addressing the three questions raised above. First, we describe research in the policy arena aimed at comprehensive management of end-to-end policy systems for privacy and security. Next we describe research in the development of cultural models and their application to multicultural policy development. Finally, we describe research on collaborative decision support aimed at advancing intelligent agents to support human actors. In each of the three sections, we focus on illustrating the connections between the specific topic of the section and the overall research questions above.

POLICY MANAGEMENT

Policies which address security and privacy are pervasive parts of technical and social systems. There is a general notion of policy-based systems as those whose behavior is guided by rules of the general form "If *condition* then *action*". Collections of rules are considered policies, and policies can be developed for various aspects of system behavior. Organizations have policies covering proper conduct to protect the safety of people and effective use of resources. Information technology systems have policies which govern who can access what resources aimed at protecting the integrity and confidentiality of the information and resources. Individuals have policies guiding their behavior towards others formed with the intention of guiding how they live their lives. These policies might be expressed in text (common in organizations), code (common for IT systems), or might be implicit (common for individuals). Such policies might be seen as including high level guidance (e.g., "to insure a safe mission") and more specific operational rules (e.g., "don't fire warning shots").

In previous research, (Karat et al. 2005) and (Karat et al. 2006) report several commonalities across a wide range of policy types in technical and social systems. First, high level policies, generally expressed in human language, are refined into operational rules while attempting to keep the intent of the high level policies. This process is difficult as the policies are often subject to differences in interpretation or context. Second, the existence of multiple, possibly conflicting, policies must be accommodated. This process is also difficult, as comparison across policies requires detailed understanding of the meaning of each policy rule and is rarely straightforward. For human or technology systems there is a resulting gap, sometimes referred to as the ***gulf of execution*** (Norman 1988) between human intentions and technology capabilities. We believe developing approaches to closing the gulf of execution would be valuable in many domains. For example, most organizations store sensitive business and personal data in heterogeneous server systems. They do not have a unified way of defining or implementing security and privacy concerns regarding the storage and use of that data throughout their organization. Changing legal requirements, social pressures and technologies are making these issues increasingly critical to organizations and society at large.

While we are interested in policies in many areas, the research reported here was focused on security and privacy policies. There were several reasons for this focus. First, there are a growing number of strict security and privacy audit and compliance requirements for healthcare, banking/finance, and government. This creates a practical need for improving the management of such information through policy-based systems. Second, there are similarities between aspects of privacy policies and aspects of security policies. Specifically, rules for a major component of security policies, namely, access control rules that define who can have access to what resources, are nearly identical to generally accepted rules which are used in the formation of privacy policies (OECD 1980) This similarity in rule structure leads us to focus on bringing together research rooted in security policy analysis with research in privacy policy authoring and implementation. Third, access control is a major topic for many types of systems and we see it as a central component of not just security and privacy, but of information handling in general. Our principal research objective has been to create an integrated privacy and security policy management framework which builds on the commonalities between the two and encompasses end-to-end solutions for use across heterogeneous configurations covering all data. This includes mechanisms and tools for supporting policy authoring, analysis, enforcement, and auditing.

Extending this, we also see similarities between the management of privacy and security policies and some important aspects of mission plans.

While such plans are not likely to consist solely of "sets of rules", they do include descriptions of intents and guidance for achieving the intent. In places where this guidance maps on to rules of the general condition/action form, we think that research in security policy management is highly relevant to authoring and managing such plans. In areas where policies consist of guidance at higher level than simple rules, we find that there is still much in policy management research that is relevant. One of the challenges in creating any effective policy management framework is to enable the logical progression from the high level statement of policy intentions to low-level implementable policies with sufficient and accurate detail to govern system behavior as intended (see Figure 1).

In our research, we had previously formulated a layered policy model with similarities to that in Figure 1 to provide a basis for reasoning about dynamic security and privacy policies. Figure 2 illustrates that these abstraction and transformation models are key enablers in providing end-to-end mechanisms for adapting system behaviors to meet high-level user-specified security and privacy goals through the enforcement of low-level controls in distributed computing systems. These models provide a basis for automated policy processes including policy specification, verification, transformation, deployment, and auditing.

Policy specification requires more than a simple policy editing tool that enables users to express policies and perform syntax analysis on them. Policies can interact with each other, often with undesirable effects, and a policy author needs to be made aware of such interactions between policies as well as possible resolutions of these issues. Furthermore, policies are usually deployed in distributed systems where it is likely that a policy author may have a partial view of

Figure 2. A model of policy transformation from high-level specification through deployment

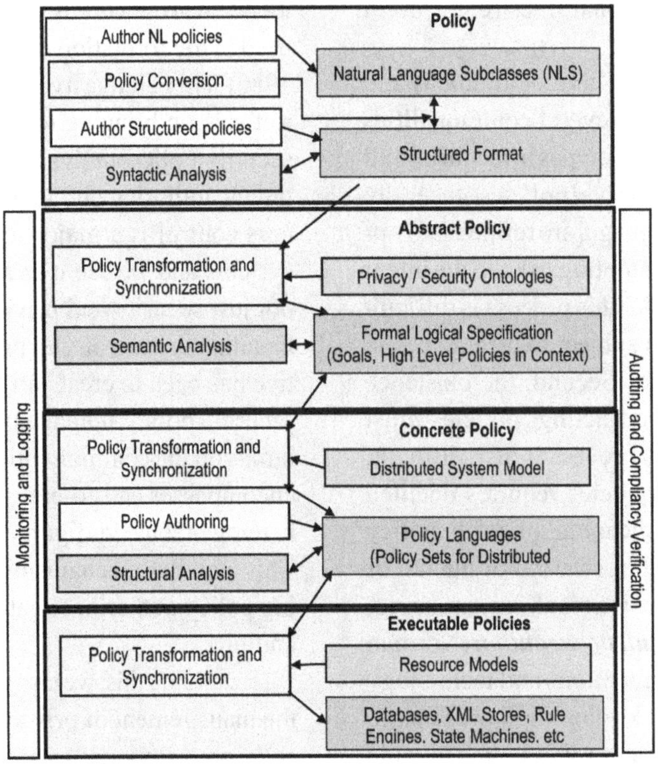

the entire system. Multiple authors may write policies applicable to the same set of resources, and processes must identify issues across these authors' policies. An end-to-end policy management framework must also provide for the full range of necessary refinements or transformations of policies from the stated human intention through system execution and monitoring.

It is important to develop technologies for effectively conveying the content of policies to the authors of the policies as well as the end-users who interact with those policies. End-users may require explanations of policy decisions, in order to understand and possibly correct situations in which requests for access were denied when the data user expected it to be allowed. Thus, technologies to help authors and end-users have been found to fall into three classes: *policy presentation* technologies, which convey the content of policies; *policy explanation* technologies, which explain policy decisions; and *processing technologies*, which provide mechanisms for conveying options, eliciting the individual's choices, and negotiating around policies. We have chosen to begin with the first two of these classes. The technologies are briefly described in the sections below. Policy negotiation is an area of future work for this project.

POLICY PRESENTATION

Approaches to policy presentation generally include two stages: first, finding a language in which to represent a policy; and second, developing a means to present the policy to users. Languages for representing policies may be natural or formal languages. Natural language representations of policies may be prone to the contextual problems of prose in which author and reader might not share a common context or understanding of the terms in the policy. Thus any natural language policy can be ambiguous or inconsistent (just as poorly written programming code is problematic). One approach to making natural language representations clear and consistent is to constrain the language by providing a fixed vocabulary in which to author a policy, in which the meaning of each term is carefully defined and known by all users. Systems which author rules by selecting terms from categories take this structured list approach. Another approach is to allow natural language authoring, and augment the policy with definitions for terms used by the author. For example, the SPARCLE system developed at IBM provides both structured list and constrained natural language methods for authoring (Karat et al. 2005), (Karat et al. 2006). SPARCLE transforms the policies into formal languages, maps the policy elements onto system elements for automated enforcement, and draws on the decision logs for auditing. This tool provides a logical and verifiable link from policy specification to the compliance audits of enforcement decisions. Another approach to making natural language representations clear is to standardize their organization and layout, as the multi-layered notice approach does.

Formal languages for representing policies remove the ambiguities and inconsistencies of natural language, allow for consistent presentation of many different policies to users, and can allow for automated comparison of different policies to each other. Examples of formal languages for representing access control and privacy policies include the Enterprise Privacy Authorization Language (EPAL) and the eXtended Access Control Markup Language (XACML).

There have been several approaches to presenting policies to users. The multi-layered approach to presenting natural language privacy policies provides a consistent, comprehensible structure to policy presentations so that users find it easy to search for specific information in a policy. The PrivacyFinder, a search engine that ranks websites by how protective of a consumer's personal data their P3P policies are, provides Privacy Reports that present P3P policies in sentences that have been vetted in user tests for consistency and readability. PrivacyFinder can also represent P3P policies graphically, by showing a graphical meter that indicates how well a website's P3P policy matches a user's privacy preferences. While the PrivacyFinder's graphical representation of P3P policies aggregates the details of a P3P policy into one simple graphical meter, IBM's SPARCLE system provides a simple visualization of policy coverage in a user configurable matrix format with policy content presented for the key policy elements (Karat et al. 2005). For enterprise systems, there is an expandable grid visualization that provides a meta-view with drill-down capability (Reeder et al. 2007), (Reeder et al. 2008). Simplified, aggregate policy representations may be most appropriate for customers, while detailed overviews are often needed by requestors in order to understand policy decisions in specific situations.

Desirable characteristics of good policy presentation technologies include:

- The ability to convert the difficult machine readable policies into human understandable policies and vice versa, and
- The ability to provide high level views of policies and access to further levels of detail as desired.

POLICY EXPLANATION

Policy explanation technologies help users to understand the decisions made by privacy and security enforcement mechanisms. When a requestor has tried to access data to which they believed they should have access, it is helpful to understand why that access was denied. When an access is denied, it may be that there is an error in a policy. Therefore, explaining why access was denied may enable a requestor to work with administrators to work out errors in policies. While there has been a fair amount of research in providing explanations in rule based systems (e.g., Clancey 1997), there has been little work to date to connect this to policy management systems. We see this as an important area for further research in the development of end-to-end policy management systems.

POLICY EXAMPLE

One of the goals of this research is to understand the important aspect of an organization, the culture within which it exists, and particular situations in which policies are created, transformed, and executed. In research to understand the needs of large civilian organizations (Karat et al. 2005) such as healthcare providers and financial institutions, researchers have recognized that differences in the industries and their associated cultures lead to differences in the types of policies that they would write and the degree to which they would be enforced. For example, there are often privacy policy rules in the healthcare domain that state that only a patient's doctor can view his or her medical record. However, if there is an emergency situation and another doctor is on the scene, then they recognize that the doctor on the scene must have access to those medical records, though perhaps with special logging and review of the access after the fact. In the financial domain, policy authors were adamant that there were no exceptions to their privacy policy rules. Either a person in a particular role is allowed access or not. If an individual thinks they need access to data, they may need an adjustment to their system defined roles or make a case for a change to the existing policy.

Table 1 shows the policies that might be written at the different layers of a policy model. In these examples, both organizations start with a policy rule defining when a practitioner within that domain can access data. In the case of healthcare, that practitioner might be a doctor while in the financial example it might be a loan officer. Because the healthcare community understands that in an emergency situation, any doctor might need access to a patient's medical record in order to save a life, a second policy is added to allow this access with the obligation that the access is tagged for a special review. In the financial domain, no cases that require this kind of exception exist.

CULTURAL MODELS

Policies specify "if-then" rules to guide behavior, as illustrated in Table 1. The sets of rules comprising policies are based on higher-level policy intentions that stem, at least in part, from culturally-shared and tacitly held concepts, causal beliefs, and values. Making such elements explicit will help to understand the intent behind the rules for a cultural group, as well as to determine the importance of policy conflicts. In the following sections we will describe the Cultural Network Analysis (CNA) approach for modeling culture (Seick et al 2010). CNA provides a principled approach for eliciting, analyzing, and representing the shared knowledge that a cultural group uses to conceptualize a task domain, including interrelated concepts, causal beliefs, and values. We will then discuss how the CNA approach can be used to inform the design of systems that support policy management in coalition operations.

From a cognitive view, the term *culture* refers to mental models, and other contents of the mind, that are shared by members of a population over a period of time (D'Andrade 1995). The term *cultural model*, in turn, can be employed both to describe the concept of culturally shared mental models, and to describe an external representation of these models that is constructed by a researcher. Mental models are people's intuitive theories and explanations about how things work (Klein et al. 2003). From our standpoint, mental models comprise interrelated networks of concepts, causal beliefs, and values. Mental models enable people to form expectations and understanding. Klein et al. identified mental models as critical mental representations for supporting decision making and other macrocognitive functions. Mental models are naturally domain specific since they pertain to the workings of particular artifacts and natural processes. Research has pointed to the existence of mental models that pertain to a variety of physical, psychological, and social domains. Mental models for a particular domain are considered "cultural" precisely to the extent that their elements are shared among members of a population (Garro 2000). Figure 3 provides a notional example of a simplified cultural model that could hypothetically form the basis for the healthcare privacy policy from Table 1.

Figure 3 depicts a number of ideas using circles, lines, and color. These ideas include simple concepts such as safety and privacy, represented as circles. It also includes causal ideas such as the concept that informal information sharing (chit-chat) about patient information reduces privacy. These are represented as lines in the figure, with +/- indicating the direction of the causal relation. Finally, Figure 3 portrays ideas of desired

Table 1. Policy content at different levels of abstraction

Layer	Healthcare Privacy Policies	Financial Privacy Policy
Specification Layer	1. Doctors can access patient records if the doctor is the patient's primary care physician, 2. Doctors can access patient records if it is an emergency situation and if the access is tagged for review.	1. Loan officers can access a customer's financial records if the loan officer had been officially assigned the customer's loan application.
High level abstract layer	1. (user category == doctors) and (action == access) and (data == patient record) and (doctor id == primary care physician id) 2. (similar to above)	1. (user category == loan officers) and (action == access) and (data == customer's financial records) and (loan officer id == customer's loan officer's id)

Figure 3. Hypothetical cultural model underlying healthcare privacy policy

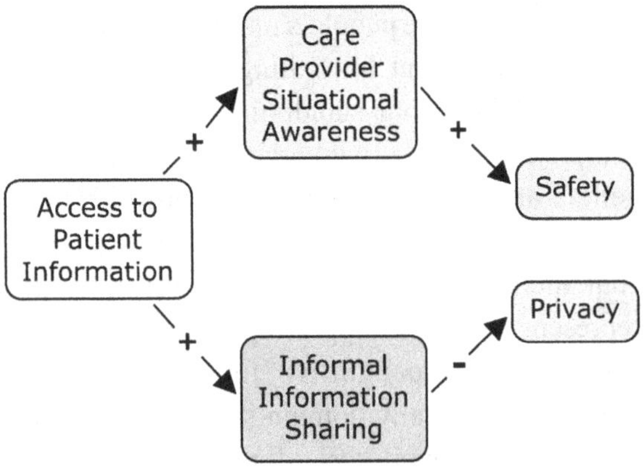

states or value using color, as well as a logical flow across desired states. Safety is a good thing, something one should have. Informal information sharing is negatively valued, at least in this specific context.

Cultural models describe and represent how the world is understood by the members of cultural groups. Thus, they provide an important basis on which to specify or evaluate policy rules for those groups. In the next section, we briefly describe a collection of methodologies for cultural modeling. We refer to this collection of methods as cultural network analysis (CNA), in light of the premise that culture itself is defined by networks of causally-interconnected concepts, causal beliefs, and values that are shared by members of a cultural group (Seick et al. 2010) .

CULTURAL NETWORK ANALYSIS

Cultural Network Analysis (CNA) refers to a collection of methodologies for building cultural models. CNA includes methods to elicit and analyze the mental models of a sample of individuals within the population, measure the degree to which elements of the mental models are shared across individuals, and finally to represent the culture.

A fully-developed, analytical cultural model represents the statistical distribution of mental models for a particular cultural group and domain. CNA builds on a synthesis of conceptually related methods for knowledge elicitation, analysis, and representation that stem from the diverse fields of naturalistic decision making, cognitive anthropology, cognitive psychology, and decision analysis. CNA offers a comprehensive, end-to-end approach for cultural modeling that is currently lacking in these individual disciplines.

Cultural Network Analysis includes techniques for exploratory and confirmatory analyses. In Exploratory CNA, semi-structured interviews are used to elicit participants' concepts and mental models with little presupposition regarding the elicited contents. In confirmatory CNA, structured interviews and field experiments are used to test hypotheses relying on systematic data that is more amenable to statistical analysis. Confirmatory CNA makes it possible to capture and represent subtle differences, such as the degree of consensus in a belief, or precise value trade-offs, which in turn makes it possible to specify policy rules that capture nuances of intent.

Exploratory CNA

Mental models are explanations about how things work. Hence, perhaps the simplest strategy for eliciting a mental model is to ask respondents to, "describe everything you know about X." The approach can be expanded to include probes intended to capture more aspects of the objects or processes under investigation. An important consideration in eliciting knowledge within social-cognitive domains is the difficulty inherent in discussing abstract concepts. One approach to achieving some concrete grounding in such abstract domains is to elicit specific incidents on the topic, and use them to tease out clues to participant's mental models.

Confirmatory CNA

One issue with purely qualitative approaches to the development of cultural models is the lack of transparency or consistent guidelines in what knowledge was deemed sufficiently shared to include in the model. (Strauss and Quinn 1997) state, "At what point in the continuum of sharedness we decide to call a given schema 'cultural' is simply a matter of taste," (p 122). Computational approaches are required for consolidating the qualitative discoveries about culturally shared mental models, and further analyzing and representing their distributions within and between populations.

Mixture modeling and Bayesian influence diagrams are two tools that can be usefully employed to meet those needs. Mixture modeling can be used as a statistical technique for assessing agreement in knowledge and beliefs among groups of respondents (Mueller et al. 2007). Mixture modeling as applied to cultural analysis starts out by defining a statistical likelihood model (i.e., a generative model) by which we assume data and errors arise. It then asks the questions, "How many groups of people with shared ideas generated the observed data." Along with the ability to test whether a consensus exists among a set of respondents, the procedure can also determine if multiple shared ideas exist, and identify the different groups or clusters of respondents.

Model Representation

Cultural Network Analysis uses influence diagrams as a default approach to representation in order to accomplish the following:

1. Provide a standard pictorial form that shows the concepts and causal linkages in a manner that can be readily digested by end users who need to routinely comprehend cultural models in varied domains
2. Permit a direct means of representing the statistical distributions of cultural knowledge, rather than just the shared knowledge
3. Yield representations in a useful form for developers of intelligent systems
4. *Provide a representational foundation for making culturally sensitive machine-based evaluations of the quality of work processes and outcomes*

Application

Using CNA, we can build cultural models that inform the design of multinational collaborative systems, such as for coalition planning or policy development (Rasmussen et al. 2008). With respect to coalition policy development, the application would depend on the collaborative decision process that coalition partners adopt. As mentioned in the introduction, policies that span coalition partners could be developed in a more or less integrative fashion. A highly integrated multinational policy development process could use cultural models from each group that represent otherwise tacit concepts, causal beliefs, and values that form the base assumptions and high-level intent to be negotiated prior to determining rules for the specification layer. In the least integrative case, exemplified in Figure 1, cultural models provide important inputs about the values and higher-level rationale for rules that enables systems to determine the importance and ultimate resolution of low-level rule policy conflicts. In addition, differences in cultural models can translate into design considerations with respect to a policy management system in the following ways:

1. The opportunities for user interaction afforded by the system
2. The kind of information made available to users at different stages of the policy authoring and validation process
3. *The support for creating, defining and editing different types of policies*

4. *The ease with which users can assess whether or not a given policy meets certain cultural criteria for quality.*

COLLABORATIVE DECISION SUPPORT

Team decision making is a bundle of interdependent activities that involve gathering, interpreting and exchanging information; creating and identifying alternative courses of action; choosing among alternatives by integrating the often different perspectives of team members; and implementing a choice and monitoring its consequences. Team decision making in coalition operations is especially challenging since coalition teams typically involve a number of units drawn from different national forces, units that are assembled on an ad hoc basis, which have not necessarily worked together before, and work under different operational regimes, with different policies and within different cultural contexts. Moreover, as the situation unfolds, teams could benefit from automated support that is adaptive to changing circumstances in the environment, such as changing mission goals or loss of teammates. While previous work in teamwork theory has focused on describing ways in which humans coordinate their activities, there has been little work on identifying interaction patterns and information flows of ad hoc and decentralized teams in time stressed and dynamic environments (Cost et al. 2001), (Lenox et al. 2000:1), (Lexon et al. 2000:2), (Sycara and Lewis 2004).

We envision agent assistance in order to achieve what we shall call culturally-sensitive policy and trust interoperability. The concept of policy interoperability means that different coalition partners with different policies should be able to collaborate and create feasible plans and execute missions effectively despite their cultural and policy differences. Agents can assist in various ways. Agents can have access to data bases of policies of coalition members, and proceed to analyze and identify conflicts in policies. The agents can proceed to evaluate whether these conflicts may affect planning and execution of given coalition missions and provide suggestions for policy conflict resolution and reconciliation. The conflict resolution that the agents suggest can be informed by cultural models that have been identified using Cultural Network Analysis, thus making the suggested resolutions culturally appropriate. The creation of these agents will be informed by past policy representation and inference techniques (Vasconcelos et al. 2007).

Another general way agents can support policy interoperability is by enforcing policies during communication and collaborative decision making of the human team mates. In other words, agents could act as censors. In this role, agents can delete any unallowable information sharing among coalition members thus enforcing the policy that prohibits sharing.

The cultural and ethnic differences of coalition members engender mistrust. Agents could assist the collaboration by promoting trust among human teammates and adequately managing the differing levels of trust between coalition members. It is widely accepted that trust within human teams is a key factor in ensuring high performance (Cost et al 2000). When considering roles for agents in teams, it is important to consider the issue of trust within the team, as this affects the activities agents can and may be required to carry out and the ways in which they must behave. Trust can be defined as the extent to which humans or agents believe that other team members can and will perform actions as requested, provide accurate and truthful information and adhere to team protocols. We distinguish between teams with high trust, where human or agent team members have a high level of trust in each other, and limited trust, where human or agent team members may not trust some or all of the other team members.

The notion of trust within a team also captures the beliefs team members have about each other's propensity for altruism, i.e. that team members will act for the greater benefit of the team even if it requires individual sacrifice. In a high trust team, agents will refrain from behaving opportunistically in pursuing their individual goals when they have knowledge about other team members' intentions and plans. We will investigate how trust and information confidentiality issues influence communication between agents supporting team activities and between human team mates and how information and arguments from specific sources are weighed during reasoning. Agents can support trust among team mates by generating and presenting convincing arguments and explanations (Oren et al. 2006).

One of the most challenging but interesting questions is, given a mission description, a cultural model and a trust model, can a policy management system be generated so that policies are culturally sensitive and allow effective mission planning and execution? To address this challenge is the ultimate goal of our collaborative research.

DISCUSSION

While we have not yet integrated cultural and collaborative models into the policy management framework presented above, our discussions have shed light on how to proceed. To date, development of policy analysis function has focused on algorithms which carry out formal analysis at the different levels of policy specification illustrated in Figure 2 (Chapter 8). For example, policies can be analyzed for conflict, which occurs when two or more policies have overlapping conditions but specify incompatible actions. While we suggest that aspects such as culture and collaboration are important elements of policies, they have not been included in the previous consideration of policy analysis precisely because we have lacked models of how culture and collaboration aspects of policy might be captured and then reasoned about. That is, we need to identify the situations in which an individual policy or sets of policies might present a potential cultural or collaborative problem. Our work here is aimed at elevating these elements in the consideration of policy by specifically asking how such models need to be extended for the purpose of policy analysis.

We see the work moving forward in a number of ways. First of all, we expect that the incorporation of cultural models will be of most value in helping to surface differences in the way that cultural assumptions are incorporated by different coalition members into their policies. Our analysis of collaboration across cultures suggests that trust is an important concept in policy creation, and that different cultures will reflect trust different assumptions concerning trust in the formation of policies. For example, in research involving different organizations, the amount of information a policy will call for sharing with others can vary significantly. Hospitals and financial institutions have very different policies for access to sensitive information which result from differences which we might model as a part of their culture. These differences reflect attitudes about how much different parties can be trusted with sensitive information and how much harm or benefit can result from policy violation in different contexts. We can see similar differences in a military context when we examine effective policies for information sharing in US and UK military organizations.

Secondly, we expect that resolution of issues raised through a cultural-based model analysis will require human collaboration. While analysis might detect possible problems suggested by the model, we do not believe that the models will be general enough to suggest successful resolutions of identified issues in all cases. Negotiation over policy differences will require support for human dialog, and mechanisms for providing this support can be informed by models of collaboration in the appropriate context. The mechanisms provided for this in policy management tools might itself be guided by various models of human collaboration and problem solving. For example, where research can point to information needs of collaborators in the specified context, policy negotiation tools can utilize this knowledge to assist decisions concerning appropriate policy changes in the field (e.g., allow one time access to information from a specific sensor) or to help negotiate policy differences before deployment (e.g., when policy authors from two organizations are trying to resolve differences between national and coalition policies). We do not believe that turning these ideas into technology for either of these forms of culturally informed policy management support is a trivial task and that it the models for culture and collaboration need further development and testing. However, we do think that a course which proceeds from attempts to apply base culture and collaboration models to policy management is the right direction toward the goal.

This chapter described efforts of three research teams to explore the intersection of policy, culture, and collaborative decision making, and through this to extend a model of policy management that incorporates aspects of cultural differences in policy making along with aspects of collaboration in coalition activities. This paper reports on the rational for the importance of such collaborative research, key research questions, and a three-phase research approach of theoretical work, integration of disparate research findings, and development of a proposed integrated model. The theoretical work led to a view of how culture impacts planning and how this result might be adapted to a model of culture in policy management. Specifically, it suggests that there might be different formulations of policy that are considered as desirable in different cultures, and that policy analysis might include culture as an element of analysis and might provide feedback to policy authors based on a cultural model. Similarly, collaboration research points to the importance of communication in group problem solving. For policy management this points to the importance of creating channels of communication when joint policies are being developed and negotiated. We believe that additional empirical research could provide data on the characteristics of a hypothesized 'common policy formulation' function in a policy management framework (i.e., a 'culturally intelligent common policy formulation function'). The data could provide evidence on whether policy management in coalition operations can be strengthened by incorporating cultural analysis and collaborative decision making variables. Integration of cultural models of commonalities and differences that capture coalition partner's understanding of the process for generating 'acceptable' (i.e., high quality) policies and the decision making criteria that analysts use to qualify policies as acceptable, could lead to better planning and policy tools.

Additional research could investigate how to design policy management systems that are usable and acceptable among collaborators from different cultures in dynamic collaborative contexts. The target is a usable and cross-culturally accepted policy management tool that employs a 'common policy formulation' function. This function would use domain-specific cultural models in developing policy analysis algorithms to predict cultural acceptability of policies, flag potential culturally-contentious policies, and identify to whom they might be contentious. The policy tool might provide culturally-sensitive suggestions on how to resolve policy 'disputes' and might provide suggestions on how to best negotiate contentious policies with different coalition stakeholders. The cultural-informed policy tool might also be leveraged to provide guidance on best practice and risk for policies. An important factor in making progress in this area is to make the integration of the models an explicit research goal. Up to this point we have collaborated in exploring the ideas presented here in an effort to address the question of how such technologies might be used by people in coalition contexts. However, advancing this vision requires additional focus not just on developing the three models, but on how the models integrate. Fitting the pieces together requires each effort to expand the topics that it considers as relevant to the development of the models.

This fundamental research in culturally-informed policy technology with decision support contributes valuable tools for coalition operations. The collaboration between different teams provides the opportunity for multidisciplinary views on the problem space. The rationale and initial research reported in this collaboration may motivate and inform other collaborations within the ITA research project.

REFERENCES

Clancey, W. J. (1997). *Situated Cognition: On Human Knowledge and Computer Representations*. Cambridge, UK: Cambridge University Press.

Costa, A., Roe, R., & Taillieu, T. (2001). Trust within teams: The relation with performance effectiveness. *European Journal of Work and Organizational Psychology, 10*(3), 225–244. doi:10.1080/13594320143000654

D'Andrade, R. G. (1995). *The development of cognitive anthropology*. Cambridge, UK: Cambridge University Press.

Garro, L. C. (2000). Remembering what one knows and the construction of the past: A comparison of cultural consensus theory and cultural schema theory. *Ethos (Berkeley, Calif.), 28*(3), 275–319. doi:10.1525/eth.2000.28.3.275

Karat, C., Karat, J., Brodie, C., & Feng, J. (2006). "Evaluating interfaces for privacy policy rule authoring," *Proceedings of the Conference on Human Factors in Computing Systems – CHI 2006*, ACM Press, pp. 83-92.

Karat, J., Karat, C., Brodie, C., & Feng, J. (2005). Privacy in information technology: Designing to enable privacy policy management in organizations . *International Journal of Human-Computer Studies, 63*(1-2), 153–174. doi:10.1016/j.ijhcs.2005.04.011

Klein, G., Ross, K. G., Moon, B. M., Klein, D. E., Hoffman, R. R., & Hollnagel, E. (2003), Macrocognition. IEEE: Intelligent Systems, May/June, 81-85

Lenox, T., Hahn, S., Lewis, M., Payne, T., & Sycara, K. (2000:1). "Task Characteristics and Intelligent Aiding", Proceedings of the 2000 IEEE International Conference on Systems, Man, and Cybernetics, October 8-11, Nashville, TN pp. 1123-1127.

Lenox T., Hahn, S., Lewis M., Payne T. and Sycara, K. (2000:2). "Agent Based Aiding for Individual and Team Planning Tasks", *IEA 2000/HFES 2000 Congress.*

Mueller, S. T., Sieck, W. R., & Veinott, E. (2007). *Cultural Metrics: A Finite Mixture Models Approach* (Technical Report DAAD19-01-2-0009). Fairborn, OH: Klein Associates Division of ARA.

Norman, D. (1988). *The design of everyday things*. New York: Double Day.

Oren, N., Norman, T. J., & Preece, A. (2006). Loose lips sink ships: A heuristic for argumentation. *Proceedings of the Third International Workshop on Argumentation in Multi-Agent Systems*, OECD- Organization for Economic Co-operation and Development, (1980). *OECD Guidelines on the Protection of Privacy and Transborder Flow of Personal Data.* Paris, France, Accessed at www.oecd.org/home/.

Rasmussen, L. J., Sieck, W. R., & Smart, P. R. (2008). US/UK Mental Models of Planning: The Relationship between Plan Detail and Plan Quality. In: *NATO RTO HFM-142 Symposium on Adaptability in Coalition Teamwork*, April 2008, Copenhagen, Denmark.

Reeder, R., Bauer, L., Cranor, L., Reiter, M., Bacon, K., How, K., & Strong, H. (2008). "Expandable Grids for Visualizing and Authoring Computer Security Policies," In Proceedings of the Conference on Human Factors in Computing Systems – CHI 2008, ACM Press.

Reeder, R., Karat, C., Karat, J., & Brodie, C. (2007). Usability Challenges in Security and Privacy Policy-Authoring Interfaces. In . *Proceedings of INTERACT, 2007*, 141–155.

Strauss, C., & Quinn, N. (1997). *A Cognitive Theory of Cultural Meaning*. Cambridge, UK: Cambridge University Press.

Sukthankar, G., Sycara, K., & Giampapa, J. A. (2008). "A Model of Human Teamwork for Agent-Assisted Search Operations", In *Proceedings of the NATO Human Factors & Medicine Panel Symposium on Adaptability in Coalition Teamwork*, Copenhagen, Denmark.

Sycara, K., & Lewis, M. (2004). Integrating Agents into Human Teams . In Salas, E. (Ed.), *Team Cognition* (pp. 203–233). Erlbaum Publishers.

Vasconcelos, W. W., Kollingbaum, M. J., & Norman, T. J. (2007). Resolving Conflict and Inconsistency in Norm-Regulated Virtual Organizations. Procs. of the International Joint Conference on Autonomous Agents and Multiagent Systems (AAMAS'07). Honolulu, Hawai'I, USA.

Chapter 10
The Network-Extended Mind

Paul R. Smart
University of Southampton, UK

Paula C. Engelbrecht
University of Southampton, UK

Dave Braines
Emerging Technology Services, UK

Michael Strub
Dstl Porton Down, UK

Cheryl Giammanco
Human Research & Engineering Directorate, USA

ABSTRACT

Whereas the traditional view in cognitive science has been to view mind and cognition as something that is the result of essentially inner, neural processes, the extended cognition perspective claims that at least some human mental states and processes stem from complex webs of causal influence involving extra-neural resources, most notably the resources of our social and technological environments. In this chapter the authors explore the possibility that contemporary and near-future network systems are poised to extend and perhaps transform our human cognitive potential. They also examine the extent to which the information and network sciences are relevant to their understanding of various forms of cognitive extension, particularly with respect to the formation, maintenance and functioning of extended cognitive systems in network-enabled environments. Their claim is that the information and network sciences are relevant on two counts: firstly, they support an understanding of the mechanisms underpinning socially- and technologically-mediated forms of cognitive extension; secondly, they serve to guide and inform engineering efforts that strive to enhance and expand our cognitive capabilities. The authors discuss the relevance and applicability of these conclusions to current and future research exploring the contribution of network technologies to military coalition operations.

DOI: 10.4018/978-1-61520-855-5.ch010

INTRODUCTION

The traditional view in the sciences of the mind sees the human brain as occupying a rather special place in the material fabric associated with the realization of human mental states and processes. One only has to flick through the pages of any contemporary text on cognitive neuroscience to appreciate the considerable dominance of what one might call the 'neurocentric view'. And it is a view that is reinforced by (and reflected in) a steady stream of brain imaging studies, many of which claim to have isolated the neuroanatomical basis of some aspect of our everyday psycho-cognitive functioning. The traditional view thus sees human mental states and processes as the direct product of what the brain does. It claims that the machinery of the mind is housed largely within the head, and that to understand more about our cognitive profile we need to understand more about how the brain works. Eventually, it is claimed, we will have a complete theory of human cognition, and within this theory the brain will occupy centre-stage.

The validity of this neurocentric, or intracranial, perspective has recently been challenged by those who embrace situated, embodied or distributed approaches to cognition (Clark, 1999; Haugeland, 1998; Hutchins, 1995a; Pfeifer & Bongard, 2007; Robbins & Ayded, 2009). Such approaches challenge the notion that mind and cognition are solely internal (neural) phenomena by emphasizing the role played by extra-neural and extra-bodily factors in shaping the profile of much real-world cognitive processing. One view that is perhaps maximally opposed to the internalist or individualistic conception of the human mind (the notion that the mind is the result of purely internal processes) is the thesis of the extended mind (Clark & Chalmers, 1998). This view explicitly endorses the idea that the human mind is not solely the product of what the brain does and that the boundaries of the human mind are not necessarily co-extensive with the biological boundaries of the brain. Instead, the claim is that much of the machinery of the human mind extends beyond the brain to encompass a much larger nexus of extra-neural (and sometimes extra-organismic) resources. According to the extended mind perspective, human mental states and processes are not always in the head; they can sometimes extend beyond the brain to encompass aspects of the external technological and social environment.

Claims about the distributed or extended nature of human cognition are commonplace in the scientific and philosophical literature (Clark, 1997, 2003, 2008; Clark & Chalmers, 1998; Dennett, 1996; Haugeland, 1998; Hollan, Hutchins, & Kirsh, 2000; Hurley, 1998; Hutchins, 1995a; Kirsh, 1996, 2006; Norman, 1993; Wilson, 1994; Wilson & Clark, 2009). But what do such claims really amount to when we consider the potential impact of network systems and technologies on our current cognitive profiles? And what role do the information and network sciences play when it comes to understanding socially- and technologically-mediated forms of cognitive extension? One thing is relatively clear: it is that as we move into an era of pervasive computing and ubiquitous network access, much of our material world is becoming infused with greater computational potential, both for ourselves and the social collectives of which we are a part. If we want to understand the opportunities (as well as the hazards[1]) for cognitive transformation in this new era, we need to have theories and approaches that are capable of operating at the interfaces of the engineering, cognitive and social sciences. It is our claim, in this chapter, that the information and network sciences are a vital source of such theories and approaches; they are suitably poised to advance our understanding of the mechanisms underpinning socially- and technologically-mediated forms of cognitive extension.

Recognizing the contribution of the wider social and technological environment to cognitive processing (at both the individual and collective level) is of particular relevance in military coalition environments. Such environments are often conceptualized in terms of multiple interconnected networks (i.e. networks of networks) that subtend the human, technological and informational domains. Such networks interact in complex, non-linear ways throughout the course of coalition operations, and the challenge for military coalitions is often to coordinate the structure and activity of these networks in ways that meliorate cognitive performance. The ability of a specific coalition element to respond in an adaptive and intelligent manner is, for example, often based on the broader ability of the coalition formation to properly create, encode, select, retrieve, transform and communicate information-bearing structures (representations), and such information manipulation processes often need to be sensitive to the structure of existing communication and social networks. Notions of distributed and extended cognition have a special role to play here because they focus attention on the fundamental interdependencies between specific cognitive performances and the wider webs of social and technological scaffolding in which such performances take place. In this chapter we aim to show why distributed and extended approaches to human cognition are relevant to our understanding of the inter-relationships between coalition networks and cognitive processing at both the individual and collective levels.

The structure of this chapter is as follows. The section entitled 'Externalism and Extension: A Brief History' provides an introduction to externalist approaches to the human mind. It reviews the key arguments associated with two forms of externalism, namely content externalism and vehicle externalism. Both of these forms of externalism raise doubts about the philosophical and scientific integrity of, what might be called, internalism (the idea that mind and cognition can be understood solely by focusing on internal, intra-cranial states-of-affairs). This leads on to a discussion about notions of cognitive extension in the section entitled 'Cognitive Extension'. Cognitive extension has been introduced using a number of real-world examples in the literature. These include long multiplication (see Wilson & Clark, 2009), ship navigation (Hutchins, 1995a), academic paper writing (Clark, 1997), puzzle solving (Kirsh, 2009; Kirsh & Maglio, 1994; Maglio, Matlock, Raphaely, Chernicky, & Kirsh, 1999), and the process of artistic creation (see Clark, 2001). In this chapter we introduce the notion of cognitive extension using a 'simple' non-cognitive example, namely the process of spider web building behaviour. This example is intended to show how a collection of capabilities that is ostensibly the product of a centralized neurological resource (the spider's nervous system), actually turns out, on closer inspection, to involve a variety of more far-flung forces and factors. Spider web weaving thus emerges as an example of what has, in the literature, been dubbed 'non-trivial causal spread' (Wheeler, 2005; Wheeler & Clark, 1999), a feature that characterizes many cases of environmentally-extended cognition. After presenting the case for network scientific approaches to extended cognitive systems in the section on 'Extended Cognitive Systems', we then present the extended mind thesis in the section entitled 'The Extended Mind'. An extended mind can be thought of as a particular kind of extended cognitive system, one that relies on the more or less permanent coupling of a human agent with cognitively-potent technological add-ons. The section entitled 'The Web-Extended Mind: A Thought (Provoking) Experiment' extends the discussion about the extended mind thesis and applies it to putative cases of cognitive extension involving the World Wide Web. In this case, we engage in a thought experiment regarding the close coupling of a human agent with near-future Web-based technologies. The thought experiment gives rise to a number of issues regarding potential shifts in our conception of ourselves as cognitively- and epistemically-bounded agents. Issues relating to socially-extended cognition (i.e. cases of cognitive extension involving other human agents) are reviewed in the section entitled 'Socially-Extended Cognition', and this is followed by a discussion of the kinds of cognitive extension that are likely to be encountered in military coalition contexts in the section entitled 'Extended Cognitive Systems and Military Coalitions'. A number of defence-related research programs, including the new Network Science Collaborative Technology Alliance (CTA), feature research that is highly relevant to some of the issues raised in this chapter, and an overview of such programs is provided in the section on 'Relevant Defence-Related Research Programs'. Within the same section we briefly present work within the joint U.S./U.K. International Technology Alliance (ITA) research program, which specifically seeks to explore a number of issues related to cognitive extension in network environments. Such work, we suggest, can be seen as the intellectual lynchpin that connects work in many other research programs, such as the Network Science and Cognitive Neuroergonomics CTAs and the Tactical Human Integration with Networked Knowledge Army Technology Objective (THINK ATO). In the 'Conclusion' section we summarize the main arguments motivating a consideration of the information and network sciences to our understanding of network-mediated forms of cognitive extension. The section also reiterates the main points of relevance regarding cognitive extension research and military coalition operations.

EXTERNALISM AND EXTENSION: A BRIEF HISTORY

Historically, cognitive science has embraced a particular view of the mind, one which sees human mental states and processes as largely the product of inner, neural mechanisms. Human mental states and processes, the view maintains, are essentially realized by physical mechanisms inside the head of human subjects, and thus the mechanistic boundaries of the human mind are roughly co-extensive with those of the biological brain. This particular view of the mind (which, following Wilson and Clark (2009), we will refer to as individualism) maintains that the human mind can be studied and understood independently of any reference to the external environment. It essentially advocates what Jerry Fodor (1980) once referred to as 'methodological solipsism', the idea that the cognitive sciences can limit their study to the individual, effectively bracketing off the world in which the individual is embedded. On the individualist view, cognition is something that is wedged between perception (on the input side) and action (on the output side), constituting the filling of what Susan Hurley (1998) refers to as a 'cognitive sandwich'. The individualist conception recognizes the role of the wider environment as an input/output space for cognitive processes, but it does not afford any constitutive role for extra-organismic elements in those processes. Despite a recent emphasis on situated (Robbins & Ayded, 2009), distributed (Hutchins, 1995a) and embodied (Clark, 1999; Pfeifer & Bongard, 2007) approaches to cognition, internalism is still apparent in the sciences of the mind[2]. In addition, it is likely that some form of neurocentric individualism best captures our contemporary 'common-sense' notions about the material origins of the human mind. As Noë (2009) points out:

"We live in a time of growing excitement about the brain…Perception, memory, our likes and dislikes, intelligence, morality, whatever – the brain is supposed to be the organ responsible for all of it." (Noë, 2009; pg xi)

Doubts about the integrity of individualism first arose during the 1970s in the work of Hilary Putnam (Putnam, 1975) and Tyler Burge (Burge, 1979). The predominant concern was that individualism failed to adequately account for the content or *meaning* of mental representations. Putnam (1975) thus argued that mental states could not be individuated in accord with the constraint of individualism because the content of a mental representation may be determined by facts external to the individual, specifically the environmental or historical location of an individual. It is only by referencing these external facts that the meaning of an inner representational item can be discerned. This form of externalism, which has been referred to as taxonomic externalism (Wilson, 2000, 2004) or content externalism (Rowlands, 2006), is clearly contrary to the main thrust of the individualist thesis. It asserts that although intentional mental states exist as internal (e.g. neural) states of an individual, they are not (in virtue of their content) supervenient[3] on purely internal (intra-individual) factors.

Despite its appeal to physical, social and historical factors, the form of externalism just described (content externalism) is still largely committed to an internalist perspective about the location of the physical structures associated with mental states and processes. Even though the content of mental representations are deemed to depend on the external environment, the physical vehicles of cognition[4], it is claimed, are still likely to be situated within the head of the individual. An alternative, and more radical, claim is that even the physical vehicles of cognition need not be restricted to the internal realm. Instead, so the claim goes, the vehicles of both mental states and mental processes are perfectly able to *extend* beyond the head into the external world. Mind, and the cognitive processes that constitute it, sometimes extend into the physical and social environment of the individual human agent.

This second form of externalism (which we will refer to as vehicle externalism) goes by a variety of names, including locational externalism (Wilson, 2000, 2004), active externalism (Clark & Chalmers, 1998), vehicle externalism (Hurley, 1998; Rowlands, 2006), environmentalism (Rowlands, 1999), and the extended mind (Clark & Chalmers, 1998). What unites all these terms is a theoretical commitment to the idea that the physical boundaries of a cognitive system should not be assumed to coincide with the traditional biological boundaries of skin and skull. Instead, cognition is seen as something that is often an environmentally-extended process. To fully understand human cognition, it is not enough to focus on the inner states of the individual, for such a focus reveals only a partial picture of cognitive processing. Instead, we need to look beyond the individual, to understand the way in which cognition is situated and embedded within a larger nexus of physical and social influences.

The notion of vehicle externalism, as just described, is something that will occupy us for the remainder of this chapter. However, before we embark on that discussion, it is important to point out that claims about vehicle externalism are largely orthogonal to those of content externalism. Content externalism is a theory about how the content of (inner) mental representations supervenes on facts about the historical and environmental location of an individual. Vehicle externalism, in contrast, does not concern itself with how representational vehicles acquire the meaning or content they do; it is primarily a theory about the nature and interaction of the vehicles themselves. In advocating vehicle externalism we are essentially committing ourselves to understanding how the physical, social and technological environment contributes to the material realization of specific states and cognitive performances; we are relatively less concerned with how those states and performances acquire their specific contents.

COGNITIVE EXTENSION

The claims of vehicle externalism have a somewhat radical sounding flavour to them[5], but the notion that the physical vehicles of cognition are not restricted to the inner, neural realm, is a notion that is perfectly compatible with the claims of both physicalism and functionalism (see Braddon-Mitchell & Jackson, 2007). And although the dominant view in artificial intelligence research (at least in the last century) was guided by predominantly individualistic and internalist conceptions of the mind, there is nothing in the bedrock claims of classical cognitivist theory (Newell, 1980; Newell & Simon, 1976; Pylyshyn, 1984) that necessarily binds intelligence to internally-situated mechanisms[6]. In spite of this, the claims of vehicle externalism are largely counter to our common-sense intuitions about the human mind, and this often results in a deep scepticism about the tenability of the core claims. To make both the claims of the vehicle externalist thesis clearer, and to invite a principled consideration of the relevant ideas, it helps to start with the simple (and in our case non-cognitive[7]). Therefore, this section begins with a simple, but powerful, demonstration of how intelligent behaviour can emerge from the delicate interplay of forces and factors that extend beyond the neural realm. In subsequent sections, we expand on this initial case study and show how vehicle externalist views can be applied to more complex forms of 'human-level' problem-solving.

Web Construction

The web of the garden cross spider (*Araneus diadematus*), like that of most orb web spiders, is a compelling example of how our initial intuitions about the problem-solving potential of a seemingly simple bio-computational system (in this case an insect nervous system) can founder in the face of real-world performance. The spider's central nervous system is composed of a number of ganglia (collections of neural tissue), of which the most prominent are the supraesophagal and subesophagal ganglia. These ganglia serve to implement and coordinate the majority of the spider's sensorimotor functions. The total number of neurons in the central nervous system is small, about 30,000 neurons in the case of the orb-web spider *Argiope* and 100,000 neurons in the case of the larger wandering spider *Cupiennius* (Foelix, 1996). This compares with somewhere in the region of 100 billion neurons for the average human brain. Given the scale of the spider's nervous system, we might expect its behavioural capacities to be limited. And yet spiders are capable of surprisingly complex behaviours[8], of which the most well known is probably web construction. The spider's web is architecturally complex, composed as it is of multiple types of silk thread, each laid down in a specific sequence and geometric pattern. Specific types of thread need to be produced at just the right time, and the overall design of the web has to be sensitive to a number of factors including the size of the prey to be caught and the shape of the local environment (the shape made available by local branches or other supporting structures). The problem might be easier if it was possible to use visual information to guide action selection processes, but *Araneus* is practically blind and does not rely on visual information to complete the web construction process (Witt, Reed, & Peakall, 1968). The average human being, blindfolded and presented with the task of creating a complex geometric structure from multiple types of building material, might be hard pressed to match the spider's feat of engineering, and this is despite the fact that our own neural systems far outstrip the size and complexity of those possessed by the average orb web spider. The feat of web construction seems to require a capacity for judgement, decision-making and planning that is profoundly out of kilter with our expectations and intuitions about what the spider should be capable of. So how does the spider do it?

The answer seems to lie in the spider's exploitation of bodily contingencies and the power of the local environment to structure and guide action choice. A detailed ethological examination of web spinning behaviour suggests that spiders are sensitive to certain bodily contingencies involving the relative positioning of their legs on certain types of silk thread (Krink & Vollrath, 1997, 1998, 1999). As the web develops, the positioning of the legs becomes a reliable cue as to what type of action needs to be executed next, as well as what type of silk needs to be produced. In essence, the web serves as "its own best model" (Brooks, 1991) of what needs to be accomplished, and the spider need only be responsive to local information concerning the structural organization of threads in the immediate vicinity of its body. At each stage of the web construction process, each of the spider's legs need only perform a local (spatial) search for the nearest thread, and, once located, the relative positioning of the legs (as well as the type of thread they are in contact with) 'represents' the web's structural status. In response to this rich body of local information, the spider need only implement locally-effective rules concerning which action to perform. And it turns out that aspects of spider web weaving behaviour can be modelled using a relatively simple (and minimal) set of rules (Krink & Vollrath, 1997, 1999). Importantly, each rule exploits facts about the spider's bodily design, and its outputs specify actions that are geared to structuring the problem space in ways that guide, constrain and simplify subsequent behaviour. The spider, it seems, distributes the computational burden associated with web spinning behaviour across a complex system that comprises its brain, body and aspects of the (self-structured) external environment[9].

So perhaps the reason we find the spider's web spinning behaviour both remarkable and mysterious (relative to its rather meagre neuro-computational resources) is because we fail to appreciate the behaviour for what it really is: a compelling example of environmentally-extended bio-morphological computation[10], one in which neural, bodily and environmental factors play representationally and computationally-significant roles. The central nervous system of the spider no doubt plays a very important coordinative role in the process of web construction, but it is only one element of a complex, environmentally extended system, and its representational resources and computational capabilities are geared not towards to the manipulation and transformation of abstract disembodied symbolic representations that occupy some inner, neural realm, but to the generation of temporally extended action sequences, actions that themselves serve to progressively structure and restructure the target problem-space in computationally- and representationally-potent ways.

The moral of this story, then, is that it is easy to be misled into thinking that intelligent action is always the sole product of neural mechanisms – that the point source of intelligent behaviour is always something that must reside in the 'head' of an agent. For what the case of web construction teaches us is that agents may often co-opt a variety of far flung forces and factors into a problem-solving routine, and not all of these forces and factors need to be biological in nature. We should not necessarily be surprised by this outcome. Evolution does not care about the material nature of problem-solving resources; it only cares about how those resources can be exploited to meet adaptive behavioural ends. Artificial evolutionary processes attest to the variety of ways in which seemingly irrelevant forces and factors may be co-opted into a design solution. Thus in using genetic algorithms to evolve real electronic circuits, Bird and Layzell (2002) managed to create an 'oscillator circuit' whose systemic oscillatory behaviour was parasitic on the radio signals being generated from a nearby computer. In essence, the evolving circuit had generated the correct oscillatory behaviour, but had done so not by creating a genuine oscillator circuit; it had solved the problem by evolving radio reception capabilities and relaying the oscillations created by nearby circuits. Such phenomena are a common feature of many evolutionary processes. Thompson, Harvey and Husbands (1996) thus argue that during the evolution of electronic circuitry:

"... it can be expected that all the detailed physics of the hardware will be brought to bear on the problem at hand: time delays, parasitic capacitances cross-talk, meta-stability constraints and other low-level characteristics might all be used in generating the evolved behaviour." (pg. 21)

What we begin to see, therefore, is that for any given problem-solving process, evolution may often assemble solutions that pay scant regard to the manner in which the problem is solved. In many cases, the nature of the solution yielded by an evolutionary process will draw on whatever resources are available to meet the representational and computational demands of the problem at hand. And the responsibility for yielding adaptive behavioural success will, in many cases, be distributed across a broad coalition of neural, bodily and environmental resources.

We thus approach the main take home message of this section. It is that, when seen in a certain light, the external environment emerges as more than just a space for sensory inputs and motor outputs; it is also poised to play an important (explanatorily-potent) role in the mechanisms by which that behaviour is realized. Intelligent behaviour, we might say, is at least sometimes realized by processing loops that *extend* beyond the neural realm and productively incorporate a variety of extra-neural resources. Some forms of behavioural intelligence are, we might say, environmentally-extended with regard to their mechanistic realization.

There is a parallel here – one that follows on nicely from the account of spiders and evolutionary processes – with Richard Dawkins' (1982) account of the extended phenotype. As part of his introduction to *The Extended Phenotype*, Dawkins (1982) encourages us to ignore the traditional biological boundary of the body and instead focus on the way in which external structures can form part of an extended system, one that is both created and maintained by specific genetic influences. From this 'extended' viewpoint, we can, he suggests, regard the spider's web as part of the spider's phenotype; it is a system that, just like the spider's body, determines the extent to which the spider's genes will be transmitted to future generations. The spider's web, when viewed through the special lens of the extended phenotype, thus emerges as a more-or-less equal partner in a complex matrix of phenotypic structures (some biological and others not) all of which are subject to evolutionary selection pressures.

But there is a deeper analogy here, one that goes beyond the level of extended phenotypes and extended behavioural mechanisms. It is the role that genes themselves play with regard to the generation of phenotypic structures. For in many ways, we suggest, the mechanisms by which genes control, regulate and contribute to the emergence of ontogenetic and cellular processes via their participation in genetic regulatory networks is directly analogous to the role played by the spider's nervous system in architecting its web. Just as the spider's web-spinning performances can seem remarkable relative to its available neuro-computational resources, so the morphological and physiological complexity of organisms can often seem surprising relative to the number of genes encoding their development (Claverie, 2001). Studies in functional genomics, for example, reveal that the number of protein-coding genes in the case of the human genome is about 20000-25000 genes (International Human Genome Sequencing Consortium, 2004), while that for the rather unsophisticated nematode worm, *Caenorhabditis elegans*, is a surprising 20,000 (C. elegans Sequencing Consortium, 1998). These results are surprising because, inasmuch as one sees genes as coding directly for specific aspects of physical form and function, one would have expected relative differences in large-scale phenotypic complexity to be reflected in large-scale differences in gene number. So how do we reconcile the apparent similarity of gene numbers in the case of *C. elegans* and *H. sapiens* with the apparent differences in physiological and structural complexity manifested by the two species?

One approach to answering this question is to emphasize the complex relationship that exists between an organism's physical structure and the genetic substrate that supposedly encodes aspects of that structure. Thus, we now recognize that genes participate in complex regulatory networks that, in addition to producing structural proteins, also serve to constrain and control the expression of specific genes via protein-based feedback mechanisms (see Kauffman, 1995). Genes do not, therefore, seem to encode directly for specific aspects of physical structure; instead, they participate in the creation of complex networks of feedback and feedforward influences that, in conjunction with other factors, contribute to much of the biological complexity that we ultimately observe. Commenting on the surprising similarity of gene numbers between species, Buchanan (2002) points out that genes encode for proteins and it is these proteins, interacting in complex webs of causal influence, that determine the differences between species. In order, to understand the real role and function of genes therefore, one needs to adopt a perspective that is specifically geared to understanding the complexity of network systems:

"To comprehend what makes us alive, and especially what distinguishes us from plants, will require insight into the architecture of this vast network; our sophistication is not due to one or another protein, but to the delicate design of the entire network." (Buchanan, 2002; pg. 16)

The analogy with the spider's web building behaviour is thus revealed. In both cases what we seem to confront is the presence of a core biological resource (neuronal or genetic) whose function it is to create networks of causal influence (some of which operate in the manner of a closed-loop feedback control system). Such networks, in conjunction with the core biological resource, realize functions whose complexity far outstrips that made possible by the initial encodings or (in the case of the brain) computational processes. To see the core biological resource as causally-relevant to the final outcome (i.e. behaviour or phenotype) of the network in question is not, of course, incorrect, but it is important to give proper explanatory weight to the role played by the networks that extend beyond the boundaries of the core resource. And it is important, in both cases, to recognize the functional contributions of the neural and genetic resources for what they really are: mechanisms to create, maintain and exploit networks of causal influence that subtend a variety of organismic and extra-organismic resources. It is not possible, we suggest, to understand the proper function and significance of the core resource (genome and brain) in the absence of this network-oriented perspective, and we certainly cannot afford to restrict our scientific attention to these resources if we ever hope to understand how higher-level phenomena (such as biological structure and intelligent behaviour) are produced. For to divest these resources of their inter-relationships with the complex networks in which they participate (and often create) is to lose sight of something explanatorily vital in our quest to understand the contribution of those resources to the target phenomena of interest. It is lose sight of the fact the functional significance of neural and genetic resources is often determined by networks that extend far beyond the neural and genetic realm.

Puzzles, Papers and Human-Level Problems

The critic will, of course, have identified a particular problem associated with the foregoing discussion, namely that, at least in the case of arachnid behaviour, we have focused on a form of intelligent behaviour that is far removed from the traditional targets of cognitive scientific enquiry (e.g. the realm of deliberative thought, planning, complex problem-solving and so on). This we accept, although it is not always clear to what extent ostensibly simple forms of adaptive behaviour should always be regarded as essentially non-cognitive in nature (see note 7). In spite of this, it is important to show how the notion of vehicle externalism can be applied to behaviours that are less controversially construed as cognitive. In the current section, therefore, we introduce a few more examples of intelligent behaviour in which human-level cognitive capabilities seem to draw on a variety of causal influences distributed across brain, body and world.

Consider first the case of multiplying two three digit numbers. A purely internalist account of how we are able to multiply the two numbers might emphasize how we first derive some symbolic encoding of the visual (or auditory) input corresponding to the two numbers. It would then invoke a computational account according to which the inner symbols are manipulated in some way so as to achieve the correct mathematical outcome. Now contrast this with what is surely a more accurate (ecologically-realistic) picture of how we implement long multiplication in the real-world. This alternative picture involves the active manipulation of external symbols in such a way that the kind of problem confronting the biological brain is profoundly simplified. In place of purely inner computational operations we see a pattern of perception-action cycles in which single digit numbers are compared and intermediate computational results are stored in an external medium using (e.g.) pen and paper. This example, described in Wilson and Clark (2009), is a case of what we might call environmentally-extended computation or 'wide computationalism' (Wilson, 1994). It takes what is, ostensibly, an inner cognitive capability (an ability to do long multiplication) and shows how crucial aspects of the problem-solving process can be (and usually are) delegated to aspects of the external environment. Importantly, the human agent in this situation emerges as a cognitive agent that (by virtue of culturally-scaffolded educational regimes) is able to make best use of a number of external props, aids and artefacts in order to meliorate problem-solving. Such melioration often occurs as a result of the way in which physical actions are used to structure and restructure aspects of the local external environment. In most cases, the result of the environmental restructuring is to radically simplify or transform the kind of problem-solving process in which the biological brain must engage.

Moving beyond the case of long multiplication, we encounter a number of cases where real-world action has been accorded an important role in enabling human subjects to navigate complex (and perhaps otherwise intractable) problem domains (Kirsh, 2009; Kirsh & Maglio, 1994; Maglio et al., 1999). David Kirsh (1995), for example, suggests a mechanism by which we are able to achieve success in the game of Scrabble[11]. Cast as a purely internal process, the cognitive demands of Scrabble seem considerable, but our problem-solving performances in the real-world often circumvent these overheads by relying on physical actions that simplify the kind of problem we are confronted with. Thus in playing Scrabble, we typically engage in a process of active manipulation of the Scrabble tiles so as to construct spatial orderings and configurations that work in concert with the pattern matching and pattern completing capabilities of the human brain. Some initial (perhaps random) spatial orderings serve to prompt the recall of specific word candidates, and these can then be evaluated and extended by further letter juxtapositions and spatial configurations. What is important here, as elsewhere, is to recognize the important and powerful role that physical action and environmental structure plays with regard to the larger problem-solving process. In the case of Scrabble and other problem domains (see Kirsh, 2009), Kirsh and Maglio (1994) suggest that certain types of action play key roles in enabling us to solve the problem in question. They refer to such actions as *epistemic actions*. These are actions that enable us to make information available[12] in ways that meliorate some aspect of our problem-solving performances. And it is epistemic actions, Clark (2008) suggests, that occupy centre-stage in discussions about how extended cognitive systems are brought into existence on the back of our active physical engagement with the external world:

"... epistemic actions, I want to suggest, are paramount among the ways in which bodily activity yields transient but cognitively crucial extended functional organizations." (Clark, 2008; pg. 70)

As a final example of extended cognition in action (!), consider the process of writing an academic paper or report, such as the one that confronts you now. One view as to how we generate such artefacts might emphasize the role of purely inner resources in contributing to fully-formed thoughts, which are then serialized as words on paper. But this, of course, is seldom, if ever, how real academic texts get written. For better or worse, what generally tends to happen is that we start by writing down a few fragmentary thoughts and ideas, and these then prompt further thoughts and ideas. As the paper emerges, a variety of external resources, such as text and papers, often themselves heavily annotated with notes and marginalia, are continually consulted. As Clark (1997) argues:

"[the text] does not spring fully formed from inner cogitations. Instead, it is the product of a sustained and iterated sequence of interactions between my brain and a variety of external props. In these cases, I am willing to say, a good deal of actual thinking involves loops and circuits that run outside the head and through the local environment. Extended intellectual arguments and theses are almost always the products of brains acting in concert with multiple external resources. These resources enable us to pursue manipulations and juxtapositions of ideas and data that would quickly baffle the un-augmented brain." (pg. 207)

Note that what is important here is the way in which some of the environmentally-extended processing loops are deemed to be *constitutive* of the thought processes giving rise to the finished article. Thinking, on this view, is not something that occurs solely within the head; it is also something that can be spread across a variety of extra-neural and, indeed sometimes, extra-corporeal resources. Thinking, as with other types of cognitive processing, is sometimes literally extended into the world outside the head.

Extended Cognitive Systems

Our aim in this section has been to highlight the way in which some forms of intelligent behaviour seem to depend on the interaction of a variety of resources, including body morphology, environmental structure, and neural processing. In fact, all of the examples presented in this section are examples of what has been called 'non-trivial causal spread' (Wheeler, 2005; Wheeler & Clark, 1999). This is something which occurs whenever we encounter a phenomenon, which we initially took to be the product of a well-demarcated system, in fact turns out to involve the exploitation of more far-flung forces and factors. Whenever we have a case of non-trivial causal spread, we also have a case of explanatory spread; i.e. a relative expansion of our explanatory frameworks to account for the phenomenon in question. Such spread seeks to give explanatory weight to factors that we initially supposed were causally-irrelevant with respect to some target phenomenon. In cases where the target phenomenon is a cognitive process, then it makes sense to see the causally-active physical vehicles of the process as extending beyond the inner, neural realm. And, inasmuch as we equate the boundaries of a cognitive system with the physical limits of the mechanisms that comprise that system's cognitive processing routines, then cognition is, at least sometimes, not bounded by the traditional boundaries of skin and skull; it emerges as something that is perfectly able to extend beyond the head and seep into the world.

Of course, in order to make this radical-sounding claim stick, we need to do adequate justice to the notion that patterns of causal influence and dependence are sufficient to warrant a readjustment of cognitive system boundaries. It is not enough to claim that an external resource becomes part of the system simply because it exerts a causal influence on some aspect of system processing. What is needed is a clear understanding of *when* environmentally-situated cognition becomes a case of genuine cognitive extension. When, in other words, does some external tool or resource become incorporated into an agent's cognitive processing routines?

There are a number of ways to approach this problem (see Haugeland (1998) and Clark (2007b) for two related, but subtly different, accounts), but much clearly rests on the extent of functional integration between the candidate component and the larger system. We tend to recognize a functionally-unified system, we suggest, when the various components of that system participate in the realization of some goal or purpose it is the system's job to achieve. What seems to be important then in the case of cognitive extension is that we confront a set of distinct components (brain, body and worldly elements) that are connected together in such a way that their functional inter-operation makes them part of a functionally-integrated (yet internally differentiated) whole. In other words, what seems to be important is the specific way in which the components cooperate in the processing and exchange of information for the purposes of accomplishing some specific task or objective, a task that we typically identify as the responsibility of a specific agent (in most cases, an individual human agent). What makes something a part of an extended cognitive system, we claim, relates to the details of the functional connectivity and patterns of information flow and influence that characterize the inter-operation of the various system components. It is in precisely this sense that we conceive of an extended cognitive system as consisting in a network of heterogeneous elements, each of which makes a specific functional contribution to the shape and profile of the cognitive performances manifest by the larger system[13].

Given this characterization of an extended cognitive system as a coordinated pattern of information flow and influence between networked components, it should be obvious why we see a role for the network and information sciences as contributing to our understanding of extended cognition. For such sciences are ideally poised to inform our understanding of how various heterogeneous components can interact in highly complex, nested and non-linear ways in order to realize cognitive functions. In addition, such a role is perfectly commensurate with the role to which such sciences are already being applied in the areas of neuroscience, economics, ecology, cellular biology, organizational analysis and epidemiology (Barabasi, 2002; Buchanan, 2002; Watts, 2003). Few would argue, we suspect, with the claim that network sciences are relevant to the project of understanding how large-scale neuronal ensembles are able to give rise to cognitively-interesting phenomena[14]; our claim is simply that the analytic targets of network science will often have to encompass a much broader range of resources when it comes to understanding the profile of much (but not necessarily all) real-world cognition. In this respect, the application of network science to extended cognitive sciences is perfectly compatible with existing research efforts in the information and network sciences; it simply extends the traditional focus of analysis to a much broader range of material resources.

One might, of course, be inclined to point out that the extended networks we see in the case of extended cognitive systems are not like those we encounter in conventional forms of network scientific analyses, especially those focused on the neural domain. The networks associated with an extended cognitive system seem to include a broad range of disparate elements (brains, bodies, and external artefacts), and this makes such networks unlike those that are the typical focus of neuroscientific enquiry. Doesn't the heterogeneity of elements within such networks mitigate against network-based analysis, and shouldn't we perhaps try to understand the capabilities of the neural sub-systems independently of the other, bio-external, components.

We reject this claim for a number of reasons, not least because it is unclear whether the capabilities and performance profile of an extended cognitive system can be understood by a strategy of piecemeal decomposition and componential analysis (see discussion on emergent capabilities in the section entitled 'The Web-Extended Mind: A Thought (Provoking) Experiment'). Moreover, the heterogeneity of extended cognitive networks is, in our view, a reason why we should embrace network- and information-based scientific approaches. The components that may comprise an extended cognitive system are indeed wildly disparate and various. They may include simple textual cues and prompts, or they may involve specific cognitive artefacts, such as slide rules, compasses, and so on (see Hutchins, 1995a). In some cases, the external technological resource may participate in computational processes independent of the human agent (e.g. mobile devices or decision support systems), or the resource may not even be technological in nature (it may, for example, be a another human agent – see the section entitled 'Socially-Extended Cognition'). Such heterogeneity merits and perhaps even necessitates the analytic techniques and conceptual theorizing of disciplines whose empirical targets are those of patterns of information-based flow and influence in materially-abstract functional organizations. The information and network sciences are ideally poised to provide this kind of abstract, functional analysis of extended cognitive systems.

Another reason why we suggest the information and network sciences are relevant to the study of extended cognitive systems relates to the fact that we are not always solely interested in analysis. Part of our interest in understanding extended cognitive systems is to be able to engineer new systems, or at least engineer environments and resources in which cognitively-relevant mergers, interactions and alliances can be established. What we need to understand, as engineers, are the kinds of technologies that are apt for integration and incorporation into existing and sometimes novel cognitive routines. Some of this is, of course, the focus of existing and well-established scientific disciplines, such as the disciplines of Human-Centered Technologies and Human-Centered Computing (Norman, 1993, 1998). But in the case of our current profile of technological innovation and development, the sciences that deal with patterns of network-mediated interaction and influence have a special relevance. This is precisely because ours is an era in which information and communication networks, as well as a host of networked multimedia devices, are both pervasive and increasingly intertwined with our daily problem-solving practices and routines. If we are to exploit the power and potential of these new network-enabled environments then we need tools, techniques and ways of thinking that are inherently sensitive to the features of network systems. It is precisely for this reason that the information and network sciences are relevant to our effort to understand and engineer network-mediated forms of bio-technological intelligence.

THE EXTENDED MIND

The previous section highlighted the way in which certain types of intelligent behaviour and cognitive processing seem to include (as wholes do their proper parts) mechanisms that extend beyond the traditional biological borders of skin and skull. The specific claim was that, under at least some conditions, we are warranted in seeing cognition as, quite literally, extending into the extra-organismic environment. The argument as currently presented, however, might be seen as applying to a narrow subset of mental states and processes, relative to those that we typically associate with a human mind. In accounting for much of the behaviour of both ourselves and others we typically make reference to a set of common-sense, mentalistic terms (such as belief, desire, hope, fear, and so on), and these are seen as playing a genuine explanatory role in psychologically-interesting patterns of behaviour. Thus my action to retrieve a beer from the fridge is explained in terms of my 'desire' to drink a beer and my 'belief' that a beer could be found in the fridge. It is this kind of intentional characterization (the ascription of intentional mental states) that helps us make sense of (to understand) patterns of human behaviour – it enables us to gain a predictively and explanatorily potent toehold on patterns of behaviour that would otherwise be psychologically unintelligible to us. So the question that arises in the case of cognitively-extended systems is whether the notion of cognitive extension gains any purchase in the more ethereal domain of folk-psychological discourse (the strategy of explaining human behaviour with respect to mental states, such as belief and desire). Can the notion of cognitive extension, as currently presented, be extended to account for the mental states that are posited as causally-relevant to the psychological understanding of our everyday patterns of behaviour? Can we, in other words, extend the case of an environmentally-extended cognitive system to the more general case of an environmentally-extended mind?

It is here (perhaps not surprisingly) that the philosophical waters begin to run deep. Perhaps the most lucid and influential account of why we should take notions such as extended belief states seriously is provided by Clark and Chalmers (1998) in their classic paper, 'The Extended Mind'. Clark and Chalmers (1998) ask us to imagine two individuals: Inga and Otto, both of whom are situated in New York City. Inga is a normal human agent with all the usual cognitive competences, but Otto suffers from a mild form of dementia and is thus impaired when it comes to certain acts of information storage and recall. To attenuate the impact of his impairment on his daily behaviour, Otto relies on a conventional notebook which he uses to store important pieces of information. Otto is so reliant on the notebook and so accustomed to using it that he carries the notebook with him wherever he goes and accesses the notebook fluently and automatically whenever he needs to do so. Having thus set the stage, Clark and Chalmers (1998), ask us to imagine a case where both Otto and Inga wish to visit the Museum of Modern Art to see a particular exhibition. Inga thinks for a moment, recalls that the museum is on 53rd street, and then walks to the museum. It is clear that in making this episode of behaviour intelligible (or psychologically transparent) to us Inga must have *desired* to enter the museum, and it is clear that she walked to 53rd street because she *believed* that that was where the museum was located. Obviously, Inga did not believe that the museum was on 53rd street in an occurrent sense (i.e. she has not spent her entire life consciously thinking about the museum's location); rather, she entertained the belief in a dispositional sense. Inga's belief, like perhaps many of her beliefs, was sitting in memory, waiting to be accessed as and when needed.

Now consider the case of Otto. Otto hears about the exhibition, decides to visit the museum and then consults his notebook to retrieve the museum's location. The notebook says the museum is on 53rd street, and so that is where Otto goes. Now, in accounting for Otto's actions we conclude, pretty much as we did for Inga, that Otto *desired* to go to the museum and that he walked to 53rd street because that is where he *believed* the museum was located. Obviously, Otto did not believe that the museum was on 53rd street in an occurrent sense (Otto has not spent much of his life constantly looking at the particular page in his notebook containing museum-related facts); rather, he entertained the belief in a dispositional sense. Otto's belief, like perhaps many of his beliefs, was sitting in the notebook, waiting to be accessed as and when needed.

Clark and Chalmers (1998) thus argue that the case of Otto establishes the case for a form of externalism about Otto's states of dispositional believing. The notebook, they argue, plays a role that is functionally akin to the role played by Inga's onboard bio-memory. If this is indeed the case then it makes sense to see the notebook as part of the material supervenience base for some of Otto's mental states, specifically his states of dispositional belief (such as those involving museum locations). The main point of the argument is to establish a (potential) role for external artefacts in constituting the physical machinery of at least some of our mental states and processes. If, as Clark and Chalmers (1998) argue, the functional contribution of an external device is the same as that provided by some inner resource, then it seems unreasonable to restrict the material mechanisms of the mind to the inner, neural realm. It seems possible, at least in principle, for the human mind to occasionally extend beyond the head and into the external world.

Such claims are, understandably, disconcerting, and it is important that we understand the precise nature of the claim that is being made. One immediate cause for concern relates to the notion of functional equivalence between the inner (e.g. bio-memory) and outer (e.g. notebook) contributions. If we allow any form of externally-derived influence to count as part of the mechanistic substrate of the mind, then doesn't this cast the mechanistic net too widely? Don't we end up confronting cases that are so blatantly counter-intuitive that they undermine the very notion of the mind as a proper focus of scientific and philosophical enquiry? Consider, for example, the case where two people have a conversation on the bus. Does this mean that their respective minds have merged into one integrated whole? And what about cases where we have some very loose coupling with an external information source, say the kind of access we have to information in a conventional textbook? Clearly, not all of the technologies or external resources that we encounter are apt to engage in the kind of bio-technological hybridization envisioned by the extended mind hypothesis. As Clark (1997) argues:

"There would be little value in an analysis that credited me with knowing all the facts in the Encyclopaedia Britannica just because I paid the monthly installments and found space for it my garage" (pg. 217).

Similarly, we suggest, it would be foolish to equate my personal body of knowledge and beliefs as co-extensive with the informational contents of the internet simply because I have an internet-enabled mobile phone. What, then, are the conditions under which we count a set of external resources as constituting part of an environmentally-extended mind? In answering this question, Clark and Chalmers (1998) embrace a particular set of criteria, ones that appeal to the accessibility, portability, reliability and trustworthiness of the external resource. The criteria are that:

1. "...the resource must be available and typically invoked" (Clark, in press-b). **[Availability Criterion]**
2. "...any information...retrieved from [the non-biological resource must] be more-or-less automatically endorsed. It should not usually be subject to critical scrutiny (unlike the opinions of other people, for example). It should be deemed about as trustworthy as something retrieved clearly from biological memory" (Clark, in press-b). **[Trust Criterion]**
3. "...information contained in the resource should be easily accessible as and when required". (Clark, in press-b) **[Accessibility Criterion]**

Clearly, such criteria serve to guide and constrain our intuitions about the kind of bio-artifactual and bio-technological couplings that are relevant to the formation of an extended mind. And they do so precisely because they delimit the range of situations under which we recognize the capabilities engendered by an external resource as being (most plausibly) that of a specific individual (or agent). In other words, what is important about the various criteria Clark and Chalmers (1998) propose is that they ensure that the capacities of an environmentally-extended, bio-technologically hybrid system are most plausibly seen by external observers (and perhaps by the agent themselves – see below) as the capacities and features of a particular agent. As Wilson and Clark (2009) suggest:

"We properly expect our individual agents to be mobile, more or less reliable, bundles of stored knowledge and computational, emotional and inferential capacities. So we need to be persuaded that the new capacities enabled by the addition of the notebook are likewise sufficiently robust and enduring as to contribute to the persisting cognitive profile we identify as Otto the agent. The bulk of Clark and Chalmers' (1998) work was an attempt to isolate and defend a specific account of the conditions under which we would be justified in identifying such an extended mind." (pg. 67).

What Wilson and Clark (2009) are suggesting here, we think, is not that the conditions cited in Clark and Chalmers (1998) (the conditions of trust, reliability, portability and so on) are necessary for *all* forms of cognitive extension. Instead, they are suggesting that the conditions apply in the specific case of the extended mind, and perhaps even here – although Wilson and Clark (2009) do not explicitly state it – they are really only relevant to the specific case of dispositional beliefs. What we seem to confront then is a set of what might generally be referred to as *coupling conditions*, conditions that determine when we are and when we are not justified in identifying cases of cognitive extension that apply to the realm of folk-psychological theorizing. In all cases of cognitive extension, we claim, what is important is a particular pattern of temporally fine-tuned information flow and influence within a networked ensemble of diverse resources. This network constitutes the mechanistic substrate of an extended *cognitive* system whenever the objective of that system, or the task in which it is engaged, is recognizably cognitive in nature (see note 7). However, this networked ensemble need not be permanent in nature. It can be a one-off organization that is assembled for the purposes of a specific cognitive task, or it can be a temporary but repeatable organization that is assembled to deal with an intermittent or periodically-occurring task (see Wilson & Clark, 2009). When the organization is more permanent, we approach the kind of conditions under which we count the external resource as constituting part of the material supervenience base associated an agent's daily patterns of psychologically-interesting behaviour. These are precisely the kind of conditions under which we are justified in seeing the emergence of an environmentally-extended mind.

THE WEB-EXTENDED MIND: A THOUGHT (PROVOKING) EXPERIMENT

Clark and Chalmers' (1998) original presentation of the extended mind thesis relies on a thought experiment involving a simple augmentative resource – a conventional notebook. It is perfectly correct and appropriate to ask whether this notebook is actually the kind of resource that could (in virtue of the kinds of human-artefact interaction it supports) fulfill the conditions for an extended mind. And, in fact, it is not clear that any actual notebook currently carried by a human agent could fulfill the criteria of portability, accessibility, reliability and so on, to the extent required. For all that, however, the main point of the notebook case was to highlight the mere possibility of an extended mind – it was not meant to suggest that most cases of notebook use actually result in genuine case of cognitive extension. But now that the notions of cognitive extension and the extended mind have been fleshed out, we can dispense with such technologically low-grade examples and focus our attention on the role played by the rich variety of emerging technologies and resources that we see in today's hi-tech environment, most of them relying, in one form or another, on complex networks of information exchange, distribution and transformation. To what extent do ubiquitous modes of network-mediated information access, as well as portable devices and wearable computers, contribute to the technological realization of extended cognitive systems and the possibility of environmentally-extended minds?

To pursue this notion in the context of our own research we have posited an extension to the original thesis of the extended mind. The thesis is called the thesis of network-enabled cognition (Smart, Engelbrecht, Braines, Hendler, & Shadbolt, 2008) (or more recently the thesis of the network-extended mind), and it makes a specific claim about the role of network systems in constituting some parts of an extended computational system, one that is capable of implementing cognitive operations and contributing to the realization of certain contentful mental states. The thesis is as follows:

Thesis of the Network-Extended Mind: The technological and informational elements of large-scale information and communication networks can, under certain circumstances, constitute part of the material supervenience base for (at least some of) an agent's mental states and processes.

Clearly, one of the things to be assessed in evaluating this thesis is whether the kinds of technologies and resources that are being made available as a result of recent research and development in the electronics and computer science domains are sufficiently well-suited to meet the kind of criteria that Clark and Chalmers (1998) insist are important. In some of our recent work we have examined this claim with regard to our (currently) best example of a large-scale networked information environment, namely the World Wide Web (Smart, Engelbrecht, Braines, Strub, & Hendler, 2009). What emerges from this analysis (see also Smart et al., 2008) is that, in many cases, the general trend of technological evolution is suitably well-aligned with the kind of criteria proposed by Clark and Chalmers (1998). Thus, in terms of concerns about portability we highlighted the fact that the current-state-of-the-art in mobile computing devices has already given us devices that are at least as portable as the conventional notebook in Clark and Chalmers' (1998) discussion. Moreover, in terms of the accessibility of information content, it is significant, that the focus of many research and technology efforts, particularly in the context of the World Wide Web, are geared towards improving user access to online information. Work of particular note here includes the development of natural language question-answering systems (Lopez, Pasin, & Motta, 2005; Tablan, Damljanovic, & Bontcheva, 2008), user-friendly semantic information browsers (schraefel et al., 2005), the use of sub-vocalization techniques to support Web navigation (Jorgensen & Binsted, 2005) and the use of intelligent forward caching and data charging mechanisms to mitigate download delays and the effects of intermittent network connectivity (Cherniack, Franklin, & Zdonik, 2001). New technologies in the field of wearable computing are also likely to enhance our access to information. Mobile device eyewear systems[15], for example, display information directly to a user's visual field using conventional eyewear equipment (e.g. spectacles). Some of the applications envisioned for this new technology include location-aware social network services, real-world visual overlays for environment navigation, battlefield situation awareness displays, and immersive virtual reality systems for education and entertainment. Such systems tend to reduce the cost of information access[16], and, we argue, they introduce new ways in which network-accessible information content can be co-opted into the information processing loops of cognitively-extended agents.

It is also important to note (and this is where our philosophical interests start to converge with our own scientific research programs) that as we move forward into an era of next-generation Web technologies, we are witnessing a move away from document-centric modes of information encoding to more data-centric modes. Document-centric modes of information encoding are those typically encountered on the conventional Web where task-relevant information is often embedded in resources such as Web pages, often surrounded by (in many cases) irrelevant or redundant information. Think about the problem of accessing factual information from a web-accessible resource, such as Wikipedia. Even if the delays associated with document retrieval (i.e. downloading) and presentation are resolved, the user is still confronted with the onerous task of surveying the document for relevant information content. In most cases this requires the user to scroll through the web page and process large amounts of largely irrelevant content in order to identify the small amount of information that is actually needed. This is a very inefficient means of information access. Even if the user tries to isolate specific information items for use on multiple occasions they cannot do this without reliably fixing the physical location of the information (perhaps by copying the required information to a local resource[17]).

What is important for the emergence of network-extended minds, we suggest, are flexible modes of data integration, aggregation and presentation, in conjunction with an ability to gear information retrieval operations to suit the task-specific needs and requirements of particular problem-solving contexts. Such capabilities are being progressively unleashed by new approaches to data representation and information access on what is (presently) the core technological infrastructure of the conventional World Wide Web. Thus notions such as the Semantic Web and Linked Data[18] initiatives (Berners-Lee *et al.*, 2006; Berners-Lee, Hendler, & Lassila, 2001; Shadbolt, Hall, & Berners-Lee, 2006) countenance an approach to data modelling and representation that is largely independent of specific presentational formats or usage contexts. Commenting on the relationship of the Semantic Web to the conventional Web, Berners-Lee et al (2006) write:

"The SW [Semantic Web] tries to get people to make their data available to others, and to add links to make them accessible by link following. So the vision of the SW is as an extension of Web principles from documents to data." (pg. 18)

This shift of emphasis (from linked documents to linked data) is, we suggest, an important milestone in enabling the kind of selective data integration, aggregation, and filtering that undergirds the emergence of cognitively-extended systems and the mechanistic realization of extended mental states.

To make this vision a little more concrete, we present a thought experiment involving a near-future case of Web-mediated information access in the context of a fully interactive (in the sense of extended Web 2.0 capabilities), linked data web environment. Imagine that our future (in our case cognitively unimpaired) human agent is equipped with a mobile networked device (a mobile phone will do), an information presentation device (such as the aforementioned mobile eyewear devices, or the memory aids being developed by the Memory Glasses project at the Massachusetts Institute of Technology[19]), and a means of controlling information access and navigation in a simple and effective manner (for the sake of argument imagine an advanced form of the electromyographic, electroencephalographic and electrooculographic interfaces being developed by a variety of academic and commercial organizations (Mason, Bashashati, Fatourechi, Navarro, & Birch, 2007; Nicolelis, 2001; Pfurtscheller, Scherer, & Neuper, 2007; Stix, 2008))[20]. Thus equipped, our future agent is able to retrieve information from the Web, on demand, in a manner that is delicately geared[21] to shaping, influencing and constraining ongoing sequences of thought and action. Our subject could, for example, be guided as to the location of interesting spatial targets by the use of simple geo-registered directional indicators overlaid onto the visual field. Our subject would not, therefore, have to rely on bio-memory to recall facts, such as the location of the Museum of Modern Art, because location-aware services would retrieve and present this information in a way that would serve to guide ongoing behaviour. Similarly, imagine that our subject has an interest in baseball and that baseball facts and figures are continually posted on the Web in a form that permits flexible forms of retrieval, combination, aggregation and inference (e.g. using the Resource Description Framework[22] or Web Ontology Language[23]). In this situation, our subject would be able to retrieve any piece of baseball-related information, on demand, in a manner that is robustly and continuously available. What, we might wonder, would our scientific, social and (indeed) subjective intuitions be in such a situation? Would it be appropriate for us to say that the subject pretty much 'knows' everything that there is to know about baseball, at least in terms of the information that is posted on the Web. If this claim seems profoundly implausible or inappropriate to you, think for a moment about what it is that determines what you think you already know. What seems to determine whether we know or do not know something is not the fact that we are continuously, consciously aware of relevant facts and figures. What seems to count is more the kind of access we have to the relevant information, the fact that when we need to recall the information it is there, easily (and sometimes not so easily) made available to us by our bio-memory systems. But need our bodies of personal knowledge be so reliant on biologically-based modes of information storage? What if our access to externally-located information was just as reliably, easily and continuously available as the access provided by our own bio-memories? It this case, it seems, there is no principled reason to suggest that the external information would not count as part of your own personal body of knowledge and dispositional beliefs. As Clark (2003) argues:

"..it sometimes makes both social and scientific sense to think of your individual knowledge as quite simply whatever body of information and understanding is at your fingertips; whatever body of information and understanding is right there, cheaply and easily available, as and when needed." (pg. 42).

If this is indeed what it means to know something, then the epistemic implications of our future contact with network systems and resources could be significant. For in such situations the boundaries of what we know seems to be limited only by the accessibility we have to various sources of environmental information, and if that information consists in the sum total of human knowledge, as stored in some large-scale networked space, then the epistemic limits of the network-extended mind are of a scale and potential that surpasses anything we have yet seen in the course of human history. What might be the long-term effect of such a cognitively-extended system on our familiar notions of knowledge-guided competence? And what might be the effect of such forms of epistemic contact on our core notions of who and what we are?

One thing that is worth considering at this point is that a potential shift in our notions of knowledge-guided competence, as applied to other agents, might also be accompanied by a correlative shift in our own subjective impressions of ourselves. In order to make this idea intelligible, think for a moment about the light in a refrigerator. If we did not know better we might be inclined to say that the light in the refrigerator is always on. Indeed, whenever we open the door to check whether the light is on, the light is, in fact, on. It seems to us as though the light is continuously lit because it is lit whenever we choose to look at it. In a similar vein, our sense of the detail in a visual scene may be attributable to the fact that the details of the scene are always made available to us whenever we try to look for them (see Myin & O'Regan, 2009). Our sense of 'seeing all the detail' in a visual scene is not necessarily because all aspects of the scene are explicitly represented[24]; rather, our conscious experience of seeing all the detail stems from the fact that we can continually visit and revisit all aspects of the scene (by moving our head and eyes) whenever we feel the need to do so. Arguments such as this form part of an influential theory of our conscious experience (Noë, 2004; Noë, 2009; O'Regan & Noë, 2001), which emphasizes how our subjective perceptual experiences are dependent on an implicit knowledge of sensorimotor dependencies (knowledge or expectations concerning the effect of movement or change on sensory stimulation). The claim that we want to make here is that this approach to accounting for our conscious experience may also be relevant in accounting for what we 'feel' or 'sense' we do and do not know. In this case, our sense of what we know would be guided by our ability to make knowledge and information available whenever we choose to do so or are required to do so. In a way that is similar to our sense of the detail in a visual scene, we sense that we know something because the thing that is known can be easily accessed and co-opted into ongoing problem-solving sequences whenever it needs it to be so. The claim is that if, by virtue of our experiences, we come to learn that certain bodies of information and knowledge (perhaps past experiences) can be easily accessed at will, then we will genuinely *feel* as if though those bodies of knowledge and information are part of us, that they are part of *our* personal body of knowledge and experience.

This touches on an issue that we address in other work (Smart, O'Hara, Engelbrecht, Giammanco, & Braines, in press), namely the extent to which 'our' memories can be externally-located and perhaps even externally-manipulated. What we suggest is that our memories need not always be in the head, and if they are not in the head then they can be manipulated in a variety of ways. This notion of manipulation touches, of course, on the classic studies in false memory research (e.g. Loftus, 1997), but the implications are somewhat broader here. What we suggest is that if our personal memories are partially constituted by what is outside the head, then we open up opportunities for radical forms of re-personalization, experiential reprogramming and memory configuration. In the extreme case, imagine if your sense of what your memories are is partially constituted by your access to various sources of external information. Now imagine that if, after some head trauma (or perhaps deliberate neurological intervention), you lose all your bio-based memories. Now your memories are entirely constituted by what is made available to you by your external cognitive aids. If you wished, you could have someone manipulate the information contents of those aids and give you, what is in effect, a new set of memories!

Whether such claims can be substantiated or not is something that only future research and engineering efforts can address. For now, the main point of our argument is to highlight the mere logical possibility of some forms of cognitive extension, and to explore their implications in terms of our cognitive capabilities and potential. Such forms of cognitive extension are not necessarily contingent on any radically new forms of science fiction style neural rewiring or neural interfacing technology, as has been proposed in other parts of the scientific literature (e.g. Stix, 2008). They are, potentially at least, part and parcel of the current trend towards increasingly intimate forms of bio-technological merger with our best network-enabled devices and network-accessible information stores. They are, we might say, intermediate stopping points *en route* to our network-enabled cognitive destiny.

What this section has intended to show is that the notion of the extended mind is an important and powerful thesis when it comes to understanding the potential impact of new network-enabled technologies on our core notions of cognitive capability and knowledge-guided competence. The point is not lost on those who embrace an extended mind perspective. Thus Clark (2008) argues:

"...as we move toward an era of wearable computing and ubiquitous information access, the robust, reliable information fields to which our brains delicately adapt their inner cognitive routines will surely become increasingly dense and powerful, perhaps further blurring the boundaries between the cognitive agent and her best tools, props and artifacts." (pg. 41)

As engineers interested in technology-mediated modes of cognitive augmentation, we can and should strive to support the emergence of systems that meet the criteria for cognitive and mental extension. The philosophically-derived coupling conditions, in this case, provide a rough set of criteria as to the required performance characteristics of putative mind-extending technologies. Such criteria obviously need to be supported by future empirical studies regarding the specific kinds of information access that are required to motivate a shift in our social, scientific and (perhaps) subjective tendencies regarding the intentional characterization of behaviour. But the conditions clearly do provide a set of useful targets for future requirements analysis and requirements-driven technology development.

And what of the role of network scientific analyses and network-theoretic approaches in supporting the emergence of network-extended minds? We saw in the case of extended cognitive systems that network scientific analyses were merited by virtue of their potential to shed new light on the emergence, maintenance and operation of circuits supporting cognitive extension. Such merits are equally applicable when it comes to understanding the contribution of information and communication networks to network-extended minds. This is so even though the nature of the external resource (e.g. network-enabled device or network-accessible information resource) may be somewhat more dynamic and invested with greater computational potential as compared to the traditional kind of cognitive artefacts featuring as part of extended cognition/extended mind accounts. Another reason to embrace the network and information sciences in relation to the thesis of the network-extended mind concerns the contribution such approaches make with respect to the development and configuration of new network-enabled artefacts and networked environments. Given the potential for our minds to become partially constituted by external technological resources, it is surely important that we seek to design those technologies so as to deliver the best profile of cognitive performance capabilities and, in the case of our adversaries, limitations.

There is a one particular sense in which network-theoretic approaches are perhaps crucial to our understanding of the role of new technologies in building network-extended minds. It concerns the way in which the integration of external resources into a hybrid cognitive processing routine (one that straddles the biological and technological realm) sometimes results in the emergence of capabilities and competences that are not reducible to those of the constituent parts. In some of our most compelling cases of cognitive extension the incorporation of an external resource does merely result in the augmentation or enhancement of some well-established ability; it engenders entirely new forms of cognitive processing capability. One only has to think of the impact that written and spoken forms of language have had on our cognitive profile (see Clark, 2008; chapter 3) to appreciate the extent to which our cognitive potential can be transformed following certain forms of cognitive merger and integration. And it is here that network science plays another potentially significant role. For network science, as a specialized branch of complexity science, is concerned with themes of self-organization, emergence and systems-level thinking. This makes it ideally poised to deal with cases in which we cannot understand the abilities of an extended cognitive system by a process of piecemeal decomposition and additive reassembly. As Wilson and Clark (2009) point out:

"To understand the integrated operation of the extended-thinking system, created, for example, by combining pen, paper, graphics program, and a trained mathematical brain, it may be quite insufficient to attempt to understand and the combine the properties of pens, graphics programs, paper, and brains." (pg 73).

The reason why this is inappropriate is for the same reason that we do not try to understand the cognitive capabilities of the human brain by exclusively focusing our analysis on the processing potential of individual neurons. Instead, areas of scientific study like cognitive neuroscience embrace the principles of systems-level analysis and modelling. They do this precisely because the capabilities of large neuronal ensembles are not those of the individual elements that comprise the ensemble. We properly recognize, in this case, that the phenomena of interest – the ones concerning cognitive processing capabilities – emerge at the systems level. So too, when it comes to cases of cognitive extension. We should not necessarily assume that we can study the elements of the extended cognitive system in isolation from the complex webs of causal influence and informational exchange that effectively couple them into functionally integrated systemic wholes. For, in many of our most compelling cases of cognitive extension, the capabilities of the whole cannot be understood by a simple strategy of componential analysis. Network science, as a specialized branch of complexity science, should be at the heart of our effort to understand the actual and potential capabilities of network-based bio-technological organizations.

SOCIALLY-EXTENDED COGNITION

The discussion so far has focused on how external, technological resources may become integrated into extended cognitive systems centered on individual human agents. However, this discussion overlooks an important aspect of human cognition – the fact that it is often embedded in complex networks of social influence and interaction. What is the relationship between technologically-mediated forms of cognitive extension and forms of cognitive extension in which the external resources consist of other human agents?

In unravelling the notion of socially-extended cognition there are a number of different perspectives that might be taken[25]:

1. The first is that within a large-scale information and communication network environment we might see a variety of socially-derived information resources as contributing to individual forms of cognitive extension. This notion of extended cognition is, at best, a weak form of socially-extended cognition. It emphasizes the role that social interactions and collaborations play with respect to the development and maintenance of external, shared resources; however, the resources in question are little more than virtual surrogates, or stand-ins, for more direct forms of social contact and communication.
2. A stronger form of socially-extended cognition sees other agents as directly constituting the supervenience base for individual forms of cognitive extension. In this case, an individual human agent (X) would become so tightly coupled with another human agent (Y) from the perspective of some cognitive processing routine that Y would come to constitute part of the machinery associated with X's cognitive profile. Both would essentially become integrated into a single cognitive system. Whether the right kind of coupling relationship between the agents

could ever, in practice, be established is unclear, but some theorists, such as Tollefsen (2006) seem favourably inclined to such a view.

3. The strongest from of socially-extended cognition is what might be called the group mind or collective mind thesis. The idea here is that a group of human agents is so organized (with regard to the flow of information and influence between them) that the group itself becomes the bearer of genuine mental states.

Clearly, the third of these is the most contentious option, and few theorists seem inclined to support it[26]. Rather than try to review or progress the philosophical debate surrounding this issue, our aim, in the current section, will be to highlight a number of issues and observations that we see as most relevant to the future study of network-mediated cognitive processing involving multiple human agents.

Firstly, we suspect that the best way of thinking about socially-extended cognition is in terms of the role that contemporary and near-future network systems might have in coordinating the thoughts and actions of a group of problem-solving agents. One way of thinking about this is to consider our earlier claim that many episodes of externally-directed cognitive processing or intelligence are at least partly constructive in nature (recall the role of genes or a spider's 'brain'[27] in actively creating structures that subsequently contribute to much of the complexity we observe at the phenotypic or behavioural level). Can something like this vision be applied to the socio-cognitive realm, the realm where cognitive processes are distributed across a network of interacting human agents?

One way in which the notion might be unpackaged is by drawing attention to the way in which many cognitively-potent external resources are the creative result of the collective actions of multiple individuals. Thus, consider the mechanisms that lie at the heart of termites' abilities to construct termite mounds. Much of this ability seems to rely on the use of stigmergic processes (Bonabeau, 1999; Bonabeau, Dorigo, & Theraulaz, 1999), processes that serve to progressively structure and coordinate collective action via the presence of simple external cues. As one termite drops a mud-ball, it leaves a pheromone marker that encourages other termites to deposit mud-balls nearby. As the collection of mud-balls increases in size, so specific architectural structures begin to emerge as the result of collective, pheromonally-mediated behaviours (see Camazine et al., 2001; chapter 18). The key point about such examples of collective intelligence and self-organization is that they show how the collective actions of multiple individuals can serve to progressively structure the environment (or at least a key problem-solving resource) in a way that meliorates some aspect of individual or collective problem-solving[28].

Perhaps in the World Wide Web context, systems like Wikipedia are already good examples of this. Such systems highlight the role that networks (in this case physical, communication networks) play in enabling individual contributions to assemble complex resources that subsequently constrain, influence and shape the profile of individual (and perhaps collective) thought and action. Sometimes when we are confronted with such resources we are enabled to pursue cognitive goals that would be difficult, if not impossible, to accomplish by ourselves. An illustrative example of this may be the way in which scientific open access initiatives[29], in conjunction with global information networks, serve to facilitate creative insight and intellectual progress in the domain of scientific endeavour. As Stevan Harnad (1999) rightly notes, the Web allows us to accomplish something akin to 'scholarly skywriting' – scientific theories, thoughts, ideas, experimental results, and sometimes data, are made available in ways that are increasingly accessible to fellow academics and scientific colleagues. It is almost as if the outputs of scientific and intellectual enquiry were written in the sky for all to see.

One idea that we want to canvass here is that the key virtue of this mode of information distribution and dissemination is that it effectively establishes linkages between ideas, thoughts and concepts that would otherwise have been too widely separated to be linked by agents engaged in individual forms of reason-constrained thought and inference. Imagine, for example, that many of the scientifically-interesting ideas which we are capable of entertaining are the nodes in a complex network whose linkages correspond to the individual transitions in a reason-respecting chain of thought. Paths through this network of ideas would then correspond to the intellectual arguments or theses that flow from some set of initial ideas, assumptions or observations. Such models, while perhaps faithful depictions of the inference chains of classical expert systems, seem congenitally ill-suited to capturing much of our human potential for creativity and insight. Perhaps this is because such models overlook the fact that our intellectual excursions are not limited to logically-constrained trajectories through a space of scientific ideas (an idea space); instead, at least in some cases, we are able to jump around in this space by virtue of our exposure to the thoughts and ideas of others. This, in conjunction with our ability to combine slow, deliberative forms of rational thought with a capacity for analogical reasoning and abstract pattern matching, enables us to effectively form conduits or shortcuts[30] to distant parts of the idea space, parts that would have been too distant or disconnected to be linked by individual (and socially-unaided) modes of exploration and search. The vision, then, is one of networks enabling individuals to exploit and benefit from mechanisms of collective search, establishing new trails through a space of ideas, some of which may, on occasion, result in discontinuous steps forward in scientific thinking, innovation and discovery.

In addition to the role of networks in supporting the collective creation of cognitively-potent artefacts and resources, there is a body of empirical research that draws close attention to the role of network structures in influencing collective problem-solving abilties. This research seeks to illuminate the specific role that factors like network topology play in enabling groups of problem-solving agents to make effective decisions and discover optimal solutions to problems. In one study involving human subjects, Mason, Jones and Goldstone (2005) explored the effect of different network topologies (e.g. small-world, random, full-connected, etc.) on the ability of groups of people to correctly guess a randomly selected number between 1 and 100. On each trial of the experiment, subjects attempted to guess the target number and were provided with feedback about the correctness/accuracy of their own response, as well as the responses of their immediate neighbours in the network (i.e. the human subjects to which they were directly connected). Mason et al (2005) found that when subjects were given simple problems involving a single target number, the fully connected networks were most effective in enabling groups to collectively settle on the correct solution. However, when the problem was more complex and involved a three-peaked payout function (one optimal solution and two sub-optimal solutions), the networks with the longest average path lengths were the most effective in enabling groups to find the optimal solution. Summarizing these results, Goldstone, Roberts and Gureckis (2008) conclude:

"Problem spaces requiring substantial exploration may benefit from networks with mostly locally connected individuals. The problem with fully connected networks is that everybody ends up knowing the same information, and they thereby become too like-minded, acting like a single explorer rather than like a federation of independent explorers."

Similar results to these have been reported by Lazer and Friedman (2007), who conducted studies with synthetic agents, again using different network topologies. Their results suggest that when agents are dealing with complex problems, the more efficient the network is at disseminating information, the better the short-run performance of the system (relative to network structures that are less efficient at disseminating information). However, as the performance of the system is monitored across time, those network structures that are less efficient at disseminating information are able to deliver better performance outcomes. In essence, the more efficient networks are better at solving problems under heavy time-constraints; however, when temporal considerations are not so important, the less efficient networks are able to deliver better long-term performance outcomes.

We thus encounter strong support for the claim that networks supporting rapid information dissemination (small-world and fully-connected networks) are more suitable for what might be called simple or 'high-tempo' problems. This contrasts with the case where the problem to be solved is more difficult and can be tackled at a more leisurely rate. In this case, more locally-connected network structures may be preferable. The reason for sub-optimal performance (at least on difficult problems) in groups connected by low average path length networks (e.g. small-world networks) seems to centre on the group's tendency to prematurely settle on sub-optimal solutions – to be drawn into sub-optimal solution outcomes on the basis of initial shared information. Such results are of potential relevance to a number of findings in the social psychological literature. These include the phenomena of groupthink (Janis, 1982), production blocking (Diehl & Stroebe, 1987) and the common knowledge effect (Stasser & Titus, 1985)[31], all of which seem to be characterized by a group's inability to find optimal solutions based on some form of precipitant interaction or early information sharing.

The empirical results of Mason et al (2005) and Lazer and Friedman (2007) are important because they highlight two things about the role of networks in socio-cognitive processing. Firstly, the suitability of a particular network structure to enable a group of problem-solving agents to reach an optimal solution outcome may depend on both the nature of the task in which the group is engaged as well the structure of the solution landscape. Secondly, the differential effectiveness of the network structure in supporting certain group-level outcomes may be accounted for by variables, such as the rate of information dissemination, that depend as much on the dynamic, time-variant functional connectivity of the network, as they do its static, structural characteristics. In respect of this latter issue, note that just because a network structure supports rapid information dissemination this does not mean that the actual flow of information through the network must be necessarily rapid. Agents or nodes within the network can effectively modulate the speed with which information is transmitted by selectively ignoring information, or by only intermittently processing information (in fact this was precisely one of the manipulations employed by Lazer & Friedman (2007)). In human networks, there are clearly a variety of factors that might contribute to this rate of information spread. These include things such as the tendency to hoard information, willingness to cooperate, vulnerability to copying/transmission errors[32], and trust. Also, of course, in situations involving mobile ad hoc networks many nodes may be expected to have only occasional and intermittent connectivity, and this may effectively impede the spread of information between all network nodes.

The studies on socio-cognitive processing in group situations highlights the relevance of network scientific approaches to our understanding about how to analyze and engineer network environments so as to best support collaborative problem-solving and decision-making. There is clearly much more work to be done here, but one thing does seem relatively clear at this early stage: it is that the kinds of information and network-theoretic approaches we advocated in the case of individual forms of cognitive extension (i.e.

extended webs of information flow and influence spun around an individual human agent), are readily applicable to the study of systems in which the webs of information flow and influence subtend multiple agents. Whether one wants to refer to such systems as socially-extended cognitive systems, or group minds, is, to our mind at least, largely irrelevant (although much may depend on whether we recognize some higher-level agency to which cognitive states and processes can be readily ascribed – see note 26). What seems important is that cognitive processing can take place in group situations, and that much of it can be supported by features of the network structure that acts to mediate group interactions. In such situations, the tools, principles and techniques of information and network science are just as relevant to our ultimate understanding of the cognitive capabilities of social organizations as they are to our understanding of extended cognitive systems involving individual human agents.

EXTENDED COGNITIVE SYSTEMS AND MILITARY COALITIONS

In considering the possibility of cognitive extension in military coalition environments, we can discern two distinct ways in which cognition may be extended beyond the bounds of individual human agents. One of these forms of cognitive extension is centered on the individual human agent. It sees the cognitive capabilities of the human agent as, in part, realized by complex webs of information flow and influence between a variety of inner (biological) and outer (social, technological and informational) resources. This is the form of cognitive extension that is most commonly encountered in the philosophical and scientific literature, and it is the form of cognitive extension that has occupied us for most of the current chapter. There is, however, a second way in which cognitive processes may be extended beyond the biological borders of specific individuals. This is the form of cognitive extension that we encounter in cases of distributed cognition research (Hutchins, 1995a, 1995b; Tribble, 2005). It emphasizes the way in which cognitive processes inhere in the complex webs of information processing that connect multiple human agents with a variety of non-biological props, aids and artefacts. This form of multi-agent cognitive extension can be discriminated from individual forms in virtue of the emphasis placed on the larger socio-technical system in which much of the relevant cognitive processing is deemed to occur. Thus, while individual forms of cognitive extension focus on the individual human agent as the target system of interest, the distributed cognition movement tends to see the larger socio-technical system as the relevant unit of cognitive analysis. Relative to this larger system, the activities of individual human agents form part of a complex web of coordinated computational activity, one that serves to propagate and transform representations in ways that ultimately lead to coherent patterns of system-level behavior.

These two forms of cognitive extension are, we suggest, highly relevant to our understanding of coalition-based cognitive abilities. Although the multi-agent form of cognitive extension might, at first glance, seem more interesting and relevant from the perspective of military coalitions (not least because it affords an opportunity to see *entire* military coalitions as functionally integrated cognitive systems), we suggest that both forms of cognitive extension are, in fact, important foci of philosophical and scientific attention. The reason for this is that we see the global effectiveness of a military coalition as dependent (at least in part) on the cognitive capabilities of both individual soldiers and the wider socio-technical systems in which these soldiers are embedded. Increases in cognitive productivity at the individual level, as produced by cognitive extension, may be magnified many times once such human-centered extended systems are combined and integrated into larger webs of collective cognitive processing. In both cases (of individual and collective cognitive processing), our understanding of how to create, configure and maintain multiple types of networks in ways that best serve the information processing objectives of the larger coalition organization is of paramount importance.

In this section we review the opportunities and challenges for cognitive extension in military coalitions, focusing exclusively on the two forms of cognitive extension identified above. The section on 'Human-Centered Cognitive Extension' reviews issues and research associated with cognitive extension at the level of the individual soldier or warfighter; the section on 'Coalitions as Extended Cognitive Systems' explores the opportunities for cognitive extension at the level of entire military coalitions (or at least significant elements thereof).

Human-Centered Cognitive Extension

The notion of human-centered cognitive extension is simply the notion of cognitive extension that has occupied us for much of the current chapter. It is the idea that the physical machinery underpinning at least some of the cognitive capabilities of an individual human agent need not necessarily reside in the head of the human agent. In understanding how to support cognitive extension at the individual human level, we have argued that we should focus on the nature of the relationships between the human agent and network-accessible information resources. Thus, in order for human-centered extended cognitive systems to emerge, we need to ensure that the appropriate channels of information flow and influence are established between the human agent and the surrounding nexus of cognitively-relevant social, technological and informational scaffolding. One of the most important issues here concerns the *bi-directional* exchange of information between individual soldiers and other (non-biological) elements of the extended cognitive system. In particular, we need to ensure that the information provided by some external resource is sufficiently poised to guide response selection and response execution processes in adaptive and intelligent ways. Furthermore, the biological elements of the soldier-centered cognitive hybrid need to be appropriately interfaced with the non-biological elements such that the hybrid system can function as a single functionally-integrated cognitive whole. What this means, in practice, is the deployment of technologies that work in concert with the human agent – technologies that are sensitive and responsive to aspects of human psycho-biological functioning, and which are capable of adapting their functional profile to meet the problem-solving goals and objectives of the larger hybrid system. Research programs such as the DARPA-funded Augmented Cognition program and the recently announced Cognition and Neuroergonomics CTA both boast scientific and technology development goals that are directly aligned with these requirements.

One problem that seems particularly pertinent to the possibility of human-centered forms of cognitive extension concerns the way in which human agents are enabled to play an active role in the retrieval, structuring and transformation of information from non-biological sources. Thus, recall that in many cases of cognitive extension (see the earlier section on 'Cognitive Extension') what we seem to encounter are feedback-loops that involve the active manipulation of an external resource by a core biological agent. Recall, also, the thought experiment discussed in the section on 'The Web-Extended Mind: A Thought (Provoking) Experiment'. This thought experiment was intended to provide a vision of the impact of near-future technologies on our traditional notions of knowledge-guided competence at the scientific, social and (perhaps) subjective levels. But of all the technological elements described as part of that thought experiment, one element emerges as particularly problematic with respect to the current state-of-the-art. This is the way in which information retrieval operations (from bio-external media) are initiated and controlled by the human agent. In the thought experiment, we discussed the use of complex sensor devices that were sensitive to minute patterns of muscular or neural activity. However, the current functionality of such devices is limited, and it is not always clear that they can be used to good effect across different situations. In the military context, for example, soldiers are typically engaged in high intensity physical activity, and such activity interferes with the controlled and deliberate expression of both muscular and neural response profiles. Ongoing work within the DARPA-funded Augmented Cognition program, as well as the forthcoming Cognition and Neuroergonomics CTA, may help to address some of these issues (see Relevant Defence-Related Research Programs'), but, in the meantime, what other strategies might we pursue in order to support the retrieval and presentation of information in ways conducive to the emergence of network-extended minds?

One potentially relevant line of research here concerns the attempt to support context-aware modes of just-in-time information retrieval (Bahrami, Yuan, Smart, & Shadbolt, 2007; Rhodes & Maes, 2000). This research seeks to proactively present relevant information by monitoring specific aspects of the task or environmental context. Complementing this research effort is work in the ITA program that seeks to monitor and infer mission status information on the basis of both physical and contextual cues (Poltrock, Handel, Bowyer, & Waggett, 2008). Importantly, once we are able to detect features of the problem-solving context, we are able to proactively disseminate information to individual agents in ways that supports the effective realization of individual and collective problem-solving goals. Clearly, our ability to dynamically configure the physical network in a way that supports this mode of context-sensitive information distribution is of vital importance (see 'Coalitions as Extended Cognitive Systems'), as is our ability to create and exploit representations of (e.g.) coalition plans (Mott & Hendler, 2007) that could be used to control information distribution and adapt communication network topologies.

One concern in relation to network-mediated forms of information retrieval and presentation involves the notorious problem of information overload. In this sense, network access is both a boon and a burden. It is a boon inasmuch as it creates new opportunities for situation awareness and improved decision-making, but it is a burden inasmuch as it runs the risk of overwhelming the capacity of the individual human agent to adaptively exploit available information in the context of ongoing decision-making processes. There are a number of lines of research that might be pursued here. One strategy is to rely on the aforementioned mechanism of context-sensitive information retrieval to limit the amount of information that is presented to a user (Bahrami et al., 2007). Another is to rely on alert and notification systems that can be tailored to a user's specific goals, interests and concerns (Smart, Russell et al., 2009). There is also an important body of research that concerns the use of subliminal[33] cuing techniques to influence behavioural output (DeVaul et al., 2005). Such techniques are important because they provide a route to behavioural influence that does not involve conscious processing.

One question that we should ask in light of these ongoing research efforts is the extent to which the various technological add-ons, changes in information accessibility and so on, are genuinely enhancing or augmenting the cognitive capabilities of a particular human agent. The answer to this question is perhaps not quite as straightforward as it might initially seem, especially since there is nothing in the bedrock claims of the extended mind thesis to suggest that all cases of cognitive extension need to be uniformly beneficial from a performance perspective. Indeed, some commentators have suggested that network technologies may have a somewhat negative impact on human cognitive processing (e.g. Carr, 2008; Greenfield, 2003). Carr (2008), for example, bemoans the apparent impact the Web is having on his cognitive capabilities:

"As the media theorist Marshall McLuhan pointed out in the 1960s, media are not just passive channels of information. They supply the stuff of thought, but they also shape the process of thought. And what the Net seems to be doing is chipping away at my capacity for concentration and contemplation. My mind now expects to take in information the way the Net distributes it: in a swiftly moving stream of particles. Once I was a scuba diver in the sea of words. Now I zip along the surface like a guy on a jet ski" (Carr, 2008; pg 57).

Clearly, we should not assume that such anecdotal reports provide any insight into the Web's true effects on human cognitive functioning[34]. Nevertheless, the cautionary flavour of Carr's (2008) commentary is well taken, and, pending further research, we should perhaps be somewhat cautious of the kind of bio-technological unions we make ourselves susceptible to.

Aside from the potential negative effects of cognitive extension on human cognitive performance, it is not always clear that technologically-mediated forms of cognitive extension should always be considered augmentative, even when the presence of such technologies seems to bolster cognitive performance. The reason for this, we suggest, is that the boundaries of the extended cognitive system are not the same as the boundaries of the individual human cognitive agent, and, inasmuch as the cognitive performances in question are attributed to the extended cognitive system (rather than the cognitive agent), it may be inappropriate to regard the capabilities of the human agent as significantly altered by the emergence of environmentally-extended cognitive circuits. Here we see an potential tension with regard to notions of cognitive agency and the physical machinery that supports cognitive processing. When the mechanisms supporting a particular cognitive performance extend beyond the biological boundaries of an individual human agent, then we arguably confront a genuine case of cognitive extension. However, it is not always clear, in such cases, that the cognitive capabilities of the larger, mechanistically-extended cognitive system should always be equated with those of the individual, biologically-bounded, human agent. Something along these lines may underlie the apparent confusion in the philosophical and cognitive scientific literature concerning the augmentative status of a number of cognitive technologies. Thus, while many commentators talk of external resources acting to augment or enhance human memory, Hutchins (1995b) suggests that we should not see such resources as enhancing the memory of individual human agents *per se*; rather, we should see the augmented capabilities as those of a new human-technology hybrid system. For example, in discussing the way speed bugs[35] contribute to memory functions in an airplane cockpit, Hutchins (1995b) argues:

"Individual pilot memory has not been enhanced; rather, the memory function has now become a property of a larger system in which the individual engages in a different sort of cognitive behavior... To call speed bugs a 'memory aide' for the pilots is to mistake the cognitive properties of the reorganized functional system for the cognitive properties of one of its human components. Speed bugs do not help pilots remember speeds; rather, they are part of the process by which the cockpit system remembers speeds." (pg. 282-283)

Such views serve to remind us that issues of cognitive extension cannot necessarily be divorced from ones of cognitive agency. In attempting to understand the extent to which the cognitive capabilities of agents are enhanced (or undermined) as a result of particular bio-technological mergers, we may need account for how the boundaries of specific cognitive agents are identified and how the cognitive capabilities of those agents get ascribed.

Coalitions as Extended Cognitive Systems

In addition to seeing cognitive extension as something that can take place at the individual, human agent level, it is also possible (on occasion) to see much larger systems, comprising multiple agents and material artefacts, as extended cognitive systems. The distributed cognition movement, for example, seeks to account for the performance of large-scale socio-technical systems in terms of the interactions of multiple human agents with a surrounding nexus of social, technological and informational resources (Hutchins, 1995a, 1995b; Tribble, 2005). Within such systems, the cognitive capabilities of the individual human agent are important but, by themselves, they are often inadequate in terms of accounting for the real targets of scientific enquiry: the systemic cognitive properties of the larger system.

The notion that large-scale socio-technical systems may be seen (and analyzed) as cognitive systems in their own right has been championed by the cognitive anthropologist Edwin Hutchins. In his studies of both ship navigation (Hutchins, 1995a) and airplane piloting (Hutchins, 1995b), Hutchins makes it clear that the proper focus of cognitive scientific attention is not always the individual human agent; often it is the larger system of social and technological resources in which the human agent is embedded. In cases where we seek to understand the cognitive capabilities of a single human agent, of course, this perspective reduces to the case of individual forms of cognitive extension (cases where the cognitive capabilities of an individual human agent are of primary interest). But in a range of cases the cognitive system in question is composed of a larger aggregation of human agents working together in support of some common or shared goal, and it is in these cases that a consideration of entire military coalitions as extended cognitive systems seems most appropriate.

Of course, just because we confront a system in which cognitive processing is distributed across a much broader nexus of resources than is the case for individual forms of cognitive extension, this does not mean that the kind of coupling conditions we saw as relevant in the case of individual forms of cognitive extension are not equally important in the case of more distributed cognitive systems. Recall our core claims about the importance of functional integration in the case of human-centered extended cognitive systems:

What seems to be important then in the case of cognitive extension is that we confront a set of distinct components (brain, body and worldly elements) that are connected together in such a way that their functional inter-operation makes them part of a functionally-integrated (yet internally differentiated) whole. In other words, what seems to be important is the specific way in which the components cooperate in the processing and exchange of information for the purposes of accomplishing some specific task or objective, a task that we typically identify as the responsibility of a specific agent (in most cases, an individual human agent). What makes something a part of an extended cognitive system, we claim, relates to the details of the functional connectivity and patterns of information flow and influence that characterize the inter-operation of the various system components. It is in precisely this sense that we conceive of an extended cognitive system as consisting in a network of heterogeneous elements, each of which makes a specific functional contribution to the shape and profile of the cognitive performances manifest by the larger system.

Much the same can be said when we confront a large-scale distributed cognitive system. Although it is not always clear that we can talk of such distributed systems as cognitive agents (at least in the same kind of way that we talk of human beings as cognitive agents), it does seem that the same kind of functional integration and coordination of the various system components is important to the cognitive outputs of the system. And just as we advocated the use of network scientific approaches in the case of human-centered cognitive extension, so it is important, we argue, to apply the tools and techniques of the information and network sciences to the case of large-scale distributed cognitive systems, ones in which a variety of biological and technological elements cooperate in producing globally-coherent patterns of systemic behavior.

One reason why network-based analyses are important when it comes to military coalition systems is that such systems are typically seen as composites of multiple interacting and interconnected networks (i.e. a network of networks). The relationships between human agents, for example, may be seen as forming one kind of network (e.g. a social network), while the relationships between elements of the physical communication infrastructure may be seen as forming a different kind of network (i.e. a communication network). Importantly, the various types of network one sees in a coalition environment interact in highly complex ways, and we may expect the systemic cognitive capabilities of a coalition-based cognitive system to depend greatly on the adaptive alignment between the disparate networks. Inasmuch as systemic cognitive performances are influenced by the structure and dynamics of the various networks comprising a military coalition system, then such networks can, we suggest, be seen as candidate elements of the cognitive machinery for coalition-based distributed cognitive systems.

Given the interdependencies between (e.g.) information, communication and human networks in coalition environments, we expect to see an important role for studies that shed light on the adaptive configuration of such networks throughout the course of coalition deployments. One, relatively simple, example of this is the case where the we seek to change the topology of physical communication networks in order to promote the appropriate exchange and transfer of information between spatially-distributed coalition elements. Another example is the case where we seek to modify the human social network in order to bring people with different (albeit related) bodies of knowledge and expertise together in order to solve some specific problem (see Huang, Contractor, & Yao, 2008). In all cases, what seems to be important is an ability to dynamically configure the structure and activity of multiple networks so as to best support the realization of organization-level or system-level goals.

The importance of dynamic configuration and functional coordination is recognized by those working in the area of distributed cognitive systems:

"In distributed cognition, one expects to find a system that can dynamically configure itself to bring subsystems into coordination to accomplish various functions." (Hollan et al., 2000; pg. 175)

Interestingly, this is a view that is echoed by the cognitive scientist, Marvin Minsky, in his book *The Society of Mind* (Minsky, 1986). Minksy argues that the human mind can be seen as a large system of experts or agencies that are dynamically assembled together in various ways in order to accomplish specific cognitive tasks. Of course, one issue that is particularly difficult to resolve here concerns the mechanisms that support the dynamic configuration of networks, or network elements, in ways that best support the realization of cognitive goals. Ideally, network structures within military coalition environments should be capable of *automatic* modes of adaptation in order to ensure that the various elements of the coalition network are functionally aligned with respect to force-level objectives. What kind of system could support such automatic modes of adaptive configuration in coalition networks?

Perhaps one way of answering this question is to turn to biology and examine the kind of solutions that nature has derived for managing information flows in large-scale network systems. Van Essen, Anderson and Olshausen (1994), for example, posit the existence of 'control neurons' in the brain of mammalian nervous systems whose function is to regulate information flows between various neural processing resources. Such regulation constitutes, they argue, a key mechanism by which we are able to adaptively focus attention on specific subsets of environmental information and efficiently organize action output in the face of competing motor commands. Their analogy is with the division of labour in a large-scale commercial organization in which the key focus of organizational activity is not the actual generation of the final product *per se*, but rather the internal trafficking of information and materials. Perhaps, therefore, some of the insights gleaned from this, and other nature-inspired solutions to the problem of automatic network configuration, could be applied to case of network configuration in military coalition systems.

Other challenges posed by a consideration of coalition-level forms of cognitive extension arise from the cultural, linguistic and technological differences between coalition force elements. One challenge, for example, relates to the need to ensure a common (or shared) understanding of informational cues against a backdrop of community-specific interpretational biases and linguistic conventions (see Smart, Huynh et al., 2009). A lack of shared understanding may compromise the functional integration of coalition system components and thereby contribute to a breakdown in collective cognitive processing. Indeed, a sufficient level of shared understanding may be deemed as one of the factors that determines whether the coalition formation can operate as an extended cognitive system. Similar threats to functional integration stem from problems associated with information exchange, trust and technological compatibility.

RELEVANT DEFENCE-RELATED RESEARCH PROGRAMS

A number of defence-related research programs feature research that is relevant to cognitive extension in network-enabled contexts. Of course, military research has often been at the forefront of efforts to develop technologically-mediated forms of cognitive augmentation and enhancement. This is reflected in the efforts to develop sophisticated wearable computing devices as part of the Land Warrior and Future Force Warrior programs (see Ashok & Agrawal, 2003). More recently, one sees efforts to develop and exploit the kind of portable projection systems that present information directly to an individual's field of view. This work, being undertaken as part of the DARPA-funded ULTRA-Vis program[36], promises to deliver the kind of portable display systems that make network-accessible information better placed to influence ongoing sequences of thought and action.

Another strand of defence-related research concerns the attempt to augment human cognition by adaptively controlling the way in which information is presented to a user based on cognitive state information. The DARPA Augmented Cognition program, for example, seeks to monitor cognitive state using a variety of physiological sensors for the purposes of adapting display configurations and task commitments. Similar research goals are expressed by the recently proposed Cognition and Neuroergonomics CTA. This program seeks to:

"...enhance Soldier-system performance in complex operational settings by optimizing information transfer between the system and the Soldier, identifying mental processes and individual differences that impact mission-relevant decision-making, and developing technologies for individualized analyses of neutrally-based processing in operational environments." (United States Army Research Laboratory, 2008)

Such efforts promise to improve our understanding of the opportunities for cognitive mergers that involve various forms of bio-artifactual coupling. By focusing on ways to optimize the presentation of task-relevant information and adapt systems to work in concert with biological functioning these programs promise to yield valuable insights into the opportunities for cognitive extension at the level of individual human agents.

Research programs such as those mentioned above are most obviously concerned with the functioning of individual soldiers, and this makes them relevant to what we have referred to as human-centered forms of cognitive extension. There are, however, a number of research programs that seek to tackle issues related to cognitive performance in more 'distributed' contexts. One such program is the joint U.S./U.K. ITA program, which focuses on military coalition contexts and includes a specific research task devoted to issues of network-mediated cognitive extension. One strand of research in this task is to explore the relationship between network-level variables and collective cognitive performances. The studies reviewed in the section on 'Socially-Extended Cognition', specifically those of Mason et al (2005) and Lazer and Friedman (2007), attest to the importance of network-level considerations in understanding collective problem solving, but such studies possess a number of limitations that undermines their applicability to the military coalition environment. One of these limitations concerns the effect of different network structures (small-world, fully-connected, etc.) and patterns of information flow in different types of task context. The studies of both Mason et al (2005) and Lazer and Friedman (2007) focus on a particular type of task involving parallel search in a fixed solution space. But the military domain features many different types of task (e.g. hierarchical planning), and some of these tasks mandate that specific agents (or agent teams) are assigned to different aspects of a more general problem. What profile of network connectivity and information flow best supports collaborative problem-solving in these more systematically-structured task contexts? A second limitation of the studies of Mason et al (2005) and Lazer and Friedman (2007) concerns the use of fixed solution spaces. In this case, the assumption is that the correct solution is the same throughout the entire course of the experiment, but in real-world environments the most appropriate solution outcome may be linked to a dynamic and evolving situational context. As the situation changes, the nature of the correct solution may also change. In light of this, it is important to study the effects of variables like network topology in problem-solving domains that feature dynamic solution spaces. Thirdly, the studies of Mason et al (2005) and Lazer and Friedman (2007) employ static network structures whose topology does not change throughout the course of the problem-solving exercise. Such static networks are unlike those typically encountered in military coalition operations where there is an increasing move towards mobile ad hoc networks with dynamically changing topologies. In order to extend the results of Mason et al (2005) and Lazer and Friedman (2007) with respect to dynamic network environments, research in the ITA program is currently exploring the effect of dynamic changes in network topology on collective problem-solving performance in simulated agent communities (see Smart, Sycara, & Huynh, in press).

Another research program where we see a potential emphasis on coalition-level cognitive outcomes is the Network Science CTA. This CTA aims to undertake fundamental research in network science in order to understand the complex interplay between a variety of types of networks (e.g. social/cognitive, information and communication networks). Such research goals are clearly relevant to our understanding of how complex networks of informational, technological and social resources can be dynamically adapted in order to support cognitive processing at the collective, multi-agent, level.

Finally, the THINK ATO is a research program that aims to enhance warfighter cognitive performance in complex dynamic network environments. It was established as a collaborative venture between the U.S. Army Research Laboratory, the U.S. Army Research Institute, and the Communications and Electronics Research, Development and Engineering Center (CERDEC). It is geared towards making best use of available networked knowledge and information in order to support individual warfighter cognitive performance, as well as collaborative decision-making and distributed problem-solving. Clearly, such objectives coincide nicely with those of research into network-mediated forms of cognitive extension at both the individual and collective levels.

CONCLUSION

The traditional view in cognitive science is that cognition is inside the head of individual human agents. In contrast to this view, the notion of cognitive extension maintains that, at least in some situations, cognition is extended beyond the traditional biological borders of skin and skull. This latter view draws on an emerging wealth of empirical data concerning in way in which the facts of material embodiment and environmental embedding contribute to the emergence of cognitive processing routines that are distributed across the brain, body and world. In this chapter, we have suggested that notions of cognitive extension can be used to understand the transformative potential of a variety of network-enabled devices and network-accessible information resources on human cognitive processing. We have also proposed an extension to the original extended mind thesis, one that specifically caters for the potential role of network systems in extending the bounds of human cognition.

Our review of the literature relating to cognitive extension and the extended mind has highlighted a number of ways in which the information and network sciences are relevant to our understanding of extended cognitive systems. These include, but are not necessarily limited to, the following:

1. Extended cognitive systems consist of networks of information flow and influence between a variety of heterogeneous resources. Network science is well suited to assist us with the project of understanding how extended cognitive circuits operate with respect to the cognitive capabilities of the larger systemic organization.
2. Network-based approaches are merited in the specific case of network-extended minds because physical networks are at the heart of contemporary technology-mediated forms of cognitive extension. Our future attempts at engineering network-extended cognitive systems, or at least enabling them to emerge, will be dictated by our ability to develop and configure network technologies in ways that expand our human cognitive potential.
3. As a specialized branch of complexity science, network scientific approaches can help us understand the emergent capabilities of extended cognitive systems. Given the complex, nested and non-linear interactions between the components of an extended cognitive system, the capabilities of the larger system are not always guaranteed to be mere augmentations or enhancements of some existing capability; they can sometimes be entirely new forms of cognitive capability and competence

The application of network scientific approaches to both the analysis and engineering of extended cognitive systems is relevant to military coalition operations because such approaches help us understand the factors that contribute to the efficiency and quality of problem-solving processes in collaborative, network-enabled, distributed teams. By developing a better understanding of the cognitive impact of network systems on both individual and collective problem-solving, we are in a much better position to engage in interventions that enhance the cognitive power and potential of military coalition formations.

Of course, the possibility for network-mediated forms of cognitive extension is not something that is relevant just to military coalitions; the increasing ubiquity and pervasiveness of network systems motivates a more general interest in the effect of network technologies on our human cognitive potential. As Hollan et al (2000) comment:

"As we build richer, more all-encompassing, computational environments it becomes more important than ever to understand the ways human agents and their local environments are tightly coupled in the processing loops that result in intelligent action." (pg. 186)

The advent of new computing technologies and network-enabled capabilities highlights a potential milestone in our human cognitive evolution. Just as the ability to use and exploit linguistic encodings marked a seachange in our individual and collective cognitive capabilities, so the development of ubiquitous network systems, wearable computing devices and pervasive computing, presents us with unparalleled opportunities for cognitive extension at both the individual and collective levels. Ours, we suggest, are 'fishnet' minds, ones that are increasingly enmeshed in complex networks of technological, linguistic and social influence. As we learn to exploit those networks for our cognitive good or ill, so too we must cast our philosophical and scientific explanatory nets ever wider. In this way we may, at last, come to see the human mind for what it really is. Not as some immaterial spirit stuff that emerges solely from the machinations of the human brain, but as a set of physical processes that occasionally escape their cranial confines and extend out into the world.

ACKNOWLEDGMENT

Research was sponsored by the U.S. Army Research Laboratory and the U.K. Ministry of Defence and was accomplished under Agreement Number W911NF-06-3-0001. The views and conclusions contained in this document are those of the author(s) and should not be interpreted as representing the official policies, either expressed or implied, of the U.S. Army Research Laboratory, the U.S. Government, the U.K. Ministry of Defence or the U.K. Government. The U.S. and U.K. Governments are authorized to reproduce and distribute reprints for Government purposes notwithstanding any copyright notation hereon.

REFERENCES

Adams, F., & Aizawa, K. (2001). The bounds of cognition. *Philosophical Psychology*, *14*(1), 43–64. doi:10.1080/09515080120033571

Adams, F., & Aizawa, K. (2008). *The Bounds of Cognition*. Oxford, UK: Blackwell.

Adams, F., & Aizawa, K. (2009). Why the mind is still in the head. In Ayded, M., & Robbins, P. (Eds.), *Cambridge Handbook of Situated Cognition*. Cambridge, UK: Cambridge University Press.

Adams, F., & Aizawa, K. (in press). Defending the bounds of cognition. In Menary, R. (Ed.), *The Extended Mind*. Aldershot, UK: Ashgate Publishers.

Ajith, A., Crina, G., & Vitorino, R. (Eds.). (2006). *Stigmergic Optimization*. Berlin, Germany: Springer-Verlag.

Ashok, R., & Agrawal, D. (2003). Next-generation wearable networks. *Computer*, *36*(11), 31–39. doi:10.1109/MC.2003.1244532

Bahrami, A., Yuan, J., Smart, P. R., & Shadbolt, N. R. (2007). *Context-Aware Information Retrieval for Enhanced Situation Awareness.* Paper presented at the Military Communications Conference (MILCOM), Orlando, Florida, USA.

Barabasi, A.-L. (2002). *Linked: The New Science of Networks*. Cambridge, MA: Perseus Publishing.

Berners-Lee, T., Hall, W., Hendler, J. A., O'Hara, K., Shadbolt, N., & Weitzner, D. J. (2006). A framework for web science. *Foundations and Trends in Web Science*, *1*(1), 1–130. doi:10.1561/1800000001

Berners-Lee, T., Hendler, J., & Lassila, O. (2001). The Semantic Web. *Scientific American*, *284*(4), 34–43. doi:10.1038/scientificamerican0501-34

Bird, J., & Layzell, P. (2002). *The evolved radio and its implications for modelling the evolution of novel sensors.* Paper presented at the Congress on Evolutionary Computation (CEC'02), Washington, DC.

Bonabeau, E. (1999). Editor's introduction: stigmergy. *Artificial Life*, *5*(2), 95–96. doi:10.1162/106454699568692

Bonabeau, E., Dorigo, M., & Theraulaz, G. (1999). *Swarm Intelligence: From Natural to Artificial Systems*. New York: Oxford University Press.

Braddon-Mitchell, D., & Jackson, F. (2007). *Philosophy of Mind and Cognition: An Introduction* (2nd ed.). Oxford, UK: Blackwell Publishing.

Brooks, R. A. (1991). Intelligence without representation. *Artificial Intelligence*, *47*, 139–160. doi:10.1016/0004-3702(91)90053-M

Buchanan, M. (2002). *Nexus: Small Worlds and the Groundbreaking Science of Networks*. New York: W. W. Norton & Company.

Burge, T. (1979). Individualism and the mental. In French, P., Uehling, T., & Wettstein, H. (Eds.), *Midwest Studies in Philosophy* (*Vol. 4*, pp. 73–121). Minnesota, MN: University of Minnesota Press.

Butler, K. (1998). *Internal Affairs: Making Room for Psychosemantic Internalism*. Dordrecht, The Netherlands: Kluwer Academic Publishers.

C. elegans Sequencing Consortium. (1998). Genome sequence of the nematode C. elegans: a platform for investigating biology. *Science, 282*(5396), 2012-2018.

Camazine, S., Deneubourg, J.-L., Franks, N. R., Sneyd, J., Theraulaz, G., & Bonabeau, E. (2001). *Self-Organization in Biological Systems*. Princeton, NJ: Princeton University Press.

Carr, N. (2008). Is Google Making Us Stupid? *Atlantic (Boston, Mass.)*, *302*(1), 56–63.

Cherniack, M., Franklin, M. J., & Zdonik, S. (2001). Expressing user profiles for data recharging. *IEEE Personal Communications: Special Issue on Pervasive Computing*, *8*(4), 32–38.

Clark, A. (1997). *Being There: Putting Brain, Body and World Together Again*. Cambridge, MA: MIT Press.

Clark, A. (1999). An embodied cognitive science. *Trends in Cognitive Sciences*, *3*(9), 345–351. doi:10.1016/S1364-6613(99)01361-3

Clark, A. (2001). Reasons, Robots and the Extended Mind. *Mind & Language*, *16*(2), 121–145. doi:10.1111/1468-0017.00162

Clark, A. (2003). *Natural-Born Cyborgs: Minds, Technologies and the Future of Human Intelligence*. Oxford, UK: Oxford University Press.

Clark, A. (2005). Intrinsic content, active memory and the extended mind. *Analysis*, *65*(285), 1–11. doi:10.1111/j.1467-8284.2005.00514.x

Clark, A. (2007a). Curing Cognitive Hiccups: A Defense of The Extended Mind. *The Journal of Philosophy, 104*(4), 163–192.

Clark, A. (2007b). Re-Inventing Ourselves: The Plasticity of Embodiment, Sensing, and Mind. *The Journal of Medicine and Philosophy, 32*(3), 263–282. doi:10.1080/03605310701397024

Clark, A. (2008). *Supersizing the Mind: Embodiment, Action, and Cognitive Extension*. New York: Oxford University Press.

Clark, A. (in press-a). Coupling, Constitution and the Cognitive Kind: A Reply to Adam and Aizawa. In R. Menary (Ed.), The Extended Mind. Aldershot, UK: Ashgate Publishers.

Clark, A. (in press-b). Memento's revenge: the extended mind, extended. In R. Menary (Ed.), The Extended Mind. Aldershot, UK: Ashgate Publishers.

Clark, A., & Chalmers, D. (1998). The Extended Mind. *Analysis, 58*(1), 7–19. doi:10.1111/1467-8284.00096

Claverie, J.-M. (2001). Gene number: What if there are only 30,000 human genes? *Science, 291*(5507), 1255–1257. doi:10.1126/science.1058969

Dawkins, R. (1982). *The Extended Phenotype: The Long Reach of the Gene*. Oxford, UK: Oxford University Press.

Dennett, D. C. (1996). *Kinds of Minds: Toward an Understanding of Consciousness*. New York: Basic Books.

DeVaul, R., Pentland, A., & Corey, V. (2005). *The memory glasses: subliminal vs. overt memory support with imperfect information.* Paper presented at the 7th IEEE International Symposium on Wearable Computers, White Plains, NY.

Diehl, M., & Stroebe, W. (1987). Productivity loss in brainstorming groups: toward the solution of a riddle. *Journal of Personality and Social Psychology, 53*, 497–509. doi:10.1037/0022-3514.53.3.497

Fodor, J. (1980). Methodological solipsism considered as a research strategy in cognitive psychology. *The Behavioral and Brain Sciences, 3*(1), 63–73. doi:10.1017/S0140525X00001771

Foelix, R. F. (1996). *Biology of Spiders*. New York: Oxford University Press.

Gibson, J. J. (1966). *The Senses Considered as Perceptual Systems*. New York: Houghton-Mifflin.

Goldstone, R. L., Roberts, M. E., & Gureckis, T. M. (2008). Emergent Processes in Group Behavior. *Current Directions in Psychological Science, 17*(1), 10–15. doi:10.1111/j.1467-8721.2008.00539.x

Gray, W. D., & Fu, W. T. (2004). Soft constraints in interactive behavior: The case of ignoring perfect knowledge in-the-world for imperfect knowledge in-the-head. *Cognitive Science, 28*(3), 359–382.

Greenfield, S. (2003). *Tomorrow's People: How 21st Century Technology is Changing the Way We Think and Feel*. London: Penguin Books Ltd.

Harnad, S. (1999). The future of scholarly skywriting. In Scammell, A. (Ed.), *I in the Sky: Visions of the Information Future*. London: Aslib.

Harnad, S., & Dror, I. E. (2006). Distributed cognition: Cognizing, autonomy and the Turing Test. *Pragmatics & Cognition, 14*(2), 209–213. doi:10.1075/pc.14.2.03har

Haugeland, J. (1998). Mind embodied and embedded. In Haugeland, J. (Ed.), *Having thought: Essays in the Metaphysics of Mind*. Cambridge, MA: Harvard University Press.

Hinsz, V., Tindale, R., & Vollrath, D. (1997). The emerging conceptualization of groups as information processors. *Psychological Bulletin, 121*, 43–64. doi:10.1037/0033-2909.121.1.43

Hollan, J., Hutchins, E., & Kirsh, D. (2000). Distributed cognition: toward a new foundation for human-computer interaction research. [TOCHI]. *ACM Transactions on Computer-Human Interaction*, 7(2), 174–196. doi:10.1145/353485.353487

Huang, Y., Contractor, N., & Yao, Y. (2008). *CI-KNOW: recommendation based on social networks.* Paper presented at the 9th Annual International Digital Government Research Conference, Montreal, Canada.

Hurley, S. (1998). *Consciousness in Action.* Cambridge, MA: Harvard University Press.

Hutchins, E. (1995a). *Cognition in the Wild.* Cambridge, MA: MIT Press.

Hutchins, E. (1995b). How a cockpit remembers its speeds. *Cognitive Science, 19*(3), 265–288.

International Human Genome Sequencing Consortium. (2004). Finishing the euchromatic sequence of the human genome. *Nature, 431*(7011), 931–945. doi:10.1038/nature03001

Janis, I. L. (1982). *Victims of Groupthink* (2nd ed.). Boston: Houghton Mifflin.

Jorgensen, C., & Binsted, K. (2005). *Web Browser Control Using EMG Based Sub Vocal Speech Recognition.* Paper presented at the 38th Annual Hawaii International Conference on System Sciences (HICSS'05), Big Island, Hawaii.

Kauffman, S. (1995). *At Home in the Universe: The Search for Laws of Self-Organization and Complexity.* London: Penguin Books Ltd.

Kirsh, D. (1995). The intelligent use of space. *Artificial Intelligence, 73*(1-2), 31–68. doi:10.1016/0004-3702(94)00017-U

Kirsh, D. (1996). Adapting the environment instead of oneself. *Adaptive Behavior, 4*(3/4), 415–452. doi:10.1177/105971239600400307

Kirsh, D. (2006). Distributed cognition: A methodological note. *Pragmatics & Cognition, 14*(2), 249–262. doi:10.1075/pc.14.2.06kir

Kirsh, D. (2009). Problem solving and situated cognition. In Ayded, M., & Robbins, P. (Eds.), *Cambridge Handbook of Situated Cognition.* Cambridge, UK: Cambridge University Press.

Kirsh, D., & Maglio, P. (1994). On distinguishing epistemic from pragmatic action. *Cognitive Science, 18*, 513–549.

Krink, T., & Vollrath, F. (1997). Analysing Spider Web-building Behaviour with Rule-based Simulations and Genetic Algorithms. *Journal of Theoretical Biology, 185*(3), 321–331. doi:10.1006/jtbi.1996.0306

Krink, T., & Vollrath, F. (1998). Emergent properties in the behaviour of a virtual spider robot. *Proceedings of the Royal Society of London. Series B. Biological Sciences, 265*(1410), 2051–2055. doi:10.1098/rspb.1998.0539

Krink, T., & Vollrath, F. (1999). Virtual Spiders Guide Robotic Control Design. *IEEE Intelligent Systems, 14*(5), 77–84. doi:10.1109/5254.796093

Lazer, D., & Friedman, A. (2007). The Network Structure of Exploration and Exploitation. *Administrative Science Quarterly, 52*(4), 667–694. doi:10.2189/asqu.52.4.667

Loftus, E. (1997). Creating false memories. *Scientific American, 277*(3), 70–75. doi:10.1038/scientificamerican0997-70

Lopez, V., Pasin, M., & Motta, E. (2005). *AquaLog: An Ontology-portable Question Answering System for the Semantic Web.* Paper presented at the 2nd European Semantic Web Conference (ESWC 2005), Heraklion, Greece.

Maglio, P., Matlock, T., Raphaely, D., Chernicky, B., & Kirsh, D. (1999). *Interactive skill in Scrabble.* Paper presented at the 21st Annual Conference of the Cognitive Science Society, Vancouver, British Columbia, Canada.

Mason, S., Bashashati, A., Fatourechi, M., Navarro, K., & Birch, G. (2007). A comprehensive survey of brain interface technology designs. *Annals of Biomedical Engineering, 35*(2), 137–169. doi:10.1007/s10439-006-9170-0

Mason, W. A., Jones, A., & Goldstone, R. L. (2005). *Propagation of innovations in networked groups.* Paper presented at the 27th Annual Conference of the Cognitive Science Society, Stresa, Italy.

Minsky, M. (1986). *The Society of Mind.* New York: Simon & Schuster, Inc.

Mott, D., & Hendler, J. (2007). *Progress on the Collaborative Planning Model.* Paper presented at the 1st Annual Conference of the International Technology Alliance (ACITA), Maryland, USA.

Myin, E., & O'Regan, J. K. (2009). Situated perception and sensation in vision and other modalities: a sensorimotor approach. In Ayded, M., & Robbins, P. (Eds.), *Cambridge Handbook of Situated Cognition*. Cambridge, UK: Cambridge University Press.

Newell, A. (1980). Physical symbol systems. *Cognitive Science, 4*, 135–183.

Newell, A., & Simon, H. (1976). Computer science as empirical enquiry: Symbols and search. *Communications of the ACM, 19*, 113–126. doi:10.1145/360018.360022

Nicolelis, M. (2001). Actions from thoughts. *Nature, 409*(6818), 403–407. doi:10.1038/35053191

Noë, A. (2004). Action. In *Perception*. Cambridge, MA: MIT Press.

Noë, A. (2009). *Out of Our Heads: Why You Are Not Your Brain, and Other Lessons from the Biology of Consciousness*. New York: Hill & Wang.

Nofi, A. A. (2000). *Defining and Measuring Shared Situational Awareness* (No. [Final). Alexandria, VA: Center for Naval Analyses.]. *CRM, D0002895*, A1.

Norman, D. A. (1993). *Things That Makes Us Smart*. Reading, MA: Addison-Wesley.

Norman, D. A. (1998). *The Invisible Computer*. Cambridge, MA: MIT Press.

O'Regan, J. K., & Noë, A. (2001). A sensorimotor account of vision and visual consciousness. *The Behavioral and Brain Sciences, 24*(5), 939–973.

Paul, C. (2004). *Morphology and Computation*. Paper presented at the From Animals to Animats 8: Proceedings of the Eighth International Conference on Simulation of Adaptive Behavior, Los Angeles, CA.

Paul, C. (2006). Morphological Computation: A Basis for the Analysis of Morphology and Control Requirements. *Robotics and Autonomous Systems, 54*, 619–630. doi:10.1016/j.robot.2006.03.003

Pfeifer, R., & Bongard, J. (2007). *How the Body Shapes the Way We Think: A New View of Intelligence*. Cambridge, MA: MIT Press.

Pfurtscheller, G., Scherer, R., & Neuper, C. (2007). EEG-based brain-computer interface. In Parasuraman, R., & Rizzo, M. (Eds.), *Neuroergonomics: The Brain at Work*. New York: Oxford University Press.

Poltrock, S., Handel, M., Bowyer, H., & Waggett, P. (2008). *A dynamic model of mission context.* Paper presented at the 2nd Annual Conference of the International Technology Alliance (ACITA'08), London.

Putnam, H. (1975). The meaning of "meaning". In Gunderson, K. (Ed.), *Language, Mind and Knowledge* (pp. 131–193). Minneapolis, MN: University of Minnesota Press. doi:10.1017/CBO9780511625251

Pylyshyn, Z. (1984). *Computation and Cognition*. Cambridge, MA: MIT Press.

Rhodes, B., & Maes, P. (2000). Just-in-time information retrieval agents. *IBM Systems Journal, 39*(3-4), 685–704.

Robbins, P., & Ayded, M. (2009). *The Cambridge Handbook of Situated Cognition*. New York: Cambridge University Press.

Rowlands, M. (2006). *Body Language: Representation in Action*. Cambridge, MA: MIT Press.

schraefel, m. c., Smith, D. A., Owens, A., Russell, A., Harris, C., & Wilson, M. L. (2005). *The evolving mSpace platform: leveraging the Semantic Web on the trail of the memex.* Paper presented at the Hypertext 2005 Conference, Salzburg, Austria.

Shadbolt, N., Hall, W., & Berners-Lee, T. (2006). The semantic web revisited. *IEEE Intelligent Systems, 21*(3), 96–101. doi:10.1109/MIS.2006.62

Smart, P. R., Engelbrecht, P., Braines, D., Strub, M., & Hendler, J. (2009). *Cognitive Extension and the Web.* Paper presented at the Web Science Conference: Society On-Line, Athens, Greece.

Smart, P. R., Engelbrecht, P. C., Braines, D., Hendler, J. A., & Shadbolt, N. R. (2008). The Extended Mind and Network-Enabled Cognition (No. ITA/P12/NEC). Southampton, UK: School of Electronics and Computer Science, University of Southampton.

Smart, P. R., Huynh, T. D., Mott, D., Sycara, K., Braines, D., Strub, M., et al. (2009). *Towards an Understanding of Shared Understanding in Military Coalition Contexts.* Paper presented at the 3rd Annual Conference of the International Technology Alliance (ACITA'09), Maryland, USA.

Smart, P. R., O'Hara, K., Engelbrecht, P., Giammanco, C., & Braines, D. (in press). *Networks, Memory and the Extended Mind.* Paper presented at the 1st ITA Workshop on Network-Enabled Cognition, Maryland, USA.

Smart, P. R., Russell, A., Liang, S. F., Shadbolt, N. R., Booth, C., Briscombe, N., et al. (2009). *Using Semantic Technologies to Improve Information Exploitation in Military and Civilian Application Contexts.* Paper presented at the Knowledge Systems for Coalition Operations Conference (KSCO'09), Southampton, UK.

Smart, P. R., Sycara, K., & Huynh, T. D. (in press). *Exploring the Dynamics of Belief Propagation and Collective Problem Solving in Simulated Agent Networks.* Paper presented at the 1st ITA Workshop on Network-Enabled Cognition, Maryland, USA.

Sporns, O. (2002). Network analysis, complexity, and brain function. *Complexity, 8*(1), 56–60. doi:10.1002/cplx.10047

Sporns, O., Chialvo, D., Kaiser, M., & Hilgetag, C. (2004). Organization, development and function of complex brain networks. *Trends in Cognitive Sciences, 8*(9), 418–425. doi:10.1016/j.tics.2004.07.008

Stasser, G., & Titus, W. (1985). Pooling of unshared information in group decision making: biased information sampling during group discussion. *Journal of Personality and Social Psychology, 48*, 1467–1478. doi:10.1037/0022-3514.48.6.1467

Sterelny, K. (2004). Externalism, Epistemic Artefacts and the Extended Mind. In R. Schantz (Ed.), The Externalist Challenge: New Studies on Cognition and Intentionality (pp. 239-254). Berlin, Germany: de Gruyter.

Stix, G. (2008). Jacking into the brain - is the brain the ultimate computer interface? *Scientific American, 299*(5), 56–61. doi:10.1038/scientificamerican1108-56

Tablan, V., Damljanovic, D., & Bontcheva, K. (2008). *A Natural Language Query Interface to Structured Information.* Paper presented at the 5th European Semantic Web Conference (ESWC'08), Tenerife, Spain.

Tarsitano, M., & Jackson, R. (1997). Araneophagic jumping spiders discriminate between detour routes that do and do not lead to prey. *Animal Behaviour, 53*(2), 257–266. doi:10.1006/anbe.1996.0372

Thompson, A., Harvey, I., & Husbands, P. (1996). Unconstrained Evolution and Hard Consequences. In Sanchez, E., & Tomassini, M. (Eds.), *Toward Evolvable Hardware: The Evolutionary Engineering Approach* (pp. 136–165). Berlin, Germany: Springer-Verlag.

Tollefsen, D. P. (2006). From extended mind to collective mind. *Cognitive Systems Research, 7*(2-3), 140–150. doi:10.1016/j.cogsys.2006.01.001

Tribble, E. (2005). Distributing Cognition in the Globe. *Shakespeare Quarterly, 56*(2), 135–155.

United States Army Research Laboratory. (2008). *Cognition and Neuroergonomics Collaborative Technology Alliance: Draft Program Announcement (No. W911NF-08-R-0014).* USA: United States Army Research Laboratory.

Van Essen, D., Anderson, C., & Olshausen, B. (1994). Dynamic routing strategies in sensory, motor, and cognitive processing. In Koch, C., & Davis, J. L. (Eds.), *Large-Scale Neuronal Theories of the Brain* (pp. 271–299). Cambridge, MA: MIT Press.

Watts, D. J. (2003). *Six Degrees: The Science of a Connected Age.* London: William Heinemann.

Watts, D. J., & Strogatz, S. H. (1998). Collective dynamics of "small word" networks. *Nature, 393*, 440–442. doi:10.1038/30918

Wheeler, M. (2005). *Reconstructing the Cognitive World.* Cambridge, MA: MIT Press.

Wheeler, M., & Clark, A. (1999). Genic representation: reconciling content and causal complexity. *The British Journal for the Philosophy of Science, 50*(1), 103–135. doi:10.1093/bjps/50.1.103

Wilcox, S., & Jackson, R. (2002). Jumping Spider Tricksters: Deceit, Predation, and Cognition. In Bekoff, M., Allen, C., & Burghardt, G. M. (Eds.), *The Cognitive Animal.* Cambridge, MA: MIT Press.

Wilson, R. A. (1994). Wide computationalism. *Mind, 103*(411), 351–372. doi:10.1093/mind/103.411.351

Wilson, R. A. (2000). The mind beyond itself. In Sperber, D. (Ed.), *Misrepresentations: A Multidisciplinary Perspective* (pp. 31–52). New York: Oxford University Press.

Wilson, R. A. (2004). *Boundaries of the Mind: The Individual in the Fragile Sciences: Cognition.* New York: Cambridge University Press. doi:10.1017/CBO9780511606847

Wilson, R. A., & Clark, A. (2009). Situated Cognition: Letting Nature Take its Course. In Ayded, M., & Robbins, P. (Eds.), *Cambridge Handbook of Situated Cognition.* Cambridge, UK: Cambridge University Press.

Witt, P. N., Reed, C. F., & Peakall, D. B. (1968). *A Spider's Web: Problems in Regulatory Biology.* Heidelberg, Germany: Springer.

ADDITIONAL READING

Clark, A. (2001). Reasons, Robots and the Extended Mind. *Mind & Language*, *16*(2), 121–145. doi:10.1111/1468-0017.00162

Clark, A. (2003). *Natural-Born Cyborgs: Minds, Technologies and the Future of Human Intelligence*. Oxford, UK: Oxford University Press.

Clark, A. (2007b). Re-Inventing Ourselves: The Plasticity of Embodiment, Sensing, and Mind. *The Journal of Medicine and Philosophy*, *32*(3), 263–282. doi:10.1080/03605310701397024

Clark, A. (2008). Supersizing the Mind: Embodiment, Action, and Cognitive Extension USA: Oxford University Press.

Clark, A. (in press-b). Memento's revenge: the extended mind, extended. In R. Menary (Ed.), The Extended Mind. Aldershot: Ashgate Publishers.

Clark, A., & Chalmers, D. (1998). The Extended Mind. *Analysis*, *58*(1), 7–19. doi:10.1111/1467-8284.00096

Goldstone, R. L., Roberts, M. E., & Gureckis, T. M. (2008). Emergent Processes in Group Behavior. *Current Directions in Psychological Science*, *17*(1), 10–15. doi:10.1111/j.1467-8721.2008.00539.x

Hutchins, E. (1995). *Cognition in the Wild*. Cambridge, Massachusetts: MIT Press.

Kirsh, D. (2009). Problem solving and situated cognition. In Ayded, M., & Robbins, P. (Eds.), *Cambridge Handbook of Situated Cognition*. Cambridge, UK: Cambridge University Press.

Noë, A. (2009). *Out of Our Heads: Why You Are Not Your Brain, and Other Lessons from the Biology of Consciousness*. New York, USA: Hill & Wang.

Pfeifer, R., & Bongard, J. (2007). *How the Body Shapes the Way We Think: A New View of Intelligence*. Cambridge, Massachusetts, USA: MIT Press.

Wilson, R. A., & Clark, A. (2009). Situated Cognition: Letting Nature Take its Course. In Ayded, M., & Robbins, P. (Eds.), *Cambridge Handbook of Situated Cognition*. Cambridge, UK: Cambridge University Press.

ENDNOTE

1 Not all forms of cognitive extension are necessarily guaranteed to impact cognition in positive ways; some forms of cognitive extension may prove deleterious to the cognitive capabilities of the larger system. This issue is taken up in the 'Human-Centered Cognitive Extension' section.

2 Harnad and Dror (2006) thus state "...cognition takes place entirely within the brains of cognizers...The causes and effects stretch more distally, but not the cognition; cognition begins and ends at the cognizer's sensor and effector surfaces."

3 The notion of supervenience represents a kind of dependency relationship between sets of properties. A set of properties (X) is said to supervene on another set of properties (Y) if objects that are indistinguishable from the perspective of Y properties are also indistinguishable from the perspective of X properties (see Braddon-Mitchell & Jackson, 2007). Thus mental states (x) supervene on brain states (y) if brain states that are physically indistinguishable are associated with mental states that are also indistinguishable. The claim of content externalism is that this is not the case: the content of mental states supervenes on facts that are external to the neurophysiological details.

4 The vehicles of cognition are the physical states and processes associated with mental states and processes. The distinction between contents (as in content externalism) and the vehicles of contents (vehicle externalism) is a distinction between the content (or meaning) and the thing that has the content (or meaning). For example, the content of a written sentence is the meaning of the sentence, while the thing that has the content (or is the bearer of the content) is the sequence of written words.

5 Indeed, the notion of cognitive extension has been the subject of a lively debate in the philosophical and cognitive scientific literature. Criticisms of the extended mind thesis centre on issues of cognitive and computational control (Butler, 1998), the distinction between intrinsic and derived contents (Adams & Aizawa, 2001, 2008, 2009, in press), and worries about the vulnerability of external resources to damage and social manipulation (Sterelny, 2004). All of these concerns have been addressed by Andy Clark in a series of recent publications (Clark, 2005, 2007a; Clark, in press-a, in press-b; Wilson & Clark, 2009). Clark (2008) provides a good summary of the criticisms and associated responses.

6 Indeed, Edwin Hutchins (1995a) depicts the symbol-manipulating vision of classical cognitive science as, in fact, a vision of environmentally-situated problem-solving. According to this vision, the human agent implements a serial, symbol manipulating processing economy by virtue of his or her interaction with a variety of external props, aids and artefacts.

7 The extent to which this behaviour is, in fact, non-cognitive depends very much on one's view of what constitutes cognition. The problem is that what is and what is not a cognitive process is often determined by ostensive definition. We can therefore point to examples of cognitive processing (e.g. perceiving, reasoning, thinking and so on), but establishing precisely what it is that makes something a cognitive process is much harder. Adams and Aizawa (2001) favour a view of cognition that highlights the role of representations with 'intrinsic' as opposed to 'derived' intentionality. Unfortunately, however, it is not entirely clear what is meant by the notion of intrinsic intentionality, or when we confront representations whose content is intrinsically given. Rowlands (2006) defines a cognitive process as "one that: (i) is required for the accomplishing of a cognitive task, (ii) involves information processing, and (iii) is of the sort that is capable a yielding a cognitive state" (pg. 32). In this definition, the notion of a 'cognitive task' is defined by ostension, and the notion of a cognitive state is construed as a genuinely representational state; i.e. a state that can be seen as representational in virtue of its satisfaction of a host of additional criteria. The main problem here, of course, concerns the fact that we are still relying on ostensive definitions for the notion of a cognitive task. We also encounter problems regarding the precise conditions under which a physical state should count as one that is genuinely representational.

8 The araneophagic spiders, in particular, have been shown to engage in a variety of complex behaviours, ranging from optimal route selection (Tarsitano & Jackson, 1997) to deceptive signalling (Wilcox & Jackson, 2002).

9 Its body is (perhaps non-accidentally) designed so as to best exploit this state of affairs – you can represent quite a lot of information when your representational repertoire is sensitive to the spatial dynamics of a system comprising eight articulated appendages!

10 Morphological computation concerns the way in which the physical body of a robot or organism can be used to perform computationally-significant functions (Paul, 2004, 2006).

11 See Maglio et al (1999) for some empirical research on this issue.

12 For an extended discussion of the notion of 'making information available', particularly with respect to Gibson's (1966) theory of visual perception, see Rowlands (2006; pg. 34-40).

13 This is not to say that such contributions are always indispensable – take away the physical rotation of Tetris zoids (see Kirsh & Maglio, 1994) and the subject may still be able to make do with purely internal rotational strategies. This does not, however, detract from the fact that when external resources are available, and productively coupled into ongoing sequences of neural operations and world-involving actions, they can still become incorporated into transient systemic wholes whose purpose is the efficient realization of a cognitive task.

14 See Sporns (2002) and Sporns, Chialvo, Kaiser and Hilgetag (2004) for some applications of network science to systems-level neuroscience.

15 http://www.microvision.com/wearable_displays/index.html

16 The notion of cost is important here because empirical studies suggest that the cost of accessing information from external resources has a significant impact on whether the resource is actually used (Gray & Fu, 2004). Information access cost is typically quantified in terms of temporal considerations, but it is possible that other types of consideration (such as physical effort) may also be important.

17 Links to sections within the page will not work because Wikipedia, like most Web 2.0 applications, features dynamic content, and the physical location of specific information items is liable to change across multiple usage contexts.

18 http://linkeddata.org/

19 http://www.media.mit.edu/wearables/mithril/memory-glasses.html

20 This is probably the most problematic aspect of our discussion: how to afford access to and interaction with network systems in low-cost ways. We revisit this later on in the discussion about context-aware information retrieval (see 'Human-Centered Cognitive Extension' section).

21 In the sense that only relevant information gets presented. The mode of information presentation is also important here. In particular, it is important to avoid concerns about information overload (see the section on 'Human-Centered Cognitive Extension'). Ideally, information should be presented in the form of simple, perhaps subliminal (see DeVaul, Pentland, & Corey, 2005), cues and prompts that serve to guide thought and action in cognitively productive ways.

22 http://www.w3.org/RDF/

23 http://www.w3.org/2004/OWL/

24 See Noë (2004) for further discussion on this issue.

25 There is also, potentially at least, another option here. This is the idea that cognitive processes are distributed across the members of a group in such a way that neither individual forms of cognitive extension nor collective minds need emerge. Much of the work in distributed cognition (Hutchins, 1995a; Tribble, 2005) can perhaps be seen in this way.

26 The key problem in this debate may centre on our ability to ascribe mental constructs to groups of people – groups of people just aren't the kind of things that behave in ways that warrant thought ascription. Something similar may confront individual forms of the extended mind thesis. When we say, for

example, that an agent, in conjunction with external technologies, solves a particular problem, there is a potential mismatch between the system that is causally implicated in the expression of behaviour (the extended cognitive system) and the thing to which mentalistic constructs get ascribed (the patterns of behaviour of a specific component of the larger system, namely the human agent). Given that beliefs are what gets ascribed to patterns of behaviour, and the behaviour that is produced (at least in the case of human agents) always results from proximate mechanisms that reside in the biological realm, does this mean that we have an inherent tendency (or bias) to always perceive the biological agent as the proper target of mental state ascriptions? Do we have an inherent tendency to discount the wider nexus of extra-biological causal influences that ultimately contributes to the profile of behaviour warranting thought ascription?

27 Technically, the spider does not have a brain; its central nervous system is composed of a number of ganglia, of which the most prominent are the supraesophagal and subesophagal ganglia.

28 Within the domain of computer science, the notion of stigmergy informs many approaches to complex problems concerning optimization, coordination and self-organization (see Ajith, Crina, & Vitorino, 2006).

29 http://www.eprints.org/openaccess/

30 Such conduits may, in some sense, be akin to the short-cuts in small-world networks (Watts & Strogatz, 1998). The shortcuts reduce the path length between our current thoughts and ideas and those that are remote, or even impossible, to reach by virtue of reason-constrained forms of inference.

31 Hinsz, Tindale and Vollrath (1997) have also highlighted some of the dangers associated with a group's over-reliance on shared information. Such insights, in combination with the results reported here, should give us pause for thought when it comes to notions of shared situation awareness (Nofi, 2000) and shared understanding (Smart, Huynh et al., 2009). Inasmuch as the interventions used to enhance shared situation awareness and shared understanding depend on the sharing of common sets of information, it is important that that we do not create a situation in which group-level problem-solving abilities are undermined as a result of trying to achieve some other human factors objective.

32 Lazer & Friedman (2007) evaluated the impact of copying errors in their computer simulation studies. They report that, in the long-run, systems with high error rates in the copying process outperformed those in which copying errors were minimized. The explanation for these results seems to be the same as that proposed for the effect of network structure on performance, namely that "Error rates in copying....alter the balance between exploration and exploitation in the system, increasing the amount of experimentation but reducing the rate with which successful strategies spread" (Lazer & Friedman, 2007).

33 Subliminal in this context means a perceptual cue that is presented at a level of intensity or duration below that necessary for it to become part of conscious awareness.

34 One reason to be cautious of such claims is that the extended mind thesis obliges us to take a systems-level perspective when thinking about the capabilities of network-extended cognitive systems. Thus, just because some aspect of the psycho-cognitive functioning of an individual seems to have been altered as a result of a specific biotechnological merger, this does not mean that those capabilities (or extensions of those capabilities) are not manifest at the system

level. To put this into context, think about the role that language plays in augmenting our cognitive capabilities (see Clark, 2008; chapter 3). It may well be that human agents are increasingly delegating much of their cognitive burden to the Web, but is this really any different from the role that written and spoken forms of language already play for us? No one, we suspect, would be comfortable with the claim that we should abandon written forms of language because they undermine the (pure) cognitive integrity of the 'real' environmentally-decoupled human agent. And this is not just because individually and collectively we are better off, in a cognitive sense, for the development of writing systems. It is because such innovations are now so deeply integrated into our everyday problem-solving routines that the very notion of establishing a neat separation between the true capabilities of the human agent and their language-infected capabilities seems untenable. Many of us, we suspect, feel that linguistically-enabled capabilities are an intrinsic part of our own personal cognitive repertoire. We see language as less a form of technological enhancement and more an aspect of our own idiosyncratic cognitive profile. The long-term vision of the network-extended mind theorist is no different in this respect. The vision is that as network technologies become more permanent, reliable and accessible, so they will become increasingly integrated into our cognitive self image – our image of who *we* are and what *we* are capable of.

[35] A speed bug is an indicator that highlights specific speeds on a airspeed instrument panel.

[36] http://www.darpa.mil/ipto/Programs/uvis/uvis.asp

Chapter 11
Cultural Network Analysis:
A Cognitive Approach to Cultural Modeling

Winston R. Sieck
Applied Research Associates, USA

Louise J. Rasmussen
Applied Research Associates, USA

Paul Smart
University of Southampton, UK

ABSTRACT

The purpose of this chapter is to describe a rigorous, end-to-end methodology for modeling culture as networks of ideas that are distributed among members of a population. The method, Cultural Network Analysis (CNA), represents an interdisciplinary synthesis of techniques drawn from the fields of cognitive anthropology, cultural and cognitive psychology, naturalistic decision making, and decision analysis. CNA is used to develop cultural models for groups and populations, typically depicted as a network representation of the culturally shared concepts, causal beliefs, and values that influence key decisions. CNA can be usefully employed for a variety of applications, including the design of tools to support multinational collaborative planning and decision making, the development of situated cultural training programs, and characterizing the cognition of target audiences to support strategic communications campaigns.

CULTURAL NETWORK ANALYSIS: A COGNITIVE APPROACH TO CULTURAL MODELING

An inherent challenge in understanding behavior in other cultures rests in gathering, analyzing, and representing the relevant cultural concepts, beliefs, and values that drive decisions in those populations. In this chapter, we present *Cultural Network Analysis* (CNA) as a broad approach that aids in providing the most relevant cognitive aspects of cultural groups for decision influence. CNA comprises a collection of methodologies for eliciting, analyzing, and representing the beliefs, values, and cognitive concepts that are shared by members of cultural groups. This paper provides a detailed description of CNA, including its applications to multinational collaboration, cultural training, and strategic communications.

DOI: 10.4018/978-1-61520-855-5.ch011

The fields of psychology and anthropology have been increasingly challenged by the separation of the study of culture and the study of the mind. The interdisciplinary field of culture and cognition has emerged as a response to this challenge (Hirschfeld & Gelman, 1994; Hutchins, 1995; Nisbett, 2003; Sperber, 1985). The cognitive revolution that began in the late 1950s influenced the fields of psychology and anthropology, leading to the development of cognitive psychology and cognitive anthropology (D'Andrade, 1981; Gardner, 1984). These fields have since progressed with little interaction. Cognitive psychologists have focused on the fundamental building blocks or "architecture" of cognition, largely ignoring the effects of content. Cognitive anthropologists have focused on the content of cognition, seeking to describe and explain knowledge that is shared among members of cultural groups.

The challenge facing these fields is rooted in a twin set of ideas that are pressuring researchers within each field to reconsider the significance of the other. In cognitive psychology, cultural variations in what were previously presumed to be universal aspects of the cognitive architecture have surfaced. These finds suggest that much of the work in cognitive psychology could potentially turn out to be ethnographical, rather than architectural in nature (Nisbett, Peng, Choi, & Norenzayan, 2001). For their part, cognitive anthropologists have been faced with a growing awareness that there are widespread commonalities in cognitive organization that point to the existence of some form of cognitive architecture, and that more recent cognitive models of knowledge acquisition, organization, and change could prove useful in developing explanations for those commonalities (Boyer, 1994).

There is also a third set of developments that offer the potential to further accelerate an interdisciplinary culture and cognition program. Work in naturalistic decision making and related areas have helped to promote a growing interest in field research within cognitive psychology. This progressive movement of cognitive psychology into the field has led to the adoption of perspectives and methods that overlap significantly with those of cognitive anthropologists, yet retain a distinctly psychological emphasis on core cognitive functions, such as decision making, planning, sensemaking, adaptation, and coordination (Klein, 1998; Klein et al., 2003). Cognitive field researchers thus have a key role to play in shaping the direction of investigations into culture and cognition, with particular emphasis on research that aims to support the cultural challenges faced by domain practitioners.

CULTURE AS DISTRIBUTIONS OF KNOWLEDGE

Within cognitive anthropology, culture is typically defined as involving shared knowledge. One specific theoretical approach to culture that characterizes culture in terms of knowledge is the epidemiological view. Here, "epidemiology" is used in the general sense of describing and explaining the statistical distributions of any property within a population. Cultural epidemiology regards culture in terms of the ideas that are widely distributed throughout a population (Sperber, 1996).

The starting point from this view is to recognize that individual minds contain vast amounts of mental content. People typically use the word *idea* to refer to any content of the mind, including conceptions of how things are and of how things should be. Networks of ideas are often referred to as folk theories or *mental models*. Such networks constitute peoples' explanations for how things work, and result in judgments and decisions that influence their behaviour (Gentner & Stevens, 1983). Furthermore, the specific nature of a person's mental models depends heavily on their cultural background (Hirschfeld & Gelman, 1994). The emphasis on "ideas" or content knowledge is consistent with work in cognitive field research and naturalistic decision making that has consistently found experiences and mental models to have a primary influence on real-world decision making. The research from this community clearly identifies the contents of cognition, as opposed to cognitive processes often studied in laboratory experiments (such as working memory), as the major driving force of decisions and behaviour.

As implied by the name, mental models reside inside the heads of individuals. However, when people communicate with each other and otherwise shape their environment, their mental models leave observable traces in the form of physical artifacts and representations, including ephemeral traces such as speech and non-verbal gestures (Sperber, 1996). People who come into contact with others' external traces generate associated thoughts, and so they produce mental models that resemble one another. Mental models can spread widely throughout a population and persist for long time periods, becoming "cultural" in the sense of being shared by many of its members across space and time. Cultural models represent these shared networks of ideas.

To take a concrete example, consider Figures 1 and 2 that illustrate American and British cultural models of planning. The set of ideas represented in Figures 1 and 2 were extracted from a study comparing American and British concepts of quality plans (Rasmussen, Sieck, & Smart, in press). As shown, a cultural model of collaborative planning contains a group's common concepts as well as their shared understanding of the causal relationships between concepts, i.e. the antecedents and consequences of planning activities and their outcomes. The cultural model influences communal expectations for how planning should unfold and provides a framework for individual selection of behaviors and goals within collaborative planning situations. Consider Figure 2, for instance. It depicts a number of shared ideas among British planners using circles, lines, and color. These ideas include concepts such as "plan complexity" and "flexible execution," represented as circles. The figure also depicts shared causal ideas such as that complex plans decrease the ability to execute flexibly. These are represented as lines in the figure, with +/- indicating the direction of the causal belief. Finally, Figure 2 portrays ideas of desired states or value using color, as well as a logical flow across desired states. Flexible execution is a good thing, something a plan should support. On the other hand, a plan with many assumptions is something to be avoided.

Figure 1. U.S. cultural model of planning

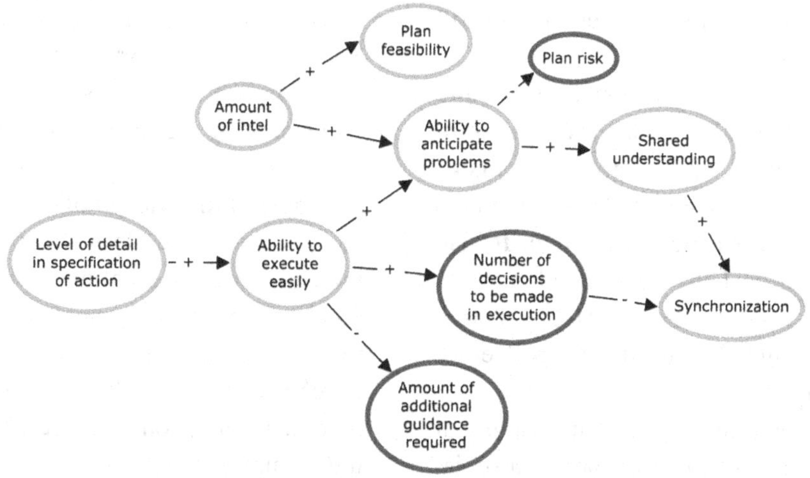

Figure 2. UK cultural model of planning

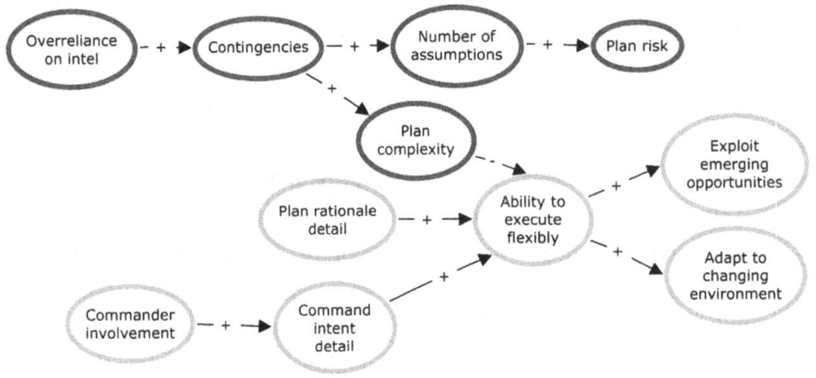

Holding this culturally-shared mental model is likely to have fairly strong consequences for how individuals from the group will decide and act in relevant situations. For example, if it is commonly held that detailed specification of intent and rationale in a plan will improve the capability of executors to adapt the plan in order to meet changing conditions, and the ability to adapt is an important value, then planning team members would be expected to focus on those components of the plan. Furthermore, multinational partners who attempt to collaborate while being guided by different cultural models may well find themselves frustrated and confused. For example, American planners who are focused on developing detailed actions so as to promote synchronization in execution may find themselves wondering why their British counterparts are still talking about goals.

At this point, it is useful to summarize and define a few related terms. First, the term *culture* refers to mental models, and other contents of the mind, that are distributed across members of a population over a period of time. It also includes the resulting behaviors and other traces that foster prolonged survival of the shared ideas by providing "habitats" for them (Berger & Heath, 2005).

Cultural group refers to a set of people that hold shared networks of ideas, whereas a social group consists of people who interact with one another. Traditionally, members of groups were connected in many different spheres, including being neighbors, engaging in the same work, and participating in the same social and religious activities. High overlap in experiences like those, clearly leads to shared ideas within a large number of domains. Hence, there was little difference between people's social groups and their cultural groups. More and more, people often identify with an increasingly wide assortment of groups that vary considerably in aspects such as purpose, size, and cohesion. Modern social groups may be best defined and described using tools such as social network analysis. Similarly, cultural groups are defined and described using cultural network analysis. Figure 3 provides an abstract representation of these two distinct levels of analysis. Figure 3 illustrates a network of people who also hold shared networks of ideas.

Generally, the size of a cultural group will depend on the *cultural domain*, that is, the kind and topic of knowledge of interest. This reflects the fact that some ideas are spread very widely among human populations, whereas others are much more narrowly held. Further, we sometimes use *cultural knowledge* in place of *culture* to refer to the networks of ideas for which there is some level of concordance among members in the cultural group.

Figure 3. Cultural networks of ideas are distributed among members of cultural groups

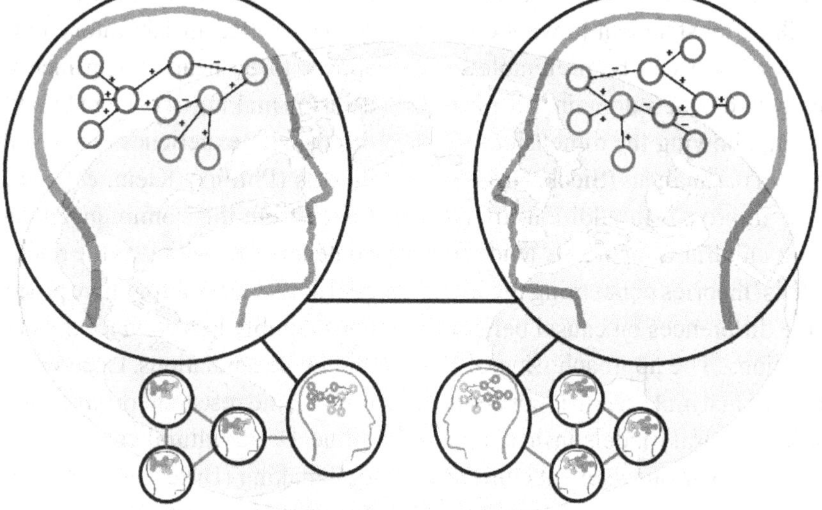

Finally, *cultural model* refers to an external representation of a culture that is constructed by a researcher. A cultural model represents a consensus of the mental models for a particular cultural group and domain.

WHY CULTURAL MODELS?

Cultural models are formal descriptions of the knowledge possessed by members of particular groups. Cultural models describe and represent how the world is understood by the members of these cultural groups. A key premise is that cultural knowledge comprises many networks of causally-interconnected ideas. These mental models become activated within particular situations to drive thinking and decision making, and can change under suitable conditions. Cultural models also seek to account for relationships between cultural knowledge and social networks, and cultural change. Cultural dynamics across social networks is especially useful for modeling shared understanding among multinational partners, as well as anticipating effects in communications campaigns.

Cultural Models vs. Cultural Dimensions

Cultural psychologists have often conceptualized culture in terms of lists of domain general, stable traits, such as individualist-collectivist value orientations (Hofstede, 2001; Schwartz, 1994). The intent of this program is to find a core set of dimensions for characterizing cultures that are important across a wide variety of domains. The motive is to provide a priori, purely analytical predictions about cultural groups that are widely applicable to many particular problems. There is some evidence at this point that general cultural dimensions may not be applicable across situations. For example, Osland & Bird (2000) point to a number of cultural paradoxes that arise in particular contexts from cultural characterization in terms of general value dimensions. One compelling reason offered for such paradoxes is that the relative importance of values varies depending on the nature of the situation. Osland & Bird refer to this phenomenon as "value trumping." From a cultural models perspective, cultural dimensions provide an initial characterization of important values that may be relevant towards understanding members of a culture in specific contexts. However, they do not provide an understanding of what values will be most salient in what contexts. This suggests that it can be preferable to begin cultural analysis of a new domain in a more exploratory fashion, allowing the dimensions of value to emerge from the analysis (Sieck, Grome, Smith, & Rababy, in press). In addition, much research involving cultural dimensions tends to rely on the scientists' theories concerning the implications of value differences on causal beliefs, other values, or actions. The approach generally results in explanations that mix scientific concepts with cultural ideas. Cultural models instead aim to directly represent the various relevant cultural ideas and their interrelations, as held by members of the cultural group.

Cultural Models vs. Cognitive Customs

A more recent trend within culture and cognition in psychology has been to move beyond knowledge contents, and study national differences in cognitive processes such as reasoning and decision making (Norenzayan, Smith, Kim, & Nisbett, 2002; J. Frank Yates, Lee, Sieck, Choi, & Price, 2002). For example, Nisbett et al. (2001) found differences between Western and South-East Asian populations in the extent to which they tended to rely on analytic and logical versus holistic and dialectical modes of thought. Yates et al. (in press) found Chinese to generate fewer arguments, and recruit a more polarized set of arguments than U. S. or Japanese participants. Corroborating these thinking modes or "cognitive customs," regarding argument recruitment, the Chinese were also consistently found to be more overconfident than U. S. or Japanese participants. This body of work is important in establishing that the mental representations studied by culture and cognition researchers should not be limited to the contents of cognition, but must ultimately include representations that govern thinking processes per se.

On the other hand, cognitive field researchers have repeatedly reported that the cognitive processes studied in laboratory experiments do not appear to be nearly as influential on real-world decision making as content knowledge in the form of episodic experiences and well-formed mental models (Phillips, Klein, & Sieck, 2004). The research from this community clearly identifies the contents of cognition as the major driving force of decisions. General cognitive processes may explain considerably less variance than content, at least in naturalistic situations. Even within laboratory settings, some researchers have shown the important influence of cultural content knowledge on decision making (Briley, Morris, & Simonson, 2000). Nevertheless, we do expect that cognitive customs will prove important, especially for understanding mechanisms underlying cultural change. Only, we expect that the research on cognitive customs could benefit from placing a greater emphasis on content. For example, content in the form of epistemological beliefs and other kinds of meta-knowledge likely contribute much to the cognitive customs of particular cultural groups, especially at the macrocognitive level. Such meta-knowledge includes folk theories about how certain macrocognitive processes function, e.g., mental models of negotiation or collaborative decision making.

KINDS OF CULTURAL DOMAINS

Mental models are naturally domain specific since they pertain to the workings of particular artifacts and natural processes. Furthermore, mental models can vary across cultures in ways that are constrained only by the domain itself and any cognitive universals that ground shared understanding across humanity (Hirschfeld & Gelman, 1994). A cultural model represents a consensus of the mental models for a particular cultural group and domain. This leads to a question about the kinds of domains for which mental models, and by extension, cultural models exist.

Most work on mental models has focused on representations in the physical domain, including folk theories of artifacts like thermostats (Kempton, 1986) and folk theories of biological systems (Atran, 1998). For example, Kempton (1986) found that some people held mental models of thermostats that included the idea that turning a thermostat to a higher setting causes a room's temperature to rise at a faster rate (a "valve model"). Others held a correct "threshold model," in which the room heats at a constant rate until the threshold is reached. Furthermore, Kempton found that those with valve models tended to adjust their thermostats on a more continual basis, whereas those with valve models tend to set their thermostats and leave them. As another example, Atran, Medin, & Ross (2005) found that cultural groups' mental models of plant/animal interactions in the rainforest were consistent with the environmental impact of those groups.

People also possess mental models that pertain to the psychological and social domains. Most fundamentally, people have theories about the workings of other people's minds (Gopnik & Wellman, 1994). People have mental models about negotiation and collaborative decision processes (McHugh, Smith, & Sieck, 2008; Van Boven & Thompson, 2003). Teammates have mental models of teamwork (Klimoski & Mohammed, 1994). Security forces have mental models of crowds (Sieck et al., in press). Just as in the physical domain, there is some evidence that mental models in the psychological and social domains guide perceptions and decisions. For example, Van Boven & Thompson (2003) found that negotiators who reached optimal settlements had mental models that reflected greater understanding of the payoffs and processes than those who did not reach optimal settlements.

Although cognitive field research and cognitive anthropology have informed each other's theory and methods, they remain distinct in terms of the domains that each pursues. For example, cognitive anthropological studies have aimed at describing and representing folk theories in domains such as kinship, subsistence, marriage, plant names, diseases, and ghosts. Cognitive field research has instead emphasized the study of mental models and experiences that support the decision making and other cognitive functions of experts working in complex, high-stakes domains.

CULTURAL NETWORK ANALYSIS

Cultural Network Analysis (CNA) refers to a collection of methodologies for building cultural models. CNA includes methods to:

- ***elicit*** the mental models of a sample of individuals within the population
- ***analyze*** the mental models in terms of their culturally-shared elements across individuals and consolidate the elements as cultural models
- ***represent*** the cultural models in accessible format for a variety of uses

CNA is an approach for building external cultural models that have been extracted from the group. CNA is based on a view of culture as comprising networks of ideas, shared to some degree within populations of investigation. CNA builds on a synthesis of conceptually related methods for knowledge elicitation, analysis, and representation that stem from the diverse fields of naturalistic decision making, cognitive anthropology, cognitive psychology, marketing, and decision analysis. None of these fields alone offers a comprehensive, end-to-end approach for cultural modeling. CNA fills that gap.

Cultural Network Analysis encompasses both qualitative, exploratory analysis, and quantitative, confirmatory analysis. In exploratory CNA, concepts and other mental model elements are extracted from qualitative sources, such as interviews and open source media (web news, blogs, email), with little presupposition regarding the elicited contents. A primary goal of exploratory CNA is to develop an initial understanding of the concepts and characteristics that are culturally relevant within the domain. Qualitative analysis and representation at this stage yield insights that can be captured in initial cultural models. Qualitative cultural models reflect the "universe" of ideas for the cultural group; that is, they seek to capture all relevant ideas mentioned. Influence diagrams are an important representation format for cultural models, as illustrated in Figures 1 and 2. Qualitative analysis may be all that is needed for some applications.

Exploratory CNA also generates hypotheses and a wealth of material for constructing structured data collection in a confirmatory CNA. Confirmatory CNA serves to test the structure of previously developed qualitative cultural models, as well as to elaborate the models with quantitative data concerning the prevalence of ideas in the population(s) of interest. In confirmatory CNA, structured interviews, field experiments, and automated semantic mining of web-based sources are used to obtain systematic data that is more amenable to statistical analysis. Statistical models used by cognitive anthropologists and market researchers are employed to assess the patterns of agreement and derive statistics describing the distribution of concepts, causal beliefs, and values. Finally, influence diagram representations of the cultural models are constructed that illustrate the statistical properties, as well as the qualitative information. Formal quantitative representation makes it possible to use cultural models in a variety of applied contexts.

EXPLORATORY CNA

Elicitation

A mental model is a person's intuitive explanation about how something works, and the explanation consists of an inter-related set of concepts, beliefs about causality, and values. Hence, in order to elicit a mental model in an interview, one needs to find ways to prompt the respondent to verbalize those key elements in great detail. Researchers from various fields have been working the issue, and considerable progress has been made regarding specific lines of probing and other techniques to tease out mental models. It is important to note that such interviews are often semi-structured, and tend to be highly dynamic, intercultural interactions. Actually conducting such interviews requires considerable skill that goes well beyond developing guides to support a particular line of questioning.

An example of mental model elicitation is provided in a study examining the cultural knowledge and understandings related to diabetes causation in a Native American community (Garro, 2000). The researchers conducted interviews following an "explanatory model framework" (Kleinman, 1978). All of the participants were members of the Anishinaabe community who had been previously diagnosed with diabetes. The researchers ensured that the following aspects of their experiences were covered in the interview:

- The cause of their illness
- Why it started and when it did
- The history of the illness
- The kinds of effects it has
- Possible and appropriate treatments for the illness

Participants were also encouraged to talk more generally about possible causes and ways of dealing with diabetes, and to answer additional related questions that arose from the responses given. Based on the results, Garro constructed a graphical outline of the culturally available understandings relevant to a cultural schema for sickness. The outline organized hierarchically the most common explanations of illness mentioned in the interviews. First, the different types of sicknesses and sickness explanations were identified. The identity and labels for these explanations were both mentioned explicitly and referenced implicitly in the different types of causes and 'treatments' mentioned. After having inferred the major types of sicknesses the causes, or perceived causes, were sorted into sickness categories. The level of detail and abstraction of the cause descriptions was dictated by the amount of information made available by informants.

In a medical illness domain like "diabetes," the general direction of peoples' subjective values regarding the disease state is probably fairly transparent (i.e. a cultural preference to avoid the disease state exists). However, that transparency does not necessarily translate to other domains, or even to various treatment options and effects within the diabetes domain. In such cases, values and objectives can be elicited directly, along with the causal beliefs that link more fundamental values with the means intended to achieve them (Gutman, 1982; Keeney, 1994). The essential idea of such "value-focused thinking" is to ask why a particular objective is important in order to elicit the more fundamental values that are anticipated in consequence. Interview approaches like this are sometimes termed "laddering" in the marketing literature (Reynolds & Gutman, 1988).

These same questioning strategies that have been used to elicit physical and biological products and processes can also be used to elicit mental models of social and cognitive functions and processes, such as collaborative planning and decision making (Rasmussen et al., in press). The data underlying the cultural models illustrated in Figures 1 and 2 were elicited using an explanatory models framework combined with a value-focused thinking approach in order to capture descriptive and prescriptive components of the mental models. In general, an explanatory framework for mental models of social-cognitive functions and processes should aim to capture the following:

- Positive/negative states and outcomes
- Conceptions of quality in process and outcomes
- Causal factors that influence the quality concepts
- Consequences of low/high quality functioning
- Artifacts, procedures and tools intended to support the natural process
- Functions of ingrained artifacts

An important consideration in eliciting knowledge about social-cognitive domains like "collaboration" or "decision making" is the difficulty inherent in discussing abstract concepts. One approach to achieving some concrete grounding in such abstract domains is to elicit specific incidents on the topic, and use them to tease out clues to participant's mental models. For example, Sieck et al. elicited incidents from Arabs in Lebanon and the US who had participated in protests as a means to gain access to Middle Eastern crowd members' understandings and expectations of how crowds work (Sieck, McHugh, & Smith, 2006). The idea is to elicit a real-lived incident from the participant, and then use that concrete example as a starting point for more abstract discussions of causality introduced through hypotheticals ("What if?" questions).

Another useful technique for getting at abstract mental models is the "Nearest neighbor method" (Klein & Hoffman, 2008). This is a useful method to use when participants may have difficulty articulating their beliefs, such as when they have an understanding and expectations about how something works, but that understanding is represented in non-verbal forms. Respondents are presented with cartoons, diagrams, other pictures, or vignettes that depict different possible variations of the mental model. The respondent selects the depiction that most closely matches their conception, and then explains where and how it does not quite fit. As with the incident-based methods, this approach can be valuable for providing some common ground between the interviewer and respondent, though it is clearly more directive than the open questioning methods.

Analysis

In contrast with elicitation, analysis of qualitative data for the specific purpose of characterizing mental models has not appeared to be developed in any great detail. In some cases, researchers seem to work individually using rather opaque processes to finally emerge with an idiosyncratic representation of a cultural model or schema. We have thus been developing our own analytical procedures with the aims of achieving a traceable, repeatable, and reliable process for extracting culturally shared mental models from text. In the most recent application of our process (described below) to some interviews we conducted with Afghans, we achieved 95.2% reliability and developed fully traceable cultural models.

The process we have developed starts by conducting an initial coding of transcripts or other data records to identify and record local concept-causal belief-value (CBV) chains. Transcripts from all participants in the cultural group are analyzed together as a unit. This step is performed by two coders working independently. The analysis team then performs a reliability check on the initial results, and holds a meeting to establish consensus codes. The reliability check is performed by tracking the conflicting codes and total number of codes. Codes are conflicting if the concepts, direction, or valence are inconsistent or incompatible. Such conflicts are resolved in the consensus meeting, and the issues discussed may include the level of abstraction/detail used to describe the CBV chains, appropriate places to divide the causal linkages, any assumptions and inferences the analysts are making. The resulting consensus descriptions are used to consolidate the local CBV codes into an overall model. This step is generally performed by one analyst. Finally, the team members perform independent reviews of the model against the data, and iteratively revise and refine until a consensus of the model is achieved. Comments and revisions to the model include flagging any concepts reflected in the consensus file that are missing from the model, pointing out inconsistencies between the consensus document and model, commenting on the overall level of abstraction of the model, and providing suggestions for reorganizing the model to make it clearer, or more succinct.

The analysis process we use relies on explicit verbalizations by the participants. Other research suggests that systematic analysis of the way people talk can offer additional insight into the nature of mental models. Metaphors are a linguistic manifestation of tacit knowledge, and hence provide a window of access to such difficult-to-verbalize thoughts. Researchers have proposed that metaphor is an indispensable part of our ordinary and conventional way of conceptualizing the world and our everyday behavior reflects our metaphorical understanding of experience (Lakoff & Johnson, 1980). A metaphor consists of the projection of one schema (the source domain of the metaphor) onto another schema (the target domain of the metaphor). Schmitt (2005) proposed a systematic approach to uncovering the origins of metaphor models, the historically-defined changes in metaphors across time, and the context-sensitivity of metaphors. These are dimensions which, according to Schmitt, are often overlooked in purely cognitive anthropological and linguistic examinations of metaphor.

As an example use of metaphor analysis for developing a cultural model, Quinn asked American husbands and wives to talk about 'marriage' and collected hours and hours of minimally guided conversations with them on that topic (Quinn, 2005). She then analyzed the data categorizing linguistic metaphors into a set of central or commonly shared conceptual metaphors that her informants used to reason about different aspects of marriage. She extracted metaphors, such as "marriage is a journey," as well as the abstract concepts and values that were being associated with each, including 'lastingness', 'sharedness', 'compatibility', and 'difficulty'. Quinn noticed that these metaphors, in isolation, did not appear to tell the whole story. She also noticed that her informants often followed the same causal chain to reach their conclusions. For example, a certain causal relationship exists between compatibility and lastingness. The resulting cultural schema that describes how Americans reason about marriage, represented as a causal chain in narrative form, is: *Marriages are successful if they last. In order to last, a marriage must be beneficial, and in order for it to be beneficial, its difficulties must be overcome, and this requires effort* (Quinn, 2005).

Graphical Representation

As with analysis, current approaches for representing cultural models appear to be relatively idiosyncratic, or even non-existent. A default approach to representation for CNA might prove quite useful, if it could accomplish the following:

1. Provide a standard pictorial form that shows the concepts and causal linkages in a manner that can be readily digested by end users who need to routinely comprehend cultural models in varied domains
2. Permit a direct means of representing the statistical distributions of cultural knowledge, rather than just the shared knowledge
3. Yield representations in a useful form for developers of intelligent systems

One such representation format that meets these requirements is an influence diagram. In an influence diagram, each node-link-node combination represents causal influence, in the sense that the value of the concept at the beginning of an arrow affects the value of the concept at the arrow's point. Fully-specified influence diagrams can also represent numerical quantities, as described in the confirmatory CNA section, but the basic structure is useful as well. Specifically, an influence diagram can present a relatively simple and useful representation of an individual's mental model of a domain that is related to key judgments and decisions that rely on that mental model. For example, Bostrom et al. provide an example of an influence diagram that illustrates an expert's mental model of radon, as related to the expert's judgment concerning risk of lung cancer (Bostrom, Fischhoff, & Morgan, 1992). Sieck and colleagues used an influence diagram to represent an expert mental model of crowd functioning, as related to key judgments of threat level and populace attitudes (Sieck et al., in press).

Likewise, influence diagrams can also be used to represent qualitative cultural models. In this case, the diagram represents the complete set of concepts and linkages for all members of each cultural group considered in the analysis. For example, Figures 1 and 2 illustrate cultural models for American and British campaign planners, respectively, each of which is related to a key judgment concerning the quality of the plan and planning process. As shown, the specific values along which such judgments are rendered were found to differ between the two cultural groups, such that the Americans rely more heavily on synchronization in judging plan quality, whereas British planners focus more on the ability of the plan to support flexible execution (Rasmussen et al., in press).

CONFIRMATORY CNA

Elicitation

Structured approaches for the elicitation of complete mental models have only recently begun serious development. Most surveys and structured questionnaires treat ideas as independent entities, and so do not provide any means for revealing their interrelated, network form. There are, however, a few studies in which causal relations between concepts have been directly assessed (Atran et al., 2005; Garro, 2000; Sieck, Smith, Grome, Veinott, & Mueller, 2009). For example, Sieck et al. selected twelve security force actions and five categories of crowd responses from an earlier qualitative cultural study in order to develop a mental models questionnaire. Each security force action was then paired once with each crowd member response. For each pair, the participant was asked whether the crowd member behavior or attitude will increase, decrease, or stay the same. The purpose of these questions was to elicit the participants' causal beliefs between security force actions and crowd member behavior. Such causal linkages form the basis of their mental models. More recently, we have begun developing questionnaires that permit the analysis of longer causal belief chains.

Analysis

One issue with purely qualitative approaches to the development of cultural models is the lack of transparency or consistent guidelines in what knowledge was deemed sufficiently shared to include in the model. Strauss and Quinn state, "At what point in the continuum of sharedness we decide to call a given schema 'cultural' is simply a matter of taste," (p 122). Structured, quantitative approaches are required for testing the qualitative discoveries about culturally shared mental models, and further analyzing and representing their distributions within and between populations. Cultural consensus theory and mixture modeling are two statistical methods that can be usefully employed to meet those needs.

Cultural consensus theory is a collection of formal statistical models designed to assess concordance in knowledge and beliefs among a set of respondents (Romney, Weller, & Batchelder, 1986). When a cultural consensus is found, it provides the consensual responses that indicate culturally shared knowledge and estimates of the strength of consensus for those responses. Individuals will also vary in the extent to which their responses agree with the consensus, and that variation is captured explicitly for each individual as a measure of "cultural competence." Cultural competence should not be confused with expertise, but rather with the degree of concordance with the culturally shared knowledge. The instigating issue that prompted development of the theory was the recognition that an anthropologist who goes in to a new culture and asks questions does not know the answers to the questions or the cultural competence of the respondents (Romney, Batchelder, & Weller, 1987). An important feature of CCT is that, assuming the data collection taps into reasonably well-shared cultural knowledge, then the number or respondents can be quite small, e.g., 10 or fewer respondents. This is important for field research, which often aims at understanding knowledge within small populations.

Cultural consensus theory has been applied to research questions in a number of cultural domains, including disease concepts and folk theories of disease processes, characterizations of alphabetic systems, national consciousness, folk theories of biology and ecology, and others. CCT can be used to analyze fixed-format questionnaires (e.g. true-false, fill-in-the-blank, multiple choice, rank order) and classification data (e.g. card sort, hierarchical taxonomy). It has also been used to analyze free-listings of concepts (Ross & Medin, 2005). CCT has most often been used to analyze data on simple concepts. However, it has also been successfully employed to analyze relationships between concepts, such as causes, consequences, and other interactions (Atran et al., 2005; Garro, 2000). This use of CCT is critical for the purposes of analyzing mental models.

Mixture modeling is a statistical technique that is growing in general popularity for a variety of uses, and provides a competing approach to CCT for cultural data analysis. Mixture models have been applied in many scientific fields, including marketing, biology, medicine, and astronomy. McLachlan and Peel (2000) provide a general description of mixture models, along with example applications.

A mixture model, or "finite mixture model," is given as a combination of different groups, each described by a distinct probability distribution. Mixture models sort through the data and group them into sets of relatively homogeneous cases or observations. For concreteness, we describe the process in an example application to market segmentation. Finite mixture modeling was used to examine whiskey usage in 2218 households (Grun & Leisch, 2007). Information on whiskey type (single malt or blend) and 21 specific brands was included in the data set. A mixture of binomial distributions was fitted to the data set, and the analysts varied the possible ("finite") number of groups between 1 and 7. Model fitting was conducted using a statistical package called "FlexMix" originally developed earlier by one of the authors (Leisch, 2004). FlexMix uses an iterative maximum likelihood procedure called the, "EM algorithm," for model estimation. The best fitting model was selected using the Bayes Information Criterion (BIC) statistic. BIC suggested that the best fit was achieved with 5 groups or segments. The largest segment was not found to consume much whiskey on the whole, used a wide variety of brands when they did partake, but avoided single malts. The largest users consisted of about 10% of the sample, and were spread across two groups best delineated by whether they consumed single malt whiskey.

Mixture modeling has also been successfully used in cultural analysis (Mueller & Veinott, 2008; Sieck & Mueller, 2009; Sieck et al., 2009). In this application, the distinct segments resulting from the analysis represent *cultural groups*, i.e., groups defined by the similarity of their ideas. Mueller and Veinott (2008) compared the technique with CCT by applying it to some of the same classic cultural data sets for which CCT was first used. The primary advantage they found for mixture modeling was in the case where CCT does not find a consensus. In that situation, one cannot tell whether there is simply no pattern to the data, or whether the data reflect multiple cultural groups. Mixture modeling can also provide a wider number of metrics for assessing the cultural groups within the population (Mueller, Sieck, & Veinott, 2007). In particular, we define here the following five cultural metrics as essential in characterizing a family of cultural models:

- Number of cultural models identified
- Consensus within a model
- Prevalence of each idea within a model
- Cultural competence of individuals
- Distinctiveness between cultural models

Although these measures stand independently of any particular statistical technique, analyses using mixture models provide quantities that can be used to assess each of these metrics.

Graphical Representation

We find influence diagrams to be useful for representing quantitative cultural models, and they provide additional information to their qualitative counterparts. When completely specified, an influence would be defined in terms of conditional probabilities, where a influences b if the probability distribution of b conditioned on a is different from the unconditional distribution of b (Howard, 1989). The full specification of influences is typically performed by experts in the domain who consult external resources and render all the probabilities themselves (Edwards, 1998; Edwards & Fasolo, 2001). This process presents a serious technical challenge to the expert and raises questions concerning the reliability and validity properties of the judgments. Fortunately, we do not need to require respondents to assess probabilities in order to develop quantitative cultural models. That is, although numerical values are incorporated in the final result, the use of influence diagrams to represent cultural models only requires that individuals be able to convey the qualitative components and directions of the influences in the diagram.

In the application of confirmatory CNA, the influence diagram represents the "culturally correct" concepts, values, and causal linkages as determined by CCT or mixture modeling for each cultural group that was found. Furthermore, the results concerning prevalence of each idea within a group are used to populate the numerical probability values in the diagram. The result in this case is a summary of not only the shared influence links across the population, but rather the full distribution of ideas, with probabilities indicating the consensus on any particular causal link (or node). An example is provided in Figure 4, illustrating a simple quantitative cultural model of Middle Eastern crowd functioning, derived using mixture modeling (Sieck et al., 2009). As shown, there is a relatively strong consensus (87%) within the represented cultural group that Americans speaking Arabic to Middle Eastern crowd members will have a positive effect on the crowd's attitudes towards the U. S. The idea that firing a warning shot will have just the opposite effect is even more prevalent within this group (93%).

APPLICATIONS

Culture is made up of contagious ideas, that is, ideas that propogate effectively and durably within a population (Sperber, 1996). Two broad objectives of research within this cultural epidemiology viewpoint are to:

- Characterize the current distribution of mental models within the cultural group
- Understand the dynamics of culture

Fundamental cultural research program seeks to address why some ideas are more infectious than others, and to explain the most widely distributed and long-lasting ideas within a population. Research for practical purposes has a slightly different focus. From a decision-making standpoint, for example, we recognize that many ideas may be pervasive but inconsequential to decisions of practical interest (Bostrom et al., 1992; Sperber, 1985). Hence, a decision-centered approach to culture and cognition begins with critical judgments and decisions that are made by members of a cultural group. Using Cultural Network Analysis, we can then study the networks of ideas that are relevant to those decisions in order to answer a host of questions, such as:

- How are networks of ideas organized in mental models?
- What is the distribution of mental models in a cultural group?

Figure 4. Example of a quantitative cultural model

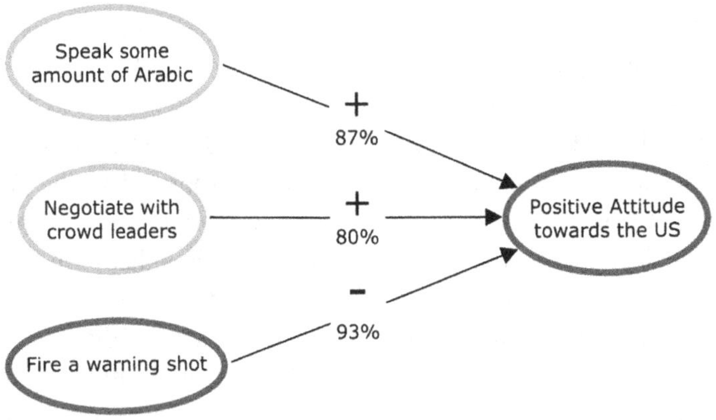

- Why are the mental models distributed in that way?
- How did the distribution get to be that way?
- How stable are those distributions?
- In what ways are the distributions changing over time?
- Why are they changing in that way?
- What makes some ideas successful in the culture?
- Why is the culture more vulnerable to some ideas than to others?

This shift in theoretical focus provides an opportunity to enhance the content of existing tools and procedures that have been developed to support multinational collaboration, but it also inspires the design of different approaches to supporting multinational collaborations altogether. Explicit representations of the distribution of implicit knowledge within two cultures can make it possible to assess differences and potential similarities between the cultures in question, including in terms of knowledge about how people should interact in order to work together effectively. The high-level goal here is to promote the development of hybrid cultures in multinational collaborative planning and decision making (Earley & Mosakowski, 2000). Hybrid cultures comprise a simplified set of shared assumptions, rules, expectations, and procedures that permit multinational teams to function effectively. Hybrid cultures develop naturally over time as teams converge on a common process of interacting. However, cultural models can be used to inform the design of tools and processes that promote the natural process.

For example, with respect to reconciling the American and British cultural models illustrated in Figures 1 and 2, we might consider a collaborative planning tool with functionality, and interfaces tailored to support distinct roles by the partners in a coherent, comprehensive coalition planning process. In this case, we could ensure that the distinct roles each national partner would be assigned align with the respective nation-specific processes. In particular, this suggests a division of labor that has the British members of the team contribute to developing the intent, plan rationale, and the logic between ends, ways, and means, whereas the Americans would focus on developing the detailed specification of actions.

Another design idea is to develop a coalition plan evaluation tool that relies on multinational metrics of plan quality. Such a tool would ideally incorporate an automated assessment of plan content using ontology representation of cultural models (Rasmussen et al., in press). The idea is to provide measures of plan quality that include value dimensions from each coalition partner (e.g. action, rationale specification). Such a tool would provide a concrete basis for discussion among planners about evaluations of plans, for example, enable them to explicitly examine and discuss the right "mix" of elements to make a good plan. Providing such online feedback to a team about what they are trying to accomplish is expected to speed up the natural process of hybrid culture building.

An explicit representation of the distribution of implicit content knowledge within a culture can also serve as a foundation for developing situated cultural training programs (Sieck, Smith, & Rasmussen, 2008). Explicit cultural models can provide a meaningful basis for inferring otherwise implicit goals and intentions, which is essential for successful communication. Whereas traditional training programs often provide a set of dos and don'ts, training the ability to infer intentions within a novel cultural knowledge structure provides a generative platform which can be used in a variety of multinational situations.

Finally, explicit representations of content knowledge within a culture can also serve as a basis for composing effective strategic communications. The models of culturally-shared content knowledge within a certain domain can serve as a basis for determining what makes for culturally meaningful messages. The CNA would allow for making predictions concerning the effectiveness of a message by providing the opportunity to assess potential unintended inferences that individuals with a certain knowledge structure might make. Specifically, in a cultural models diagram, each concept and causal belief represents an opportunity to affect a change in beliefs or concepts. Hence, such diagrams can provide a basis for determining the content of communications. Messages are created so as to affect the values of the most vulnerable concept nodes (i.e. those for which there is the least consensus) which then propagate across perceived influences to affect the values of other concepts. These effects spread through the cultural knowledge network, ultimately changing the value in overall perceptions or cognitions. With this CNA approach, information efforts focus on transmitting the most relevant information to affect conceptual change in a way that makes sense within the cultural group's understanding.

ACKNOWLEDGMENT

This research was sponsored by the U.S. Army Research Laboratory and the U.K. Ministry of Defence and was accomplished under Agreement Number W911NF-06-3-0001. The views and conclusions contained in this document are those of the author(s) and should not be interpreted as representing the official policies, either expressed or implied, of the U.S. Army Research Laboratory, the U.S. Government, the U.K. Ministry of Defence or the U.K. Government. The U.S. and U.K. Governments are authorized to reproduce and distribute reprints for Government purposes notwithstanding any copyright notation hereon.

REFERENCES

Atran, S. (1998). Folkbiology and the anthropology of science: Cognitive universals and cultural particulars. *The Behavioral and Brain Sciences*, *21*, 547–609. doi:10.1017/S0140525X98001277

Atran, S., Medin, D. L., & Ross, N. O. (2005). The cultural mind: Environmental decision making and cultural modeling within and across populations. *Psychological Review*, *112*(4), 744–776. doi:10.1037/0033-295X.112.4.744

Berger, J. A., & Heath, C. (2005). Idea habitats: How the prevalence of environmental cues influences the success of ideas. *Cognitive Science*, 29.

Bostrom, A., Fischhoff, B., & Morgan, M. G. (1992). Characterizing Mental Models of Hazardous Processes: A Methodology and an Application to Radon. *The Journal of Social Issues*, *48*(4), 85–100.

Boyer, P. (1994). Cognitive constraints on cultural representations: Natural ontologies and religious ideas. In Hirschfeld, L. A., & Gelman, S. A. (Eds.), *Mapping the mind: Domain specificity in cognition and culture* (pp. 391–411). New York: Cambridge University Press.

Briley, D., Morris, M., & Simonson, I. (2000). Reasons as carriers of culture: Dynamic versus dispositional models of cultural influence on decision making. *The Journal of Consumer Research*, *27*, 157–178. doi:10.1086/314318

D'Andrade, R. G. (1981). The cultural part of cognition. *Cognitive Science*, *5*, 179–195. doi:10.1207/s15516709cog0503_1

Earley, P. C., & Mosakowski, E. (2000). Creating hybrid team cultures: An empirical test of transnational team functioning. *Academy of Management Journal*, *43*, 26–49. doi:10.2307/1556384

Edwards, W. (1998). Hailfinder: Tools for and experiences with Bayesian normative modeling. *The American Psychologist*, *53*(4), 416–428. doi:10.1037/0003-066X.53.4.416

Edwards, W., & Fasolo, B. (2001). Decision technology. *Annual Review of Psychology*, *52*, 581–606. doi:10.1146/annurev.psych.52.1.581

Gardner, H. (1984). *The mind's new science: A history of the cognitive revolution*. New York: Basic Books.

Garro, L. C. (2000). Remembering what one knows and the construction of the past: A comparison of cultural consensus theory and cultural schema theory. *Ethos (Berkeley, Calif.)*, *28*(3), 275–319. doi:10.1525/eth.2000.28.3.275

Gentner, D., & Stevens, A. L. (1983). *Mental Models*. Hillsdale, NJ: Lawrence Erlbaum Associates.

Gopnik, A., & Wellman, H. M. (1994). The theory theory. In Hirschfeld, L., & Gelman, S. (Eds.), *Mapping the Mind: Domain Specificity in Cognition and Culture* (pp. 257–293). New York: Cambridge University Press.

Grun, B., & Leisch, F. (2007). FlexMix: An R package for finite mixture modelling. *R News*, *7*(1), 8–13.

Gutman, J. (1982). A means-end chain model based on consumer categorization processes. *Journal of Marketing*, *46*(2), 60–72. doi:10.2307/3203341

Hirschfeld, L., & Gelman, S. (Eds.). (1994). *Mapping the mind: Domain specificity in cognition and culture*. New York: Cambridge University.

Hofstede, G. (2001). Culture's consequences (2 ed.). Thousand Oaks, CA: Sage.

Howard, R. A. (1989). Knowledge maps. *Management Science*, *35*, 903–922. doi:10.1287/mnsc.35.8.903

Hutchins, E. (1995). *Cognition in the Wild*. Cambridge: MIT Press.

Keeney, R. L. (1994). Creativity in decision making with value-focused thinking. *Sloan Management Review*, *35*(4), 33–41.

Kempton, W. (1986). Two theories of home heat control. *Cognitive Science*, *10*, 75–90.

Klein, G. (1998). *Sources of Power: How People Make Decisions*. Cambridge: MIT Press.

Klein, G., & Hoffman, R. R. (2008). Macrocognition, mental models, and cognitive task analysis methodology. In Schraagen, J. M., Militello, L. G., Ormerod, T., & Lipshitz, R. (Eds.), *Naturalistic Decision Making and Macrocognition* (pp. 57–80). Aldershot, UK: Ashgate.

Klein, G., Ross, K. G., Moon, B. M., Klein, D. E., Hoffman, R. R., & Hollnagel, E. (2003). Macrocognition. *IEEE Intelligent Systems, May/June*, 81-85.

Kleinman, A. (1978). Concepts and a model for the comparison of medical systems as cultural systems. *Social Science & Medicine*, *12B*, 85–93.

Klimoski, R., & Mohammed, S. (1994). Team mental model: Construct or metaphor. *Journal of Management*, *20*, 403–437. doi:10.1016/0149-2063(94)90021-3

Lakoff, G., & Johnson, M. (1980). *Metaphors we live by*. Chicago: University of Chicago Press.

Leisch, F. (2004). FlexMix: A general framework for finite mixture models and latent class regression in R. *Journal of Statistical Software*, *11*(8), 1–18.

McHugh, A. P., Smith, J. L., & Sieck, W. R. (2008). Cultural variations in mental models of collaborative decision making. In Schraagen, J. M. C., Militello, L., Ormerod, T., & Lipshitz, R. (Eds.), *Naturalistic Decision Making and Macrocognition* (pp. 141–158). Aldershot, UK: Ashgate Publishing Limited.

McLachlan, G. J., & Peel, D. (2000). *Finite Mixture Models*. New York: Wiley. doi:10.1002/0471721182

Mueller, S. T., Sieck, W. R., & Veinott, E. S. (2007). *Cultural Metrics: A Finite Mixture Models Approach (Technical Report DAAD19-01-2-0009)*. Fairborn, OH: Applied Research Associates.

Mueller, S. T., & Veinott, E. S. (2008). *Cultural mixture modeling: Identifying cultural consensus (and disagreement) using finite mixture modeling.* Paper presented at the Proceedings of the Cognitive Science Society, Washington, DC.

Nisbett, R. E. (2003). *The geography of thought: How Asians and Westerners think differently.and why*. New York: The Free Press.

Nisbett, R. E., Peng, K., Choi, I., & Norenzayan, A. (2001). Culture and systems of thought: Holistic versus analytic cognition. *Psychological Review, 108*, 291–310. doi:10.1037/0033-295X.108.2.291

Norenzayan, A., Smith, E. E., Kim, B. J., & Nisbett, R. E. (2002). Cultural preferences for formal versus intuitive reasoning. *Cognitive Science, 26*, 653–684.

Osland, J. S., & Bird, A. (2000). Beyond Sophisticated Stereotyping: Cultural Sensemaking in Context. *The Academy of Management Executive, 14*(1), 65–79.

Phillips, J. K., Klein, G., & Sieck, W. R. (2004). Expertise in judgment and decision making: A case for training intuitive decision skills. In Koehler, D. J., & Harvey, N. (Eds.), *Blackwell Handbook of Judgment & Decision Making* (pp. 297–315). Malden, MA: Blackwell. doi:10.1002/9780470752937.ch15

Quinn, N. (2005). How to reconstruct schemas people share, from what they say. In Quinn, N. (Ed.), *Finding culture in talk*. New York: Palgrave MacMillan.

Rasmussen, L. J., Sieck, W. R., & Smart, P. (in press). What is a Good Plan? Cultural Variations in Expert Planners' Concepts of Plan Quality. *Journal of Cognitive Engineering and Decision Making*.

Reynolds, T. J., & Gutman, J. (1988). Laddering theory, method, analysis, and interpretation. Journal of Advertising Research, Feb/March, 11-31.

Romney, A. K., Batchelder, W. H., & Weller, S. C. (1987). Recent applications of cultural consensus theory. *The American Behavioral Scientist, 31*(2), 163–177. doi:10.1177/000276487031002003

Romney, A. K., Weller, S. C., & Batchelder, W. H. (1986). Culture as consensus: a theory of culture and informant accuracy. *American Anthropologist, 88*, 313–338. doi:10.1525/aa.1986.88.2.02a00020

Ross, N., & Medin, D. (2005). Ethnography and experiments: Cultural models and expertise effects elicited with experimental research techniques. *Field Methods, 17*, 131–149. doi:10.1177/1525822X04271681

Schmitt, R. (2005). Systematic metaphor analysis as a method of qualitative research. *Qualitative Report, 10*(2), 358–394.

Schwartz, S. H. (1994). Cultural dimensions of values: Towards an understanding of national differences. In Kim, U., Triandis, H. C., Kagitcibasi, C., Choi, S. C., & Yoon, G. (Eds.), *Individualism and collectivism: Theory, method, and applications* (pp. 85–119). Newbury Park, CA: Sage.

Sieck, W. R., Grome, A. P., Smith, J., & Rababy, D. A. (in press). Expert cultural sensemaking in the management of Middle Eastern crowds. In Mosier, K. L., & Fischer, U. M. (Eds.), *Informed by Knowledge: Expert Performance in Complex Situations*. San Francisco: Taylor and Francis.

Sieck, W. R., McHugh, A. P., & Smith, J. L. (2006). Use of cognitive field research methods to investigate cultural groups: The case of individual decision making in Middle Eastern crowds. In R. Sun & N. Miyake (Eds.), *Proceedings of the 28th Annual Conference of the Cognitive Science Society,* (pp. 2164-2168).

Sieck, W. R., & Mueller, S. T. (2009). Cultural variations in collaborative decision making: Driven by beliefs or social norms? In *Proceedings of the International Workshop on Intercultural Collaboration* (pp. 111-118), Palo Alto, CA.

Sieck, W. R., Smith, J., Grome, A. P., Veinott, E. S., & Mueller, S. T. (2009). *Violent and Peaceful Crowd Reactions in the Middle East: Cultural Experiences and Expectations.* Paper presented at the International Academy for Intercultural Research, Honolulu, HI.

Sieck, W. R., Smith, J., & Rasmussen, L. J. (2008). *Expertise in making sense of cultural surprises.* Paper presented at the Interservice/Industry Training, Simulation, and Education Conference (I/ITSEC), Orlando, FL.

Sperber, D. (1985). Anthropology and psychology: Towards an epidemiology of representations. *Man, 20*, 73–89. doi:10.2307/2802222

Sperber, D. (1996). *Explaining culture: A naturalistic approach.* Malden, MA: Blackwell.

Van Boven, L., & Thompson, L. (2003). A look into the mind of the negotiator: Mental models in negotiation. *Group Processes and Interpersonal Relations, 6*, 387–404. doi:10.1177/13684302030064005

Yates, J. F., Ji, L.-J., Oka, T., Lee, J.-W., Shinotsuka, H., & Sieck, W. R. (in press). Indecisiveness and thoroughness: Cultural variations in customs, values, and expectations. *Journal of Cross-Cultural Psychology.*

Yates, J. F., Lee, J.-W., Sieck, W. R., Choi, I., & Price, P. C. (2002). Probability judgment across cultures. In Gilovich, T., Griffin, D., & Kahneman, D. (Eds.), *Heuristics and Biases: The Psychology of Intuitive Judgment* (pp. 271–291). Cambridge, UK: Cambridge University Press.

Chapter 12
Conversation Analysis of Coalition Communication in Network Centric Operations

Ping Xue
Boeing Research & Technology, USA

Stephen Poteet
Boeing Research & Technology, USA

Anne Kao
Boeing Research & Technology, USA

ABSTRACT

Team verbal communication is a central component of coalition operations. Investigating the nature of (mis)communication in coalition operations, we conducted analyses based on both natural discourse data recorded during mission execution exercises, and the data collected through interviews of experienced coalition personnel. The authors' analyses shows that pragmatic aspects of language use are essential parameters to successful understanding of communication threads, which demonstrate the ineffectiveness of purely semantic-similarity-based approaches to conversational data. This analysis provides insights into the strengths and weaknesses of current computational approaches such as Latent Semantic Analysis as a methodology for analyzing conversational discourse. As a result, the authors propose a computational framework that incorporates both LSA and computational pragmatics for automating the analysis of coalition communication to support the concept of network-centric operations.

INTRODUCTION

Current major military deployment almost always involves collaborations among multiple nations with groups and team members from diverse backgrounds (Pierce 2002a, 2002b; Chiarelli and Michaelis 2005). The success of coalition operations requires highly effective and efficient communication between teams and team members. In a network centric environment in this Information Age, human communication constitutes an important dimension of the human networks, interacting with other types of networks such as communication networks and information networks. The degree of effectiveness of

DOI: 10.4018/978-1-61520-855-5.ch012

human communication will impact on the degree of the coalition interoperability, determining the effectiveness of coalition forces. However, it has been recognized that diverse backgrounds among multinational groups and team members have presented serious challenges in coalition communication (Pierce 2002a, 2002b; Chiarelli and Michaelis 2005). The challenges include frequent miscommunication among coalition teams. The related issues have been studied from various perspectives and methods including automated methods and techniques. In recent years, statistical or numerical text analysis methods such as Latent Semantic Analysis (henceforth LSA, Landauer et al. 1998) have been used to automate analysis and measurement of team communication conversation and discourse. In this chapter, we examine this human dimension of the coalition network in a network-centric context. In particular, we investigate the nature of miscommunication in coalition operations by exploring computational approaches to conversation analysis. Language is used in complex ways involving linguistic, cognitive, cultural and contextual differences. We focus on linguistic variations and variations of language use in Command and Control (C2) and planning in military coalition teams, especially among teams and members from English-speaking countries. We conduct an exploratory analysis of the discourse data resulting from a mission execution simulation commonly known as the Singapore Data.[1] We also report our analysis of data collected from interviews of UK and US military officers who had experienced miscommunications with coalition partners during multination training and operations.

Our analyses suggest that many relevant issues are largely pragmatic in nature, beyond not only lexical and grammatical differences but also "semantic" (topic) similarity of the communication content. This supports our on-going efforts to develop a computational pragmatics methodology. Departing from conventional computational approaches to language, this approach aims to understand the use of language in a social and communicative context by incorporating social, discourse and contextual information following the tradition advanced by Austin (1962) and Grice (1978, 1989), among other researchers. We believe that pragmatic dimensions such as speech acts (Austin 1962; Searle 1969) and conversation sequence patterns (Sacks 1992) are important parameters to successful understanding of the true nature of communication threads, especially the evaluation of communication effectiveness.

This chapter is organized as follows. In Section 2, we briefly review current computational approaches to the analysis of conversational data, in particular, Latent Semantic Analysis (LSA). We will discuss its general strengths and limitations from a theoretical point of view. In Section 3, we present a detailed description of the Singapore Data. In Section 4, we first provide background information about the Text Representation Using Subspace Transformation (henceforth TRUST, Booker et al. 1999) and the Starlight Information Visualization System (henceforth Starlight, Risch et al. 1999, see Figure 1). We then describe our analysis methods, and present the results of our preliminary analysis. In Section 5, we will discuss the implications of this analysis, particularly the need to incorporate pragmatic aspects of human team communication such as speech acts and conversation patterns including patterns of speech act sequences. In Section 6, we discuss the aspects that are important to a computational analysis of team communication, sketching an approach that incorporates both LSA and computational pragmatics methods. Finally in Section 7, we will conclude by outlining the issues and topics for future work.

PREVIOUS AND CURRENT COMPUTATIONAL APPROACHES

Evaluation of team communication and communication performance has largely relied on human observation and manual analysis. Earlier studies focused on the usefulness of counting various aspects of communication acts to study use cases such as aircrew performance. Orasanu (1990), and Mosier and Chidester (1991), for example, found that better-performing teams demonstrated more overt communications than did crews that performed worse. Other studies (e.g., Jentsch et al. 1995), however, have only partially confirmed the hypothesis that more communications are associated with better performance. Bowers et al. (1998) proposed analyzing patterns instead of merely counting statements or different types of statements. They performed analysis of variance on various action statements and acknowledgement utterance sequences, and found that utterance sequences of these two types correlated with team performance: poor performing teams were less likely (i) to follow action statements with other action statements; (ii) to follow air traffic statements with planning statements; and (iii) to follow uncertainty statements with acknowledgements. While these analyses have shown insightful results, manual methods often require tedious effort such as transcription and annotation, which is expensive and time-consuming. For large-scale operations like coalition operations with multinational teams and large organizations involved, it becomes impractical or impossible to depend on manual methods for conducting efficient, systematic and time-critical studies. This calls for research on automated methods to analyze team communication. The goal for an automated method is to not only be able to analyze the communication data automatically, but also to support constructing models that can enable automatically monitoring, evaluating and predicting communication process based on the analysis results.

In the past decade or so, computational conversation analysis research has been largely focused on some specific application domains. Stolcke et al. (2000) focus on speech dialog system, using a Hidden Markov Model, modeling a conversation as a sequence of utterance types with the transition between adjacent types being a probabilistic function. They work with speech data from the Switchboard corpus (Godfrey et al. 1992). The utterance sequences are often simply dialogues with highly focused topics and purposes. In addition, unlike Bowers et al. (1998) above, there are no performance measures that these dialog systems are trying to predict from the conversation.

Speech-based dialog systems often include multimodal interface using dialog text (Cox et al. 2000; Perez-Quinones 1994). These systems track and make use of contextual information. The intelligence of the system is largely encapsulated in a service logic that embodies domain knowledge about the data and/or about the environment. It should be pointed out that the domain knowledge used in these systems is quite limited and specific and, as a result, such systems are not easily adapted to another technical domain. Importantly, these systems have primarily been developed to resolve lexical ambiguities and, to a lesser extent, pronoun reference. In such systems, deictic orientation resolution largely depends on other components of the system such as the deictic gestures acquired by using the devices like a mouse or pen. The system usually does not provide a pragmatic framework for conversational communication. As a result, it is unable to capture and represent the communicative intention of language use, to derive the presupposition of the communication, or to track and manage the deictic orientation of the context in which the communication occurs. More importantly, speech dialog systems are often task-oriented and are intended to emulate humans in human computer interaction to serve specific application purposes. For coalition operations, by contrast, we need to analyze language communication so as to monitor and evaluate communication performance, especially in multinational teams.

More recent computational work utilizes LSA to extend the analysis of team communications in a number of ways. LSA is a mathematical technique for analyzing the relations between documents and the terms that occur in the documents (Landauer et al. 1998). Using a term-document matrix, it allows one to represent texts of various lengths in the form of vectors, to use linear algebra techniques like singular value decomposition (SVD) to generate a "semantic" space, and to compute the distance between document vectors in this semantic space. This provides a measure of the underlying topics represented in the compared vectors or, loosely speaking, the semantic similarity of any two texts even if they do not contain words in common. LSA allows one to perform search, to compare the utterances of different team members with each other or with the conversation as a whole, to compare the utterances of individuals or whole teams with gold standards (e.g. training documents or the commander's briefing), or to assess the topical coherence of a conversation from utterance to utterance. Importantly, it can do all these without requiring extensive dictionaries or knowledge bases or even a marked-up training sample. Given these properties, LSA has been used for a wide range of applications, and more recently it has been used for dialog act classification and discourse analysis of team conversation (Foltz et al. 2006; Gorman et al. 2003; Martin and Foltz 2004; Serafin and Eugenio 2004).

In Gorman et al. (2003), LSA is combined with a nearest neighbor classification algorithm to automate the categorization of utterances into utterance types that Bowers et al. (1998) performed manually, using the same categories. While the results of the automatic classification were not as accurate as the manual annotation, they were reasonable and considerably faster. They also noted that use of the additional "syntactic" features (i.e. the question mark "?") in addition to the LSA features improved the classification results considerably. Furthermore, they also used LSA to analyze text coherence by comparing utterances or utterance sequences with preceding utterances within a given window, with the tentative conclusion that higher performing teams had more topic agreement between distant tags than lower performing teams. Finally, they measured what they called "communication density", a sort of measure of meaningfulness, as the length of the LSA vector divided by the number of words in the utterance, and found it had not a linear but a negative quadratic effect on performance; i.e. there was an optimal communication density for high performing teams – both more and less were detrimental.

Foltz et al. (2006) extended the analysis of Gorman et al. in several ways. First, they not only predicted the utterance types for the same data that Gorman et al. used; they also used the utterance types to predict the performance (though, unlike Bowers, they apparently just used individual utterance types rather than 2-utterance sequences for the predictions). Secondly, they applied the same approach to other data sets to predict performance on a number of measures with similar encouraging results. Note that, while they speak of generalizing, what they do is apply their approach (using LSA and a nearest-neighbor classifier to assign tags to utterances) to a new set of data. They are not applying the results of training on one set of data to a new set of data, as might be suggested by their use of the term "generalization".

In summary, speech act based approaches (e.g., Stolcke et al. 2000) focus on the sequential structures in the conversation data, and have been concerned with predicting the appropriate next speech act for interactive system like automated phone reservation systems but not with analyzing the quality of the communication. LSA-based approaches (e.g., Foltz et al. 2006), on the other hand, while attempting to assess communication quality, focus on topic relatedness as the primary metric. We will argue that, while these latter approaches tackle important aspects of communication, they fail to take seriously the pragmatic structure of conversation, including speech acts and their appropriate structuring, and thus fail to address essential factors for evaluating the effectiveness of the communication. In the following sections, we examine Singapore data and demonstrate that many issues in question are indeed pragmatic in nature, beyond not only lexical and structural properties, but also what can be addressed by topic relatedness.

SINGAPORE DATA

Our analysis is conducted on the basis of the Singapore Data, which were collected in October 2006, when DARPA sponsored an experiment based on a joint exercise involving the Singapore Armed Forces and the U.S. Army Research Laboratory, Human Research and Engineering Directorate (Pierce 2007).[2] The purpose of the experiment was to simulate multinational team communication scenarios and to provide data for developing and evaluating technologies to improve collaboration and interoperability among multinational teams. The scenario involved combined intervention of US and Singaporean forces in a confrontation between two imaginary countries. Roles included, for each partner, embassy personnel, combat forces and commanders at various level, intel officers etc. In addition, the enemy (one of the two imaginary countries), the media and NGOs (non-government organizations) were also represented. There were two runs of the experiment corresponding to two different conditions: one with standalone computing systems (Singapore's Basic Mission Planner in non-collaborative mode) and only basic intergroup communication systems (e.g., face-to-face communication, text chat, email and phone calls), and the other with highly interoperable computing systems and sophisticated communication systems (e.g., CATISA in its fully collaborative mode which provided secure text chat, VoIP and emails, and a collaborative planning tool, and sense making / information sharing tools). Prior to the start of each run, the participants attended a pre-planning and training session to familiarize themselves with the communication systems and experimental contexts. Each run lasted two days, during which each person could participate in conversations through emails and/or by joining different chat rooms.[3]

Understanding of the commander's intent and mission is a main objective performance measure in the experiments. Importantly, the experiment sessions started with a briefing by the commander of the Coalition Task Force Headquarters. This experiment resulted in two sets of conversational data, with each utterance annotated with additional parameters of the communication. The two sets of conversational data, Run1 and Run2, contain 2,694 and 3,356 records of utterances respectively. In addition to a text field containing a message ranging from a single word to multiple sentences, each record contains the attributes shown in Table 1.

ANALYSIS OF SINGAPORE DATA

Trust and the Starlight System

We processed language and the related communication parameters both from a topic structure and topic dynamics point of view, examining these aspects of the team conversation data using Text Representation Using Subspace Transformation (TRUST, Booker et al. 1999), developed by the Text Mining team in Boeing Research & Technology. TRUST is an enhancement of LSA, combining methods and techniques of mathematics, statistics, high performance computing and computational linguistics, and is designed to process large amounts of text data. The analysis results are graphically displayed using the Starlight Information Visualization System (Starlight, Risch et al. 1999), developed by Pacific Northwest National Laboratory (PNNL) with TRUST as its text processing engine. TRUST provides 3 key capabilities to Starlight:

- A numeric representation of the free text data to enable display of document in topic space.

Table 1. Data attributes

SID	unique identifier
Timestamp	system time
Date	date and time that the communication occurred
Domain	the TeamViz domain; for Run1 the domain is Coalition Ops Run1, and for Run2 the domain is Coalition Ops Run2
Conversation Type	email or chat room conversation
Group	group that the message sender belongs to
Role	the role of the person that sent the communication
Broadcast	true indicates that the message was broadcast to the entire chat room[4]

- Text Summarization capability which gives users a quick at-a-glance view.
- Natural language query capability.

Using a visualization-oriented user interface, Starlight displays documents as topically organized clusters in a 3-D space, where individual documents are represented within the clusters as scatter plots (Figure 1). Documents with similar topics are represented closer together in the 3-D space. To augment the user's understanding of the display, Starlight provides brushing, annotation and various querying capabilities. Thus, Starlight coupled with TRUST and other advanced information modeling and management functionality captures information relationships in meaningful ways, and presents these relationships in intuitive graphical formats. This enables powerful forms of information visualization, access, exploitation, and understanding. Starlight is being used by over 1,000 users in the intelligence community. Here we are trying it out for the first time on a new type of text data, utterances from conversations involving multiple participants and teams.

Methods and Results

We initially hypothesized that connected conversations or threads would be by-and-large confined to a chat room, but when we tried to analyze utterances from individual chat rooms, we discovered that most of the conversations tended to be broken up with numerous gaps. This was our first indication that "Broadcast" might mean that a message was being posted on all chat rooms, rather than confined to a single chat room. We then combined all of the data for each run into a single group. We reformatted the data to meet Starlight's requirements with the original data relations completely preserved. Looking at the clusters of utterances, we discovered that one cluster seemed to refer exclusively to email. Starlight provides a means to extract the text data from a cluster for presentation in tabular form. Using this we discovered that it did in fact refer to a coherent conversational thread that spanned most of the two days. Using another of Starlight's features, we visually linked the dots representing the utterances in the 3-D space according to their temporal sequence. This provided a more or less continuous movement through the semantic space of the cluster, with occasional sudden long shifts followed within two or three utterances by returns to the vicinity of the document preceding the shift. Upon examination of the accompanying text (by expanding the dots to the text record they represent), we discovered that these shifts and returns corresponded to brief side conversations, in parallel with the main thread, and a return indicated the end of a side conversation. We found this pattern interesting and went on to explore other clusters.

Figure 1. Screen shot from analysis of Singapore data using starlight

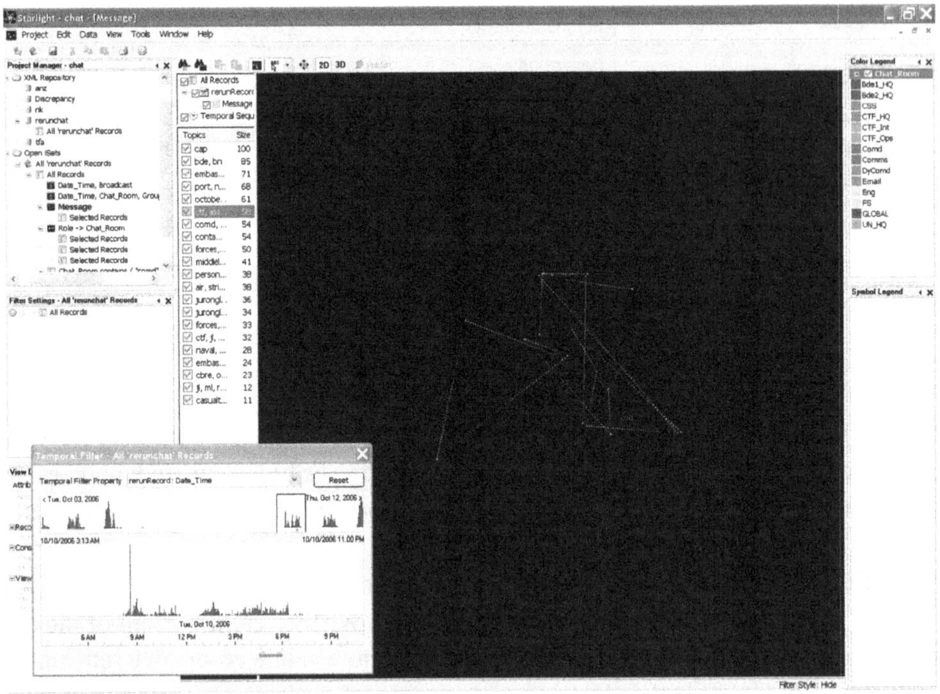

We found that the other clusters did not correspond to other chat rooms as we had expected (the email was classified as a "chat room" in the experiment) and our initial explorations had not yet revealed any clear patterns to the nature of these clusters and their relationships. Reading through the data, we discovered that many of the conversational threads spanned chat rooms, though some utterances seemed to be disconnected from the surrounding dialogue and email conversations in general were orthogonal.

Processing steps in Starlight:

- Message content clustering with TRUST
- Query on "Jurongland, Middleland, delete" for a subset of messages
- Apply temporal filter (from 4:30pm to 6:00pm on Oct. 10)
- Link Contents with temporal sequence

Significantly, we observed that, while the topic-relatedness embedded in the semantic clusters may be an important indication of the relevance of the communication content of the conversation in question, the semantic similarity of the conversation topics among teams and team members did not provide a meaningful performance measure for the level of understanding of commander's intent and effectiveness of communication. We discuss this observation in detail below.

DISCUSSION AND IMPLICATION

Our initial analysis results at least partially agree with the conclusion by previous studies (e.g., Foltz et al. 2006) that LSA based method can be used to analyze and model conversational discourse. More precisely, LSA can be a powerful tool for inferring the topics (semantic content) of a team communication and for measuring the semantic similarity of the communication topics among teams. It is a promising approach for automatically monitoring, evaluating and predicting the communication process based on the analysis of conversational texts and communication discourse.

Our analysis results also revealed the limitations of the conventional LSA approach for evaluating communication performance. LSA is used to measure the "shared interest" between CTF (Combined Task Force group) and each of the brigades. The focus of LSA is to detect topic relatedness. The longer a stretch of conversation centers around the same topic (or closely related topics), the higher the LSA scores the text will receive. Consequently, a longer sequence of sentences would receive a higher score as long as the topics of these sentences stay related. This is counterintuitive. As some of the subjects in the experiment pointed out,[5] what is important is to measure the quality not the quantity of the conversation. In our view, pragmatics is essential to the evaluation of conversation quality, and this is crucially missed in an LSA approach.

In general, LSA's strength is in finding the subtle topic relationships between pieces of text (from words to documents). It does this by, in a sense, "averaging" the contributions of all the words in a text unit. This means that, not only is it incapable of answering questions like who did what to whom, but it is also incapable of picking out the individual contributions of words and phrases. These are essential in determining presuppositions, the speech act being performed, and the subtle structuring of dialogue. Existence presuppositions can arise from words like "the" which are typically on stopword lists and ignored by LSA in computing its latent semantic space. Similarly, responses like "yes" and "no" and discourse connectives like "moreover" and "therefore" are significant in structuring conversations, but again are frequently on stopword lists in an LSA approach. Furthermore, a question such as "Has he left?" and a statement like "He has left" communicate different speech acts, but the only clue to this is the ordering of the words, which is ignored in a pure LSA approach.

This provides a clue as to why email and chat room utterances in the Singapore Data yielded such differing results. Email tended to consist of longer utterances than chat room, containing at least full sentences, while chat room utterances frequently consisted of only one or two words. The longer utterances provide richer 'context' words that LSA is able to take advantage of in building the latent semantic model. In addition, emails tended to have preceding emails attached, artificially making the topics of the various emails even more similar because they shared much of the same text. It is not surprising that the emails all clustered together in semantic space.

In addition, LSA is good at finding relationships between text units or utterances that are superficially different, but less useful at finding distinctions. This may be especially crucial in trying to assess communication failures due to subtle differences in language or in the world view and presuppositions underlying the use of language in a multicultural context. Even confining ourselves to a single cultural or organizational context, communication failures may occur even when a stretch of discourse is topically coherent. Let us use an example from Run2 to illustrate this last point.[6]

- *On 10/10/2006 at 16:49, in the UN_HQ (UN headquarters) chat room, SG_Rep (The Singapore representative) in the White Cell group asked Commander and Deputy Commander to delete a paragraph (as a revision).*

- *He also requested an acknowledgement on this.*

- *At 16:52, PMO_US (the US Political Military Officer) acknowledged this in the same chat room.*

- *At 16:54, Bde1_2IC (Brigade 1 2nd in Command) in group Bde1 stated in chat room Bde1_HQ (Brigade 1 headquarters) that there was a change in the generation situation (which most likely refers to this revision).*

- At 16:55, Bde1_2IC asked in chat room UN_HQ to confirm that there was no invasion.

- A few seconds later, PMO_US requested the attention of the Commander in chat room CTF_HQ that there was a request to delete a paragraph in the message sent out by SG_Rep.

- At 17:00, G2 in group CTF asked in chat room UN_HQ to confirm that there was an invasion.

- At 17:01, Bde1_2IC asked PMO_U and SG_Rep in chat room CTF_HQ whether that paragraph in question was still valid or not.

- At 17:02, G2 in group CTF said in the same chat room that "there seem[s] to be some missed comm[unication]".

- A few seconds later, G2 said in the same chat room that the invasion did take place. It is not clear where he got this information.

- At 17:08, Bde1Comd in chat room Bde1_HQ asked everyone to note that there was a paragraph being deleted.

- At 17:10, PMO_SG in chat room CTF_HQ asked (in non-broadcast mode) about the deletion of the paragraph.

- At 17:12, PMO_SG in the same chat room (also in non-broadcast mode) confirmed (indirectly) the deletion of the paragraph because it was a typo.

- At 17:13, after a few more related, non-broadcast chat by PMO_US, he stated (in broadcast mode) in chat room UN_HQ that there was a correction of the initial report.

- At 17:13, SG_Rep in the same chat room (in non-broadcast mode) reported the deletion and asked for acknowledgement.

- At 17:14, PMO_SG acknowledged it, and one minute later, asked (in non-broadcast mode) for a "line" saying it was due to a typo. There was no recorded reply from SG_Rep on this.

- A few seconds later, in the same chat room, he sent out a message citing SG_Rep to the effect that there are instructions to delete the text due to the typo.

- At 17:26, SG_Rep acknowledged that the deletion was due to a typo in chat room CTF_HQ.

- At 17:33, in chat room CTF_HQ, Bde1Comd said that he thought the information in that paragraph was deleted. Commander in the same chat room confirmed that (indirectly) at 17:34.

Given this thread of conversation, the topic stayed closely on the same question: whether there is indeed an instruction to delete a paragraph. The topic relatedness is very high according to LSA. This is nevertheless clearly a case of poor communication, demonstrating one of the shortcomings of a purely LSA-based approach. In order to uncover the true nature of this thread of conversation, it is necessary to analyze the pragmatics of the conversation, in particular the patterning of speech acts like acknowledgement and requests for acknowledgement and other types of expressions when uncertainty is involved. This thread of conversation started with a request for the deletion of a paragraph, and asked for acknowledgement. After the acknowledgement, the confusion or uncertainty persisted. There were more requests for clarification and reconfirmation of the same point, followed by answers and acknowledgements of the same question, for about 45 minutes. In short, while topic relatedness is an important measure of some dimension of communication content, the topic-related clusters are not a reliable indicator of the level of understanding of the commander's intent and the effectiveness of communication. On the contrary, a high degree of relatedness among the words distributed in the context might be an indicator of the opposite, namely difficulty of communication. A purely LSA-based approach is incapable of detecting the difference.

These challenges can be addressed to some extent by incorporating contextual information in two ways: (i) an LSA approach could use the commander's initial briefing or other supplementary materials (such as related military doctrines and manuals) to establish a representation that is deemed to be most closely related to the current mission; (ii) domain specific ontology could be used to further supply related topics for the LSA representation. For example, if the mission is to help clean up nuclear material pollution, all the related cleaning materials and methods should also be included. One could use an ontology in conjunction with appropriate sample data collection to allow this type of information to be highly relevant to the LSA representation. Even with these augmentations, LSA alone may not be sufficient to determine whether the communication is effective in the sense of C2 and planning. In short, communication effectiveness represents a dimension that cannot be simply measured by the degree of topic-relatedness.

VARIATIONS OF LANGUAGE USE AND LINGUISTIC PRAGMATICS

Poteet et al. (2008a, 2008b) conducted research under the International Technology Alliance Program on the issue of misunderstanding due to miscommunication, as an attempt to address two of the military coalition "hard problems" (Verma 2009) related to reducing information load and improving shared understanding. The focus of the research has been on exploring variations in language and language use between US and UK personnel due to UK/US military background and cultural differences, and contextual differences in general. Poteet et al. employed semi-structured interviews of UK and US military officers who had experienced miscommunications with coalition partners during multination training and operations. The study analyzed instances of miscommunication between UK and US military groups based on the data collected. Various types of linguistic variations and cultural differences manifested by the US and UK groups were identified. American English and British English differ in complex ways not only in terms of lexical differences but also, perhaps more importantly, in terms of language use due to cultural differences. More specifically, the study identified at least five areas where linguistic variations and cultural differences manifested by the UK and US groups led to miscommunication. These areas include use of acronyms, use of slang and colloquialisms, and use of jargon.

What is important to the discussion here is that miscommunication often results from misinterpretation of speech acts and/or inappropriate interpretation of speaker's intended meaning due to the differences between the speaker and the hearer in terms of background knowledge and understanding of the context in which the communication occurs. Language is used to do more than perform simple functions (or speech acts) such as making statements or assertions. It can be used to question, to promise, to request, to greet, to congratulate. While there is sometimes a syntactic correlate of the speech act (in English, typical questions have the first auxiliary verb before the subject and many have a question (*wh-)* word introducing the sentence), this is not always the case and often what looks like one speech act on the surface is really another speech act. For example, the apparent yes/no question "Could you pass the salt?" is actually a request. Various interrogative forms are used in English to perform speech acts other than questions. Misinterpreting the intended speech act can be more complex due to socio-cultural factors involved. For example, what was offered by a commander as an observation or statement might be interpreted by subordinates as an order; or contrariwise, a polite command commonly used in one culture may be interpreted as just a suggestion by people from another culture.

One of the most familiar examples of a speech act which is not limited to the military was reported by an interviewee who is on a US exchange in the UK. A standard UK greeting "Are you all right?" was interpreted by him as a question about his health or situation, and he initially responded accordingly. Unlike some simple lexical differences, he reported that this misunderstanding took a while to figure out. A related aspect is concerned with differences in styles of communication, which are more closely associated with cultures. For example, a cultural tendency to understatement may result in a misinterpretation and misunderstanding of the extent of a problem. Social status and context may also have impacts on interpretation. Such examples include differences in habitual inferences e.g., a simple order by a commander to "go to" a certain location suspected of being friendly to hostiles may be interpreted by one group as a command to take that area by force but by another as simply an order to go there and assess the situation.

Unlike many of the instances of lexical miscommunication, these cases were not instances of a failure to understand but rather of a misunderstanding, a communication that was believed to have succeeded on both parts, but might have in fact failed disastrously. The misunderstanding is due to knowledge of English accompanied by insufficient knowledge of English-speaking societies and cultures leading the hearer to "improvise inferential solutions – to construct the final message – based on divergent socio-cultural realities." (Garcia and Otheguy 1989, p.2). Miscommunications (or ineffective communication) like these cannot be solved by a common dictionary with standard military terminology.

In summary, from a rather different perspective with a rather different methodology, these studies (Poteet et al. 2008a, 2008b) converge with the analysis presented in Sections 4 and 5 in suggesting that many relevant issues of team miscommunication in coalition operations belong to the issues of language use and are pragmatic in nature, not just a matter of lexical or grammatical differences or something that measures of topic similarity could detect. These issues can neither be adequately addressed by conventional language processing approach, nor by standard LSA approaches purely based on topic-relatedness represented in term-document matrix. This supports our on-going efforts to develop a computational pragmatics methodology.

A COMPUTATIONAL PRAGMATICS APPROACH TO THE ANALYSIS OF TEAM COMMUNICATION

It has been long recognized that the study of language cannot be separate from the social and cultural context. Austin (1962) pointed out that 'there is no simple and handy appendage of a word called 'the meaning of the word(x)"', and warned us to take care when removing words from their ordinary usage. Searle elaborated the theory of speech acts in (Searle 1969) and other work, as part of a theory of social acts distinct from physical actions. Grice (1968) distinguished utterer's meaning from utterance's meaning, which led to his theory of conversational implicature. The important insight that pragmatics provides is that we use language not only to convey information but also to perform communicative actions such as making statement, suggestion, promise or request, or asking question, thus indicating functions of the utterances in the context (Austin 1962; Levinson 1983; Morris 1938). Linguistic pragmatics is concerned with the relations between linguistic utterances, the utterer's intention, and the appropriate (e.g. social) context, focusing on how language is used and what effect context has on the interpretation of linguistic expressions.

Although it has been widely acknowledged that the understanding of pragmatic aspects of language use is of crucial importance in many theoretical and practical areas, the current computational research on these aspects of language use have been quite primitive and have largely focused on rather specific and limited domains. This is probably because while most people tacitly understand the basic concepts of linguistic pragmatics, these concepts with their related aspects are not easy to define precisely or to implement computationally. The notion of context is one of these fundamental concepts. The term *context* refers to a wide range of factors such as the participants in the communication, the temporal and spatial parameters of the communication event, the beliefs and intentions of the participants, and other socio-cultural aspects of language use in human communication that may have effects on the use and interpretation of language. While it is very important in the study of pragmatics, the notion of context is unfortunately even less understood. Systematic investigation of context and context representations has been undertaken primarily in cognitive science and artificial intelligence fields (see e.g., Bunt 2000; Dey and Abowd 2000; McCarthy 1993). Pragmatics has also been used in linguistics and philosophy to refer to a wide range of phenomena of language use, including common ground, implicature, presuppositions, discourse structure, speech acts, and so on. Computational pragmatics is concerned with the same range of relations between linguistic phenomena and context. Computationally, each of the topics in pragmatics can be thought as an inference problem, resolving apparent ambiguity or filling in the information that is not apparently present. What has been achieved by human language technology in computational pragmatics is far from what is needed. Reference resolution, interpretation of speech acts, discourse analysis and abduction are the primary areas that have received the most attention in the area of computational pragmatics (Bunt and Black 2000; Jurafsky 2006). However, reference resolution seems to be the only aspect that has been widely applied in applications utilizing human language technology.

Given the complex and diverse aspects of the problem, no realistic single solution seems to be possible to solve the general case of pragmatics analysis. We do not attempt to address all the major aspects of pragmatics. Instead, we will initially focus on computational analysis of speech acts and their patterning in conversation. Specifically, we sketch a method that computationally analyzes speech acts of team discourse to demonstrate a computational pragmatics approach to the analysis of team communication. We show that this approach is capable of measuring and evaluating the performance of team communication. Recall that the thread of conversation that we discussed in Section 4 started with a request for the deletion of a paragraph, and asked for acknowledgement. After the acknowledgement, the confusion or uncertainty persisted. There were more requests for clarification and reconfirmation of the same point, followed by answers and acknowledgements of the same question, for about 45 minutes. Although the initial misunderstanding or confusion might be simply due to misunderstanding of some lexical items or possibly other reasons, a crucial aspect of this thread is the sequence of speech acts, i.e. a request, then an acknowledgement, followed by repeated requests for clarification and reconfirmation and repeated answers and acknowledgements, which is an indication of unsuccessful and ineffective communication independent of what the original cause was. If we would add the dimension of speech acts and other pragmatic cues in addition to the topics represented by LSA, this communication situation would be clearly revealed.

There are two distinct computational approaches to speech act analysis: logical-inference approach and probabilistic-inference approach.

While deductive reasoning can be useful in guiding logical inference and avoiding logical errors, abductive reasoning seems to be more common in human communication. Often there are not enough cues represented in the utterance in the communication that can serve as premises to allow derivation of a valid conclusion. Instead, there is a collection of cues that collectively suggest the most probable interpretation. This collection of cues can be summarized to describe the patterns and to draw inference on a probabilistic basis. The inference may not be valid but it is the inference that best explains the observed evidence. Following this line of research, we use probabilistic reasoning in our computational analysis of speech acts. More specifically, we follow the spirit of cue-based approach described in Jurafsky (2006) using surface cues including specific lexical, collocational, and syntactic features as well as cues provided by conversational patterns available in the context. A cue is a surface feature in language that is probabilistically associated with some speech act. For example, lexical cues include *wh-* words for questions, and *yes* or *no* words for responses and syntactic cues include word order such as English aux-inversion. As mentioned above, discourse connectives like *moreover* and *therefore* are significant in structuring conversations. Conversation patterns are also important cues. Certain pattern of speech act sequences may suggest the status of the communication. For example, repeated requests for and confirmation of the same information manifested in the Singapore Data is an indication of poor communication performance or even communication failure, as discussed above.

We are devising a set of speech act-like categories based on the DAMSL (Dialog Act Markup in Several Layers) architecture (Allen and Core 1997; Core and Allen 1997) to describe and code various dialog and utterance types for our computational purpose.[7] We plan to use a supervised machine learning algorithm to build probabilistic models (or classifiers) statistically trained on manually annotated corpus of conversation utterances with speech acts. Certain statistical classifiers (e.g. Naïve Bayes, Support Vector Machines, cf. Sebastiani 2002) can make predictions of the category of an utterance based solely on the features of that utterance (lexical, grammatical, punctuation). This general approach can also be extended to account for variations of speech act interpretation between English-speaking nations. Recall that there are speech act variations such as "Are you all right?", which is interpreted as a question in American English but a greeting in British English. Such variations can be intuitively coded in the training data and thus the models trained on the training data for American English and British English respectively. However, discourse structure is also an important discourse cue for identifying the relevant speech act. Often the speech act of an utterance varies depending on the utterance that precedes it. For example, the utterance "No, it isn't a success" can be a disagreement or agreement depending on whether the preceding statement is a positive statement like "it is a great success!" or a negative statement like "it isn't a success at all" (see Allwood 1995 for similar examples). In order to take sequence information into account, a different type of classifier would need to be used, such as a Hidden Markov Model (HMM, Lawrence 1989) or a Conditional Random Field classifier (CRF, Lafferty 2001).

Thus, like the work in Jurafsky (2006), our computational analysis of speech acts makes crucial use of surface cues in the context. But unlike the work in Jurafsky (2006) and other speech dialog systems which were designed to serve a small set of specific application purposes, we analyze conversation discourse in coalition operations which represent a much more dynamic environment. In addition, one of the major goals of our analysis is to evaluate team communication performance, which is irrelevant for a task-oriented speech dialog system. Recognition of the relevant speech act for an utterance provides an important basis for team communication interpretation and evaluation.

Given the discussion above and an appropriate set of the speech acts determined, we outline the basic aspects of our computational approach to the analysis of team communication in the following:

i) Use TRUST (a version of Latent Semantic Analysis) to identify topic coherence and patterns of topic change throughout a segment of conversation.
ii) Use a high level ontology of the designated domain, based on, for example, the commander's intent, to help identify the semantics of different segments of the conversation. For example, a plan to move a military unit will require certain logistics planning such as time line, destination, and assets.
iii) Identify the sequences and patterns of speech acts and other pragmatic features in conversation, especially during planning. For example, after a commanding officer gives an order for a troop movement, he usually receives a certain number of acknowledgements and a certain number of requests for clarification.
iv) Based on a conceptual framework outlining specific areas in which we expect to see differences in communication patterns across cultures developed in our field work and experiments (Poteet et al. 2008a; 2008b; Verma 2009), we will develop discourse models and correlate language features from (i), (ii) and (iii) above with them. This will help establish a range of patterns for normal communications as a baseline.
v) We will then monitor a new communication and assess it for any deviation from the baseline as an indication of possible communication failures. This can be used off-line to support more focused training or even on-line to alert participants to potential communication failures before it is too late.

CONCLUSION

In summary, many relevant issues of team miscommunication in coalition operations are pragmatic in nature and pragmatic aspects of language use are essential parameters to successful understanding of communication threads, which demonstrate the ineffectiveness of purely semantic-similarity-based approaches to conversational data. More specifically, our analysis of both Singapore Data and the data collected through interviews suggests the need for considering speech acts and pragmatics in general in analyzing team communications. There are various surface cues including lexical, syntactic, collocational, and discourse structure cues that probabilistically correlate with the speech act identity of an utterance. A statistical cue-based approach leverages these cues and provides a method for automatic speech act recognition. We have showed that the information provided by speech acts such as statement, question, request, and confirmation,[8] in addition to the topics represented by LSA, provide valuable insights for the analysis and evaluation of team communication discourse.

In recent years, conversational aspects of language use have been attracting attention from a number of research communities. Speech acts have played an increasingly important role, for example, in software agent communication languages (ACLs) such as KQML (Finin et al. 1997), FIPA (FIPA 1997) and KAoS (Bradshaw 1997). In addition to syntax and semantics, an utterance in one of these ACLs also explicitly specifies what sort of speech act is being performed, e.g. INFORM, REQUEST, and OFFER. These have their formal semantics and, in addition, must fit into some version of conversational policy, specifying constraints on how these speech acts can be used, e.g. what speech act types can follow what other speech act types (Greaves et al. 1998). This work may provide some suggestions on how to analyze conversations between people, but it is, first of all, prescriptive (the ACL designers are specifying rules for how software agents are permitted to interact) and, second, the interaction between software agents is, of necessity, considerably simpler than the interaction between humans.[9]

As noted by Jurafsky (2006), it is critical to determine the relationships between utterance types and performance effectiveness, and both Bowers et al. and Foltz et al. (2006) have indeed shown that certain types of utterance correlate to team performance. Future work will involve conversational analysis (Hutchby and Wooffitt 1997) and developing a set of speech act-like utterance types appropriate to the military planning domain and, in particular, to the kind of multi-participant conversations exemplified by chat rooms. In addition, it is necessary to investigate the patterns that these utterance types enter into, and to model the ways that can be used to indicate difficulty or failures of communication. Well constructed models, as we envision, can provide not only diagnosis for communication failures but also timely warning for communication difficulty before the communication breaks down.

In addition, this approach can be further enhanced by leveraging knowledge-based methods and publicly available lexical and knowledge resources such as WordNet (Fellbaum 1998) and domain specific ontologies. A well constructed framework of computational pragmatics with enhanced LSA techniques and natural language processing techniques is needed to enable automated analysis of communication and provide metrics for essential coalition processes such as planning, in order to facilitate coalition communication in general. Computational pragmatics is still a new frontier where many areas are waiting to be investigated and explored (Stolcke et al. 2000) A well constructed framework of computational pragmatics will serve as the basis to enable technology that can potentially provide automated measurement and feedback on coalition processes such as planning and automated evaluation of the plans and intent statements, and to facilitate coalition communication in general. More precisely, the proposed framework, if successful, will both provide a more fine-grained analysis of planning session failures to facilitate improved training and will allow on-line monitoring of US-UK coalition communication during planning, assessing it for any deviation, and providing warning for potential communication breakdowns when identified verbal communication features exceed an established threshold, advancing the state-of-the-art of the current computational approaches to team communication analysis.

Miscommunication in multinational context is a broad topic with significant impact for military coalition operations. Significant improvements in this area require that efforts be carried out using multiple approaches and from multiple angles. This paper highlights language analysis, especially using computational pragmatics as an important means for addressing transnational miscommunication problems. This approach is especially important in a networking environment, where the ability to automatically track and flag occurrences of communication problems would be advantageous. Under the International Technology Alliance Program, the authors have been collaborating with US Army Research Lab, UK Defence Science and Technology Lab, and Systems Engineering and Assessment in the UK to jointly develop additional models, methods and recommendations to improve training, and to reduce miscommunication incidences as important components of the strategy for tackling miscommunication problems. The ultimate solution to overall miscommunication problems calls for all these approaches to work in concert (Larsson et al. 2008; Poteet et al. 2008a; 2008b).

ACKNOWLEDGMENT

Research was sponsored by the U.S. Army Research Laboratory and the U.K. Ministry of Defence and was accomplished under Agreement Number W911NF-06-3-0001. The views and conclusions contained in this document are those of the authors and should not be interpreted as representing the official policies, either expressed or implied, of the U.S. Army Research Laboratory, the U.S. Government, the U.K. Ministry of Defence or the U.K. Government. The U.S. and U.K. Governments are authorized to reproduce and distribute reprints for Government purposes notwithstanding any copyright notation heron. We also would like to thank the anonymous reviewers for helpful comments on an earlier version of this paper.

REFERENCES

Allen, J., & Mark, C. (1997). *Draft of DAMSL: Dialog act markup in several layers*. Unpublished manuscript, University of Rochester. Available online at: http://www.cs.rochester.edu/research/speech/damsl/RevisedManual/

Allwood, J. (1995). An activity-based approach to pragmatics. *Gothenburg Papers in Theoretical Linguistics 76*.

Austin, J. L. (1962). *How to Do Things with Words*. Oxford, UK: Clarendon Press.

Booker, A., Condliff, M., Greaves, M., Holt, F. B., Kao, A., & Pierce, D. J. (1999). Visualizing text data sets. *Computing in Science & Engineering*, (July/August): 26–35. doi:10.1109/5992.774838

Bowers, C. A., Jentsch, F., Salas, E., & Braun, C. (1998). Analyzing communication sequences for team training needs assessment. *Human Factors, 40*, 672–679. doi:10.1518/001872098779649265

Bradshaw, J. M. (1997). KAoS: Toward an industrial-strength open agent architecture. In Bradshaw, J. M. (Ed.), *Software Agents*. Cambridge, MA: MIT Press/AAAI Press.

Bunt, H. (1994) Representing knowledge extracted from natural language dialog. In *Proceedings of PRICAI'94, International Workshop on Knowledge Engineering and Application*, Beijing.

Bunt, H. (2000). Dialog pragmatics and context specification. In Bunt, H. C., & Black, W. (Eds.), *Abduction, Belief and Context: Studies in Computational Pragmatics*. Amsterdam: John Benjamins.

Bunt, H., & Black, W. (2000). The ABC of the computational pragmatics. In Bunt, H. C., & Black, W. (Eds.), *Abduction, Belief and Context: Studies in Computational Pragmatics*. Amsterdam: John Benjamins.

Chiarelli, P. W., & Michaelis, P. R. (2005). Winning the peace, the requirement for full-spectrum operations. *Military Review*, 4–17.

Core, M., & Allen, J. (1997). Coding dialogs with the DAMSL annotation scheme. In *Working Notes of the AAAI Fall Symposium on Communicative Action in Human and Machines*, (pp. 28035), Cambridge, MA.

Cox, K., Grinter, R. E., & Mantilla, D. (2000). Using dialog and context in a speech-based interface for an information visualization environment. In *Proceedings of ACM 2000*.

Dey, A. K., & Abowd, G. D. (2000, April). Towards a Better Understanding of Context and Context-Awareness. In the *Workshop on The What, Who, Where, When, and How of Context-Awareness, as part of the 2000 Conference on Human Factors in Computing Systems (CHI 2000),* The Hague, The Netherlands.

Fellbaum, C. (Ed.). (1998). *WordNet: An Electronic Lexical Database*. Cambridge, MA: MIT Press.

Finin, T., Labrou, Y., & Mayfield, J. (1997). KQML as an Agent Communication Language. In Bradshaw, J. M. (Ed.), *Software Agents*. Cambridge, MA: MIT Press/AAAI Press.

FIPA. (1997). Foundation for Intelligent Physical Agents. FIPA97 Agent Communication Language Specification, (1.2 ed.).

Foltz, P. W., Martin, M. J., Abdelali, A., Rosenstein, M., & Oberbreckling, R. (2006). Automated team discourse modeling. In *Test of Performance and Generalization, Proceedings of the 28th Annual Cognitive Science Conference*, Vancouver, Canada.

Garcia, O., & Otheguy, R. (1989). *English Across Cultures, Cultures Across English: A Reader in Cross-cultural Communication*. New York: Mouton de Gruyter.

Godfrey, J. J., Holliman, E. C., & McDaniel, J. (1992, March). SWITCHBOARD: Telephone speech corpus for research and development. In *Proceedings of the IEEE Conference on Acoustics, Speech, and Signal Processing*, San Francisco, (Vol. 1, pp. 517-520).

Gorman, J. C., Foltz, P. W., Kiekel, P. A., Martin, M. J., & Cooke, N. J. (2003). Evaluation of latent semantic analysis-based measures of tam communications content. In *Proceedings of the Human Factors and Ergonomics Society 47th Annual Meeting*.

Greaves, M., Holmback, H., & Bradshaw, J. (1998). *What Is a conversation policy? Unpublished technical report*. Seattle, WA: The Boeing Company.

Grice, P. (1968). Utterer's meaning, sentence-meaning, and word-meaning. *Foundations of Language*, *4*, 1–18.

Grice, P. (1978). Further notes on logic and conversation. In Cole, P. (Ed.), *Syntax and Semantics 9: Pragmatics* (pp. 113–128). New York: Academic Press.

Grice, P. (1989). *Studies in the Way of Words*. Cambridge, MA: Harvard University Press.

Hutchby, I., & Wooffitt, R. (1997). *Conversation Analysis*. Oxford, MA: Blackwell.

Jentsch, F. G., Sellin-Wolters, S., Bowers, C. A., & Salas, E. (1995). Crew coordinatin behaviors as predictors of problem detection and decision making times. In *Proceedings of the Human Factors and Ergonomics Society 39th Annual Meeting* (pp. 1350-1353). Santa Monica, CA: Human Factors and Ergonomics Society.

Jurafsky, D. (2006). Pragmatics and computational linguistics. In Horn, L. R., & Ward, G. (Eds.), *Handbook of Pragmatics*. Oxford, MA: Blackwell. doi:10.1002/9780470756959.ch26

Kao, A., Poteet, S., & Xue, P. (2007). Conversation analysis of coalition communication in computational pragmatics. In *Proceedings of the First Annual Conference of the International Technology Alliance*, New York.

Lafferty, J., McCallum, A., & Pereira, F. (2001). Conditional random fields: Probabilistic models for segmenting and labeling sequence data. In *Proceedings of the Eighteenth International Conference on Machine Learning (ICML-2001)*, Williamstown, MA.

Landauer, T., Foltz, P. W., & Laham, D. (1998). Introduction to latent semantic analysis. *Discourse Processes*, *25*, 259–284. doi:10.1080/01638539809545028

Larsson, A., Solodilova-Whiteley, I., Kao, A., Poteet, S., & Xue, P. (2008, September). A multidimensional approach to studying cultural differences and coping strategies in a multinational coalition environment. In *Proceedings of the 13th InternationalCommand and Control Research and Technology Symposium.*, Bellevue, WA.

Lawrence, R. R. (1989). A tutorial on hidden markov models and selected applications in speech recognition. *Proceedings of the IEEE*, *77*, 257–286. doi:10.1109/5.18626

Levinson, S. C. (1983). *Pragmatics*. Cambridge, UK: Cambridge University Press.

Martin, M., & Foltz, P. W. (2004). Automated team discourse annotation and performance prediction using LSA. In *Proceedings of the Human Language Technology and North American Association for Computational Linguistics Conference (HLT/NAACL)*, Boston.

McCarthy, J. (1993). Notes on formalizing context. In *Proceedings of 1993 International Joint Conference on Artificial Intelligence*.

Morris, C. (1938). Foundations of the Theory of Signs. In Carnap, R., & Morris, C. (Eds.), *International Encyclopedia of Unified Science* (pp. 77–138). Chicago: University of Chicago Press.

Mosier, K. L., & Chidester, T. R. (1991). Situation assessment and situation awareness in a team setting. *Designing for Everyone: Proceedings of the 11th Congress of the International Ergonomics Association*, Paris, 15-20 July, (pp. 798-800).

Orasanu, J. M. (1990). Shared mental models and crew decision making. *Cognitive Science Laboratory Report No. 46*. Princeton, NJ: Princeton University.

Perez-Quinones, M. A. (1994). *Computational pragmatics in HCI: using dialog context in a multimodal interface*. Technical Report 20375, Naval Research Laboratory, Washington DC.

Pierce, L. G. (2002a). Barriers to adaptability in a multinational team. In *Proceedings of the 45th Human Factors and Ergonomics Society Annual Meeting*, Baltimore, (pp. 225-229).

Pierce, L. G. (2002b) Preparing and supporting adaptable leaders and teams for support and stability operations. In *Proceedings of the 11th ROK-US Defense Analysis Seminar(Manpower Policy, Session (4)*, Seoul, Korea, (pp. 97-129).

Pierce, L. G. *(2007)*. October 2006 Defense Advanced Project Agency (DARPA) sponsored joint experiment between the Singapore (SN) Armed Forces and the U.S. Army Research Laboratory, Human Research and Engineering Directorate (ARL-HRED). *Unpublished white paper.*

Poltrock, S.E., Handel, M., & Bowyer, H. (2009). *Collaborative behavior patterns and heuristic for recognizing these patterns*. Technical Report, International Technology Alliance, TA4.

Poltrock, S. E., Handel, M., Waggett, P., & Bowyer, H. (2008, September). A dynamic model of mission context. In *Proceedings of the 2nd Annual Conference of the International Technology Alliance*, London.

Poteet, S., Patel, J., Giammanco, C., Whiteley, I., Xue, P., & Kao, A. (2008a). Words Are Mightier Than Swords… and Yet Miscommunication Costs Lives! In *Proceedings of the Second Annual Conference of the International Technology Alliance*, London.

Poteet, S., Xue, P., Patel, J., Kao, A., Giammanco, C., & Whiteley, I. (2008b). *Linguistic Sources of Coalition Miscommunication*. Paper presented at NATO HFM-142 Symposium, Copenhagen.

Risch, J. S., Rex, D. B., Dowson, S. T., Walters, T. B., May, R. A., & Moon, B. D. (1999). The STARLIGHT information visualization system. In Card, S., Mackinlay, J., & Shneiderman, B. (Eds.), *Lectures on Conversation*. Oxford, UK: Basil Blackwell.

Sacks, H. (1992). *Lectures on Conversation*. Oxford, UK: Basil Blackwell.

Searle, J. R. (1969). *Speech Acts. An Essay in the Philosophy of Language*. Cambridge, UK: Cambridge University Press.

Sebastiani, F. (2002). Machine learning in automated text categorization. [CSUR]. *ACM Computing Surveys, 34*(1), 1–47. doi:10.1145/505282.505283

Serafin, R., & Di Eugenio, B. (2004, July). FLSA: Extending latent semantic analysis with features for dialog act classification. In *ACL04, 42nd Annual Meeting of the Association for Computational Linguistics*, Barcelona.

Stolcke, A., Reies, K., Cocarro, N., Shriberg, E., Bates, R., & Jurafsky, D. (2000). Dialogue Act Modeling for Automatic Tagging and Recognition of Conversational Speech. *Computational Linguistics, 26*(3), 339–373. doi:10.1162/089120100561737

Verma, D. (Ed.). (2009). International Technology Alliance in Network and Information Sciences Biennial Program Plan 2009, May 12th 2009 - May 11th 2011.

ENDNOTES

1 See Section 3 for a detailed description of Singapore Data.

2 Note: None of the authors was involved in designing or running the experiment. The resulting conversational data was made available to us for analysis. This analysis was also reported in Kao et al. (2007).

3 It should be noted that the conversations also involved other means of communication such as Voice over IP and personal conversation but only email and chat room conversations were recorded. Thus the data does not present a complete picture of the conversations.

4 This is the description of the Broadcast field that was provided to us. As we will describe below, this does not seem to correspond with the data, but we were unable to get any further clarification about this from participants in the experiment.

5 Somewhat surprisingly, the subjects seemed to be aware of much of the nature of the experiment and made a number of meta-comments about the metrics being used and the experimenters.

6 The participants of this exercise include four major groups, namely Brigade 1 (Bde1), Brigade 2 (Bde2), Combined Task Force (CTF), and White Cell. G2 stands for Army or Marine Corps component intelligence staff officer. Bde1 2IC stands for Brigade 1 2nd in Command.

7 Thus, our speech acts are more like dialog acts as discussed in Bunt (1994).

8 The most commonly referred speech acts are statement, question and request, often so called "big three"; *confirmation is* less studied, however.

9 Most these systems like speech dialog systems are task-oriented. In the context of International Technology Alliance, certain mission-oriented scenarios (such as training scenarios), where the mission is clearly set and the tactics are well defined, may be characterized as a type of task-oriented cases, where conversations might follow certain operational and communication patterns (see Poltrock et al. 2008, 2009).

Compilation of References

Abrams, Z., Goel, A., & Plotkin, S. A. (2004, April 26-27). Set k-cover algorithms for energy efficient monitoring in wireless sensor networks. In ISPN 04, Berkeley, CA, (pp. 424-432).

Adams, F., & Aizawa, K. (2001). The bounds of cognition. *Philosophical Psychology, 14*(1), 43–64. doi:10.1080/09515080120033571

Adams, F., & Aizawa, K. (2009). Why the mind is still in the head. In Ayded, M., & Robbins, P. (Eds.), *Cambridge Handbook of Situated Cognition.* Cambridge, UK: Cambridge University Press.

Adams, F., & Aizawa, K. (in press). Defending the bounds of cognition. In Menary, R. (Ed.), *The Extended Mind.* Aldershot, UK: Ashgate Publishers.

Aggarwal, C., & Yu, P. (2007). *A Survey of Synopsis Construction in Data Streams. Data Streams: Models and Algorithms.* Berlin: Springer.

Aggarwal, C., Ta, N., Feng, J., Wang, J., & Zaki, M. J. (2007). XProj: A Framework for Projected Structural Clustering of XML Documents. In *KDD Conference.*

Aggarwal, C., Xie, Y., & Yu, P. (2009). GConnect: A Connectivity Index for Massive Disk-Resident Graphs. In *VLDB Conference.*

Agrawal, D., Calo, S. B., Lee, K.-W., & Lobo, J. (2007). Issues in Designing a Policy Language for Distributed Management of IT Infrastructures. In Integrated Network Management, (pp. 30-39).

Agrawal, D., Giles, J., Lee, K., & Lobo, J. (2005). *Policy Ratification* (pp. 223–232). POLICY.

Agrawal, R., Borgida, A., & Jagadish, H. V. (1989). Efficient Maintenance of transitive relationships in large data and knowledge bases. In *ACM SIGMOD Conference.*

Ahuja, R., Orlin, J., & Magnanti, T. (1992). *Network Flows: Theory, Algorithms, and Applications.* Englewood Cliffs, NJ: Prentice Hall.

Ajith, A., Crina, G., & Vitorino, R. (Eds.). (2006). *Stigmergic Optimization.* Berlin, Germany: Springer-Verlag.

Akylidiz, I., Su, W., Sankarasubramariam, Y., & Cayirci, E. (2002, Aug). A survey on sensor networks. *IEEE Communications Magazine, 40*(8), 102–113. doi:10.1109/MCOM.2002.1024422

Alaettinoglu, C., Villamizar, C., Gerich, E., Kessens, D., Meyer, D., Bates, T., Karrenberg, D. & Terpstra, M. (1999). Routing Policy Specification Language (RPSL). *RFC 2622 (Proposed Standard)*, June 1999. *Updated by RFC 4012.*

Alberts, D. A., Garstka, J. J., Hayes, R. E., & Signori, D. A. (2001). *Understanding Information Age Warfare.* Washington, DC: CCRP.

Allen, J., & Mark, C. (1997). *Draft of DAMSL: Dialog act markup in several layers.* Unpublished manuscript, University of Rochester. Available online at: http://www.cs.rochester.edu/research/speech/damsl/Revised-Manual/

Allen, R., & Garlan, D. (1994). Beyond definition/use: architectural interconnection. In *Proceedings of the ACM Workshop on Interface Definition Languages*, New York, NY, (pp. 35–45).

Allwood, J. (1995). An activity-based approach to pragmatics. *Gothenburg Papers in Theoretical Linguistics 76.*

Alon, N., Matias, Y., & Szegedy, M. (1996). The Space Complexity of Approximating the Frequency Moments. In *ACM Symposium on Theory of Computing.*

Ashok, R., & Agrawal, D. (2003). Next-generation wearable networks. *Computer, 36*(11), 31–39. doi:10.1109/MC.2003.1244532

Atran, S. (1998). Folkbiology and the anthropology of science: Cognitive universals and cultural particulars. *The Behavioral and Brain Sciences, 21*, 547–609. doi:10.1017/S0140525X98001277

Atran, S., Medin, D. L., & Ross, N. O. (2005). The cultural mind: Environmental decision making and cultural modeling within and across populations. *Psychological Review, 112*(4), 744–776. doi:10.1037/0033-295X.112.4.744

Austin, J. L. (1962). *How to Do Things with Words.* Oxford, UK: Clarendon Press.

Bahrami, A., Yuan, J., Smart, P. R., & Shadbolt, N. R. (2007). *Context-Aware Information Retrieval for Enhanced Situation Awareness.* Paper presented at the Military Communications Conference (MILCOM), Orlando, Florida, USA.

Bandara, A. K., Lupu, E. C., Moffett, J., & Russo, A. (2004). A Goal-based Approach to Policy Refinement. In *Proceedings of the IEEE Policy Workshop.*

Barabasi, A.-L. (2002). *Linked: The New Science of Networks.* Cambridge, MA: Perseus Publishing.

Barbir, A., Murphy, S. & Yang, Y. (2006, October). Generic Threats to Routing Protocols. *RFC 4593 (Informational).*

Bar-Noy, A., & Brown, T. M, Johnson, P., Sarïoz, D., Verma, D. & Wu, C. W. (2008). More is More: the Benefits of Dense Sensor Deployment. In MASS 08.

Bar-Noy, A., Brown, T., Johnson, M. P., & Liu, O. (2007, November 9-10), Covering With Inexactly Placed Sensors. In *17th Fall Workshop on Computational and Combinatorial Geometry.* Yorktown Heights, NY: IBM T.J. Watson Research Center.

Bar-Noy, A., Brown, T., Johnson, M. P., & Liu, O. (2009, June). Cheap or Flexible Sensor Coverage, Intl Conference on Distributed Computing in Sensor Systems. In *DCOSS.*

Barros, J. H., & Rodrigues, M. R. D. (2006). Secrecy Capacity of Wireless Channels. In *IEEE International Symposium on the Information Theory.*

Bartolini, N., Calamoneri, T., Fusco, E. G., Massini, A., & Silvestri, S. (2008). A Self-deployment Algorithm for Mobile Sensor Networks. In *DCOSS 2008.* Snap and Spread.

Basagni, S., Chlamtac, I., Syrotiuk, V. R., & Woodward, B. A. (1998). A Distance Routing Effect Algorithm for Mobility (DREAM). In *Proceedings of the ACM/IEEE Mobicom.*

Becker, M. Y., & Sewell, P. (2004). Cassandra: Flexible trust management, applied to electronic health records. In CSFW, (pp. 139–154). Washington, DC: IEEE Computer Society.

Becla, J., & Lim, K. T. (2007). *Report from 1st Workshop on Extremely Large Databases.* Held at Stanford Linear Accelerator Centre, Menlo Park, CA.

Beigi, M., Calo, S., & Verma, D. (2004, June). Policy Transformation Techniques in Policy-based Systems Management. In Policy Workshop 2004, Yorktown, New York.

Belokosztolszki, A., Eyers, D. M., Pietzuch, P., Bacon, J., & Moody, K. (2003). *Role-Based Access Control for Publish/Subscribe Middleware Architectures.* Distributed Event Based Systems.

Benaloh, J. (1999). *Dense Probabilistic Encryption.* Retrieved from http://research.microsoft.com/crypto/papers/dpe.ps

Bent, G., Dantressangle, P., Vyvyan, D., Mowshowitz, A., & Mitsou, V. (2008). A dynamic distributed federated database. In *Proceedings of ACITA'08*, London.

Bergamaschi, F., Conway-Jones, D., Gibson, C., Stanford-Clark, A., Verma, D., Calo, S., et al. (2008). Policy Enabled ITA Sensor Fabric, A Distributed Framework for the Validation of Experimental Algorithms Using Real and Simulated Sensors. In Proc. POLICY.

Berger, J. A., & Heath, C. (2005). Idea habitats: How the prevalence of environmental cues influences the success of ideas. *Cognitive Science*, 29.

Berners-Lee, T., Hall, W., Hendler, J. A., O'Hara, K., Shadbolt, N., & Weitzner, D. J. (2006). A framework for web science. *Foundations and Trends in Web Science*, *1*(1), 1–130. doi:10.1561/1800000001

Berners-Lee, T., Hendler, J., & Lassila, O. (2001). The Semantic Web. *Scientific American*, *284*(4), 34–43. doi:10.1038/scientificamerican0501-34

Bias, R., & Mayhew, D. (Eds.). (2005). *Cost-Justifying Usability: An Update for the Internet Age* (2nd ed.). San Francisco: Morgan Kaufman.

Biggs, N. (1994). *Algebraic Graph Theory (Cambridge Mathematical Library)*. Cambridge, UK: Cambridge University Press.

Bird, J., & Layzell, P. (2002). *The evolved radio and its implications for modelling the evolution of novel sensors.* Paper presented at the Congress on Evolutionary Computation (CEC'02), Washington, DC.

Bisdikian, C., Branch, J., Leung, K. K., & Young, R. I. (2009). A Letter Soup for the Quality of Information in Sensor Networks. *First Int'l Workshop on Information Quality and Quality of Service* (IQ2S'09), Galveston, TX, USA, March 9.

Bisdikian, C., Kaplan, L. M., Srivastava, M. B., Thornley, D. J., Verma, D., & Young, R. I. (2009). Building Principles for a Quality of Information Specification for Sensor Information. In *Twelfth Int'l Conf. on Information Fusion* (FUSION 2009), Seattle, WA, USA, July 6-9.

Blasch, E., & Plano, S. (2005). DFIG level 5 (user refinement) issues supporting situational assessment reasoning. In Eights Int'l Conf. on Information Fusion (FUSION), Philadelphia, PA, July 25-28.

Blaze, M., Feigenbaum, J., Ioannidis, J., & Keromytis, A. D. (1999). The role of trust management in distributed systems security. In Secure Internet Programming, (pp. 185–210).

Boldyreva, A., Gentry, C., O'Neill, A., & Yum, D. H. (2007). *Ordered multi-signatures and identity-based sequential aggregate signatures, with applications to secure routing. In 14th ACM conference on Computer and communications security* (pp. 276–285). New York.

Bonabeau, E. (1999). Editor's introduction: stigmergy. *Artificial Life*, *5*(2), 95–96. doi:10.1162/106454699568692

Bonabeau, E., Dorigo, M., & Theraulaz, G. (1999). *Swarm Intelligence: From Natural to Artificial Systems*. New York: Oxford University Press.

Booker, A., Condliff, M., Greaves, M., Holt, F. B., Kao, A., & Pierce, D. J. (1999). Visualizing text data sets. *Computing in Science & Engineering*, (July/August): 26–35. doi:10.1109/5992.774838

Bostrom, A., Fischhoff, B., & Morgan, M. G. (1992). Characterizing Mental Models of Hazardous Processes: A Methodology and an Application to Radon. *The Journal of Social Issues*, *48*(4), 85–100.

Bottino, A., & Laurentini, A. (2006, December 17 - 20). Experimental results show near-optimality of a Sensor location algorithm. In *Proc. 2006 IEEE International Conference on Robotics and Biomimetics*, Kunming, China.

Botts, M., & Robin, A. (2007). *Open Geospatial Consortium Inc: OpenGIS® Sensor Model Language (SensorML)*. Implementation Specification.

Bowers, C. A., Jentsch, F., Salas, E., & Braun, C. (1998). Analyzing communication sequences for team training needs assessment. *Human Factors*, *40*, 672–679. doi:10.1518/001872098779649265

Boyer, P. (1994). Cognitive constraints on cultural representations: Natural ontologies and religious ideas. In Hirschfeld, L. A., & Gelman, S. A. (Eds.), *Mapping the mind: Domain specificity in cognition and culture* (pp. 391–411). New York: Cambridge University Press.

Braddon-Mitchell, D., & Jackson, F. (2007). *Philosophy of Mind and Cognition: An Introduction* (2nd ed.). Oxford, UK: Blackwell Publishing.

Bradshaw, J. M. (1997). KAoS: Toward an industrial-strength open agent architecture. In Bradshaw, J. M. (Ed.), *Software Agents*. Cambridge, MA: MIT Press/ AAAI Press.

Brass, P. (2007). Bounds on coverage capabilities for models of networks of mobile sensors. *ACM Transactions on Sensor Networks, 3*(2).

Brik, V., Banerjee, S., Gruteser, M., & Oh, S. (2008, September). Wireless Device Identification using Radiometric Signatures. In ACM Mobicom, Burlingame, CA.

Briley, D., Morris, M., & Simonson, I. (2000). Reasons as carriers of culture: Dynamic versus dispositional models of cultural influence on decision making. *The Journal of Consumer Research, 27*, 157–178. doi:10.1086/314318

Brin, S., & Page, L. (1998). The anatomy of a large-scale hypertextual Web search engine. In *WWW Conference*.

Broder, A. Z., Charikar, M., Frieze, A., & Mitzenmacher, M. (1997). Syntactic clustering of the web. *WWW Conference, Computer Networks, 29*(8–13), 1157–1166.

Broder, A., & Mitzenmacher, M. (2002). Network applications of bloom filters: A survey. *Internet Mathematics, 1*(4), 485–509.

Brodie, C., George, D., Karat, C.-M., Karat, J., Lobo, J., Beigi, M., et al. (2008).The Coalition Policy Management Portal for Policy Authoring, Verification, and Deployment. In *IEEE Workshop on Policies for Distributed Systems and Networks, POLICY 2008*, (pp. 247 – 249).

Brooks, R. A. (1991). Intelligence without representation. *Artificial Intelligence, 47*, 139–160. doi:10.1016/0004-3702(91)90053-M

Brown, T., Sarïoz, D., Bar-Noy, A., La Porta, T., & Verma, D. (2007, September). Full Coverage of a Region Allowing Inexact Placement of Sensors. In *Annual Conference of ITA*.

Brown, T., Sarïoz, D., Bar-Noy, A., La Porta, T., Verma, D., Johnson, M. P., & Rowaihy, H. (2007). Geometric considerations for distribution of sensors in ad-hoc sensor networks. *Proc. SPIE.*

Brugger, A., Bunke, H., Dickinson, P. J., & Riesen, K. (2008). Generalized Graph Matching for Data Mining and Information Retrieval. In *ICDM Conference* (pp. 298-312).

Buchanan, M. (2002). *Nexus: Small Worlds and the Groundbreaking Science of Networks*. New York: W. W. Norton & Company.

Bunt, H. (1994) Representing knowledge extracted from natural language dialog. In *Proceedings of PRICAI'94, International Workshop on Knowledge Engineering and Application*, Beijing.

Bunt, H. (2000). Dialog pragmatics and context specification. In Bunt, H. C., & Black, W. (Eds.), *Abduction, Belief and Context: Studies in Computational Pragmatics*. Amsterdam: John Benjamins.

Bunt, H., & Black, W. (2000). The ABC of the computational pragmatics. In Bunt, H. C., & Black, W. (Eds.), *Abduction, Belief and Context: Studies in Computational Pragmatics*. Amsterdam: John Benjamins.

Burge, T. (1979). Individualism and the mental. In French, P., Uehling, T., & Wettstein, H. (Eds.), *Midwest Studies in Philosophy* (*Vol. 4*, pp. 73–121). Minnesota, MN: University of Minnesota Press.

Butler, K. (1998). *Internal Affairs: Making Room for Psychosemantic Internalism*. Dordrecht, The Netherlands: Kluwer Academic Publishers.

C. elegans Sequencing Consortium. (1998). Genome sequence of the nematode C. elegans: a platform for investigating biology. *Science, 282*(5396), 2012-2018.

Calo, S., Wood, D., Zerfos, P., Vyvyan, D., Dantressangle, P., & Bent, G. (2009). Technologies for Federation and Interoperation of Coalition Networks. In *proceedings of The 12th International Conference on Information Fusion*, Seattle, WA.

Camazine, S., Deneubourg, J.-L., Franks, N. R., Sneyd, J., Theraulaz, G., & Bonabeau, E. (2001). *Self-Organization in Biological Systems*. Princeton, NJ: Princeton University Press.

Camtepe, S. A., & Yener, B. (2007). *Key Distribution Mechanisms for Wireless Sensor Networks: Survey*. RPI Technical Report TR-05-07.

Cardelli, L., & Gordon, A. D. (1998). Mobile ambients. In *Proceedings of the 1st International Conference on Foundations of Software Science and Computation Structure (FoSSaCS)*, (pp. 140–155). London, UK: Springer-Verlag.

Cardenes, A., Radosavac, S., & Baras, J. H. (2005). Detection and Prevention of MAC Layer Misbehavior in Ad-Hoc Networks. In *ACM Workshop on Security of Ad-Hoc and Sensor Networks (SASN)*.

Carr, N. (2008). Is Google Making Us Stupid? *Atlantic (Boston, Mass.)*, *302*(1), 56–63.

Carroll, J. (Ed.). (1995). *Scenario-Based Design: Envisioning Work and Technology in System Development*. New York: Wiley and Sons.

Casey, S. (1998). *Set Phasers on Stun: True Tales of Design, Technology, and Human Error*. Santa Barbara, CA: Aegean Publishing Co.

Chaimowicz, L., Kumar, V., & Campos, M. F. M. (2004). A paradigm for dynamic coordination of multiple robots. *Autonomous Robots*, *17*(1), 7–21. doi:10.1023/B:AURO.0000032935.30271.a5

Chakrabarti, D., Wang, Y., Wang, C., Leskovec, J., & Faloutsos, C. (2008). Epidemic thresholds in real networks. *ACM Transactions on Information and System Security*, *10*(4). doi:10.1145/1284680.1284681

Chakrabarti, D., Zhan, Y., & Faloutsos, C. (2004). R-MAT: A Recursive Model for Graph Mining. In *SDM Conference*.

Chakrabarty, K., Iyengar, S. S., Qi, H., & Cho, E. (2002). Grid coverage for surveillance and target location in distributed sensor networks. *IEEE Transactions on Computers*, *12*(51), 1448–1453. doi:10.1109/TC.2002.1146711

Chan, H., Perrig, A., & Song, D. (2006). Secure Hierarchical In-Network Aggregation in Sensor Networks. In *ACM Conference on Computer and Communication Security (CCS)*.

Chan, H., Perrig, A., & Song, D. (2008). Efficient Security Primitives from a Secure Aggregation Algorithm. In *ACM Conference on Computer and Communication Security (CCS)*.

Chau, C. K., Crowcroft, J., Lee, K. W., & Wong, S. H. Y. (2008). How to Enable Policy-based Interactions in Dynamic Wireless Networks. In *Proceedings of the IEEE Policy Workshop*.

Chau, C. K., Crowcroft, J., Lee, K.-W., & Wong, H. Y. (2008, January). *IDRM: Inter-Domain Routing Protocol for Mobile Ad Hoc Networks*. Technical Report UCAM-CL-TR-708, University of Cambridge, Computer Laboratory.

Chawathe, S. S. (1999). Comparing Hierachical data in external memory. *Very Large Data Bases Conference*.

Chen, A., Lai, T., & Xuan, D. (2008). *Measuring and guaranteeing quality of barrier-coverage in wireless sensor networks*. MobiHoc.

Chen, Y., Liestman, A., & Liu, J. (2004). Clustering Algorithms for Ad Hoc Wireless Networks. In *Proceedings of the Ad Hoc and Sensor Networks*.

Cheng, J., Xu Yu, J., Lin, X., Wang, H., & Yu, P. S. (2006). Fast Computation of Reachability Labelings in Large Graphs. In *EDBT Conference*.

Cheng, J., Xu Yu, J., Lin, X., Wang, H., & Yu, P. S. (2008). Fast Computing Reachability Labelings for Large Graphs with High Compression Rate. In *EDBT Conference*.

Cheng, M.-K., Lu, Y.-E., Vendruscolo, M., Lió, P., & Blundell, T. L. (2008). Prediction by graph theoretic measures of structural effects in proteins arising from non-synonymous single nucleotide polymorphisms. *PLoS Computational Biology*, *4*(7). doi:10.1371/journal.pcbi.1000135

Cheriyan, J., Hagerup, T., & Melhorn, K. (1996). An -time maximum-flow algorithm. *SIAM Journal on Computing*, *25*(6), 1144–1170. doi:10.1137/S0097539791278376

Cherniack, M., Franklin, M. J., & Zdonik, S. (2001). Expressing user profiles for data recharging. *IEEE Personal Communications: Special Issue on Pervasive Computing, 8*(4), 32–38.

Chiang, C.-Y., Chadha, R., Newman, S., Lo, R., & Bauer, R. (2007). Integrated Network Operations For Future Army Tactical Networks. In MILCOM'07, Orlando, FL.

Chiarelli, P. W., & Michaelis, P. R. (2005). Winning the peace, the requirement for full-spectrum operations. *Military Review*, 4–17.

Chomicki, J., Lobo, J., & Naqvi, S. A. (2003). Conflict Resolution Using Logic Programming. *IEEE Transactions on Knowledge and Data Engineering, 15*(1), 244–249. doi:10.1109/TKDE.2003.1161596

Chung, F. (1997). *Spectral Graph Theory* (No. 92: CBMS Regional Conference Series in Mathematics). Providence, RI: American Mathematical Society.

Cisco Systems. (2000). *Internetworking Technologies Handbook* (3rd ed.). Indianapolis, IN: Cisco Press.

Clancey, W. J. (1997). *Situated Cognition: On Human Knowledge and Computer Representations.* Cambridge, UK: Cambridge University Press.

Clark, A. (1997). *Being There: Putting Brain, Body and World Together Again.* Cambridge, MA: MIT Press.

Clark, A. (1999). An embodied cognitive science. *Trends in Cognitive Sciences, 3*(9), 345–351. doi:10.1016/S1364-6613(99)01361-3

Clark, A. (2001). Reasons, Robots and the Extended Mind. *Mind & Language, 16*(2), 121–145. doi:10.1111/1468-0017.00162

Clark, A. (2003). *Natural-Born Cyborgs: Minds, Technologies and the Future of Human Intelligence.* Oxford, UK: Oxford University Press.

Clark, A. (2005). Intrinsic content, active memory and the extended mind. *Analysis, 65*(285), 1–11. doi:10.1111/j.1467-8284.2005.00514.x

Clark, A. (2007a). Curing Cognitive Hiccups: A Defense of The Extended Mind. *The Journal of Philosophy, 104*(4), 163–192.

Clark, A. (2007b). Re-Inventing Ourselves: The Plasticity of Embodiment, Sensing, and Mind. *The Journal of Medicine and Philosophy, 32*(3), 263–282. doi:10.1080/03605310701397024

Clark, A. (2008). *Supersizing the Mind: Embodiment, Action, and Cognitive Extension.* New York: Oxford University Press.

Clark, A. (in press-a). Coupling, Constitution and the Cognitive Kind: A Reply to Adam and Aizawa. In R. Menary (Ed.), The Extended Mind. Aldershot, UK: Ashgate Publishers.

Clark, A. (in press-b). Memento's revenge: the extended mind, extended. In R. Menary (Ed.), The Extended Mind. Aldershot, UK: Ashgate Publishers.

Clark, A., & Chalmers, D. (1998). The Extended Mind. *Analysis, 58*(1), 7–19. doi:10.1111/1467-8284.00096

Clausen, T. & Jacquet, P. (2003). *Optimized Link State Routing Protocol (OLSR)*, RFC 3626.

Claverie, J.-M. (2001). Gene number: What if there are only 30,000 human genes? *Science, 291*(5507), 1255–1257. doi:10.1126/science.1058969

Clouqueur, T., Phipatanasuphorn, V., Ramanathan, P., & Saluja, K. K. (2002). Sensor deployment strategy for target detection. In Raghavendra, C. S., & Sivalingam, K. M. (Eds.), *WSNA* (pp. 42–48).

Clouqueur, T., Phipatanasuphorn, V., Ramanathan, P., & Saluja, K. K. (2002). Sensor deployment strategy for detection of targets traversing a region. *MONET, 8*(4), 453–461.

Cohen, E. (1997). Size-estimation framework with applications to transitive closure and reachability. *Journal of Computer and System Sciences, 55*(3), 441–453.

Cohen, E., Halperin, E., Kaplan, H., & Zwick, U. (2002). Reachability and distance queries via 2-hop labels. In *ACM Symposium on Discrete Algorithms.*

Cook, D., & Holder, L. (2007). *Mining Graph Data*. New York: John Wiley & Sons, Inc.

Core, M., & Allen, J. (1997). Coding dialogs with the DAMSL annotation scheme. In *Working Notes of the AAAI Fall Symposium on Communicative Action in Human and Machines*, (pp. 28035), Cambridge, MA.

Cormode, G., & Muthukrishnan, S. (2005). Space-Efficient Mining of Multigraph Streams. In *ACM PODS Conference*.

Costa, A., Roe, R., & Taillieu, T. (2001). Trust within teams: The relation with performance effectiveness. *European Journal of Work and Organizational Psychology*, *10*(3), 225–244. doi:10.1080/13594320143000654

Cox, K., Grinter, R. E., & Mantilla, D. (2000). Using dialog and context in a speech-based interface for an information visualization environment. In *Proceedings of ACM 2000*.

Craven, R., Lupu, E., Ma, J., Russo, A., Sloman, M., & Bandara, A. (2009). *An expressive policy analysis framework with enhanced system dynamicity* (pp. 239–250). ASIACCS.

Crawley, E., Nair, R., Rajagopalan, B., & Sandick, H. (1998). *A Framework for QoS-based Routing in the Internet* (RFC 2386). Internet Engineering Task Force (IETF).

Crowcroft, J., Hand, S., Mortier, R., Roscoe, T., & Warfield, A. (2003). *Plutarch: an argument for network pluralism*. ACM Computer Communication Review.

D'Andrade, R. G. (1981). The cultural part of cognition. *Cognitive Science*, *5*, 179–195. doi:10.1207/s15516709cog0503_1

D'Andrade, R. G. (1995). *The development of cognitive anthropology*. Cambridge, UK: Cambridge University Press.

Dalamagas, T., Cheng, T., Winkel, K., & Sellis, T. (2005). Clustering XML Documents Using Structural Summaries. In *Information Systems*. Thousand Oaks, CA: Elsevier.

Dawkins, R. (1982). *The Extended Phenotype: The Long Reach of the Gene*. Oxford, UK: Oxford University Press.

Dennett, D. C. (1996). *Kinds of Minds: Toward an Understanding of Consciousness*. New York: Basic Books.

DeVaul, R., Pentland, A., & Corey, V. (2005). *The memory glasses: subliminal vs. overt memory support with imperfect information*. Paper presented at the 7th IEEE International Symposium on Wearable Computers, White Plains, NY.

Dey, A. K., & Abowd, G. D. (2000, April). Towards a Better Understanding of Context and Context-Awareness. In the *Workshop on The What, Who, Where, When, and How of Context-Awareness, as part of the 2000 Conference on Human Factors in Computing Systems (CHI 2000)*, The Hague, The Netherlands.

Dhillon, S. S., & Chakrabarty, K. (2002). Sensor placement for grid coverage under imprecise detections. In *FUSION* (pp. 1581–1587). Iyengar.

Dhillon, S. S., & Chakrabarty, K. (2003). *Sensor placement for effective coverage and surveillance in distributed sensor networks* (pp. 1609–1614). WCNC.

Diehl, M., & Stroebe, W. (1987). Productivity loss in brainstorming groups: toward the solution of a riddle. *Journal of Personality and Social Psychology*, *53*, 497–509. doi:10.1037/0022-3514.53.3.497

Dijkstra, E. W. (1959). A note on two problems in connection with graphs. *Numerische Mathematik*, *1*, 269–271. doi:10.1007/BF01386390

DMTF. (2007). *CIM Simplified Policy Language*. CIM-SPL.

Douceur, J. (2002). The Sybil Attack. In *2nd IEEE Workshop on Peer-to-Peer Systems (IPTPS)*.

Driscoll, P. J., Tortorella, M., & Pohl, E. (2005). *Information Product Quality in Network Centric Operations* (Tech. Rep. No. DSE-TR-0516). West Point, NY: United States Military Academy, Operations Research Center of Excellence.

Earley, P. C., & Mosakowski, E. (2000). Creating hybrid team cultures: An empirical test of transnational team functioning. *Academy of Management Journal, 43*, 26–49. doi:10.2307/1556384

Edwards, W. (1998). Hailfinder: Tools for and experiences with Bayesian normative modeling. *The American Psychologist, 53*(4), 416–428. doi:10.1037/0003-066X.53.4.416

Edwards, W., & Fasolo, B. (2001). Decision technology. *Annual Review of Psychology, 52*, 581–606. doi:10.1146/annurev.psych.52.1.581

Faloutsos, M., Faloutsos, P., & Faloutsos, C. (1999). On Power Law Relationships of the Internet Topology. In *SIGCOMM Conference*.

Feigenbaum, J., Kannan, S., McGregor, A., Suri, S., & Zhang, J. (2005). On Graph Problems in a Semi-streaming Model. *Theoretical Computer Science, 348*(2).

Fejes Tóth, L. (1972). *Lagerungen in der Ebene, auf der Kugel, und im Raum (2. Auflage)*. Berlin: Springer Verlag.

Fellbaum, C. (Ed.). (1998). *WordNet: An Electronic Lexical Database*. Cambridge, MA: MIT Press.

Finin, T., Labrou, Y., & Mayfield, J. (1997). KQML as an Agent Communication Language. In Bradshaw, J. M. (Ed.), *Software Agents*. Cambridge, MA: MIT Press/AAAI Press.

FIPA. (1997). Foundation for Intelligent Physical Agents. FIPA 97 Agent Communication Language Specification, (1.2 ed.).

Fisher, M., Ellis, J., & Bruce, J. (2003). *JDBC API Tutorial and Reference* (3rd ed.). Upper Saddle River, NJ: Prentice Hall PTR.

Fjallstrom, P.-O. (1998). Algorithms for Graph Partitioning: A Survey. *Linkoping Electronic Articles in Computer and Information Science, 3*(10).

Flake, G., Tarjan, R., & Tsioutsiouliklis, M. (2003). Graph Clustering and Minimum Cut Trees. *Internet Mathematics, 1*(4), 385–408.

Fodor, J. (1980). Methodological solipsism considered as a research strategy in cognitive psychology. *The Behavioral and Brain Sciences, 3*(1), 63–73. doi:10.1017/S0140525X00001771

Foelix, R. F. (1996). *Biology of Spiders*. New York: Oxford University Press.

Foltz, P. W., Martin, M. J., Abdelali, A., Rosenstein, M., & Oberbreckling, R. (2006). Automated team discourse modeling. In *Test of Performance and Generalization, Proceedings of the 28th Annual Cognitive Science Conference*, Vancouver, Canada.

Future Combat Systems. (n.d.). Retrieved from https://www.fcs.army.mil

Gamma, E., Helm, R., Johnson, R., & Vlissides, J. M. (1995). Design Patterns: Elements of Reusable Object-Oriented Software, (1st ed., ser. Professional Computing Series). Reading, MA: Addison-Wesley.

Garcia, O., & Otheguy, R. (1989). *English Across Cultures, Cultures Across English: A Reader in Cross-cultural Communication*. New York: Mouton de Gruyter.

Gardner, H. (1984). *The mind's new science: A history of the cognitive revolution*. New York: Basic Books.

Garey, M. S., & Johnson, D. S. (1979). *Computers and Intractability: A Guide to the Theory of NP-completeness*. New York: W. H. Freeman.

Garlan, D., & Shaw, M. (1993). An introduction to software architecture. In Ambriola, V., & Tortora, G. (Eds.), *Advances in Software Engineering and Knowledge Engineering* (pp. 1–39). Singapore: World Scientific Publishing Company.

Garro, L. C. (2000). Remembering what one knows and the construction of the past: A comparison of cultural consensus theory and cultural schema theory. *Ethos (Berkeley, Calif.), 28*(3), 275–319. doi:10.1525/eth.2000.28.3.275

Gentner, D., & Stevens, A. L. (1983). *Mental Models*. Hillsdale, NJ: Lawrence Erlbaum Associates.

Ghosh, A., & Das, S. (2006). Coverage and Connectivity Issues in Wireless Sensor Networks. In Shorey, Ananda, Chan, & Ooi (Eds.), Mobile, Wireless and Sensor Networks: Technology, Applications and Future Directions. Chichester, UK: John Wiley & Sons

Gibson, D., Kumar, R., & Tomkins, A. (2005). Discovering Large Dense Subgraphs in Massive Graphs. In *VLDB Conference*.

Gibson, J. J. (1966). *The Senses Considered as Perceptual Systems*. New York: Houghton-Mifflin.

Gillies, D., Thornley, D., & Bisdikian, C. (2009). Probabilistic Approaches to Estimating the Quality of Information in Military Sensor Networks. *British Computer Society: The Computer Journal*. doi:10.1093/comjnl/bxp047.

Girko, V. L. (2005). The circular law. twenty years later. part iii. *Random Operators and Stochastic Equations*, *13*(1), 53–109.

Godfrey, J. J., Holliman, E. C., & McDaniel, J. (1992, March). SWITCHBOARD: Telephone speech corpus for research and development. In *Proceedings of the IEEE Conference on Acoustics, Speech, and Signal Processing*, San Francisco, (Vol. 1, pp. 517-520).

Goldblatt, R. (1992). *Logics of time and computation* (2nd ed.). Stanford, CA: Center for the Study of Language and Information.

Goldstone, R. L., Roberts, M. E., & Gureckis, T. M. (2008). Emergent Processes in Group Behavior. *Current Directions in Psychological Science*, *17*(1), 10–15. doi:10.1111/j.1467-8721.2008.00539.x

Goldwasser, S., & Micali, S. (1982). *Probabilistic encryption and how to play mental poker keeping secret all partial information. In 14th* (pp. 365–377). ACM STOC.

Gomes, H. (1996). *Quality Quotes*. Chicago: Irwin Professional Publishing.

Gopnik, A., & Wellman, H. M. (1994). The theory theory. In Hirschfeld, L., & Gelman, S. (Eds.), *Mapping the Mind: Domain Specificity in Cognition and Culture* (pp. 257–293). New York: Cambridge University Press.

Gorman, J. C., Foltz, P. W., Kiekel, P. A., Martin, M. J., & Cooke, N. J. (2003). Evaluation of latent semantic analysis-based measures of tam communications content. In *Proceedings of the Human Factors and Ergonomics Society 47th Annual Meeting*.

Gozdecki, J., Jajszczyk, A., & Stankiewicz, R. (2003). Quality of Service Terminology in IP Networks. *IEEE Communications Magazine*, *41*(3), 153–159. doi:10.1109/MCOM.2003.1186560

Gray, W. D., & Fu, W. T. (2004). Soft constraints in interactive behavior: The case of ignoring perfect knowledge in-the-world for imperfect knowledge in-the-head. *Cognitive Science*, *28*(3), 359–382.

Greaves, M., Holmback, H., & Bradshaw, J. (1998). *What Is a conversation policy? Unpublished technical report*. Seattle, WA: The Boeing Company.

Greenfield, S. (2003). *Tomorrow's People: How 21st Century Technology is Changing the Way We Think and Feel*. London: Penguin Books Ltd.

Grice, P. (1968). Utterer's meaning, sentence-meaning, and word-meaning. *Foundations of Language*, *4*, 1–18.

Grice, P. (1978). Further notes on logic and conversation. In Cole, P. (Ed.), *Syntax and Semantics 9: Pragmatics* (pp. 113–128). New York: Academic Press.

Grice, P. (1989). *Studies in the Way of Words*. Cambridge, MA: Harvard University Press.

Grun, B., & Leisch, F. (2007). FlexMix: An R package for finite mixture modelling. *R News*, *7*(1), 8–13.

Gupta, H., Zhou, Z., Das, S. R., Gu, Q. (2006). Connected sensor cover: self-organization of sensor networks for efficient query execution. *IEEE/ACM Transactions on Networks*, *1*(14), 55-67.

Gutman, J. (1982). A means-end chain model based on consumer categorization processes. *Journal of Marketing*, *46*(2), 60–72. doi:10.2307/3203341

Hall, D. L., & McMullen, S. A. H. (2004). *Mathematical Techniques in Multisensor Data Fusion* (2nd ed.). Boston: Artech House.

Harnad, S. (1999). The future of scholarly skywriting. In Scammell, A. (Ed.), *I in the Sky: Visions of the Information Future*. London: Aslib.

Harnad, S., & Dror, I. E. (2006). Distributed cognition: Cognizing, autonomy and the Turing Test. *Pragmatics & Cognition, 14*(2), 209–213. doi:10.1075/pc.14.2.03har

Haugeland, J. (1998). Mind embodied and embedded. In Haugeland, J. (Ed.), *Having thought: Essays in the Metaphysics of Mind*. Cambridge, MA: Harvard University Press.

He, H., & Singh, A. K. (2005). Efficient Algorithms for Mining Significant Substructures in Large Graphs with Quality Guarantees. In *ICDM Conference*.

He, T., Zafer, M., & Bisdikian, C. (2008). Detecting Transient Signals with Incomplete Observations. In MILCOM 2008, San Diego, CA, Nov. 17-19.

Heffernan, A. (1998, August). *Protection of BGP Sessions via the TCP MD5 Signature Option*. RFC 2385 (Proposed Standard).

Hero, A. O. III. (2003). Secure space-time communication. *IEEE Transactions on Information Theory, 49*(12). doi:10.1109/TIT.2003.820010

Hinsz, V., Tindale, R., & Vollrath, D. (1997). The emerging conceptualization of groups as information processors. *Psychological Bulletin, 121*, 43–64. doi:10.1037/0033-2909.121.1.43

Hirschfeld, L., & Gelman, S. (Eds.). (1994). *Mapping the mind: Domain specificity in cognition and culture*. New York: Cambridge University.

Ho, I. W., Ko, B.-J., Zafer, M., Bisdikian, C., & Leung, K. (2008). Cooperative Transmit-Power Estimation in MANETs. In *IEEE Wireless Communication and Networking Conference (WCNC)*.

Hofstede, G. (2001). Culture's consequences (2 ed.). Thousand Oaks, CA: Sage.

Hollan, J., Hutchins, E., & Kirsh, D. (2000). Distributed cognition: toward a new foundation for human-computer interaction research. [TOCHI]. *ACM Transactions on Computer-Human Interaction, 7*(2), 174–196. doi:10.1145/353485.353487

Howard, A., Mataric, M. J., & Sukhatme, G. S. (2002a). An incremental self-deployment algorithm for mobile sensor networks. *Autonomous Robots, 13*(2), 113–126. doi:10.1023/A:1019625207705

Howard, A., Mataric, M. J., & Sukhatme, G. S. (2002b). *Mobile sensor network deployment using potential fields: A distributed, scalable solution to the area coverage problem* (pp. 299–308). DARS.

Howard, R. A. (1989). Knowledge maps. *Management Science, 35*, 903–922. doi:10.1287/mnsc.35.8.903

Howard, R. D. (1966). Information Value Theory. *IEEE Trans. on Systems Science and Cybernetics, 2*(1), 22–26. doi:10.1109/TSSC.1966.300074

Hu, Y. C., & Perrig, A. (2004). A Survey of Secure Wireless Ad Hoc Routing. In IEEE Security and Privacy Magazine.

Hu, Y-C., Perrig, A. & Sirbu, M. (2004, October). SPV: secure path vector routing for securing BGP. *ACM SIGCOMM Computer Communications Review, 34*(4).

Huang, C.-F., & Tseng, Y.-C. (2005). The coverage problem in a wireless sensor network. *Mobile Networks and Applications, 10*(4), 519–528. doi:10.1007/s11036-005-1564-y

Huang, Y., Contractor, N., & Yao, Y. (2008). *CI-KNOW: recommendation based on social networks*. Paper presented at the 9th Annual International Digital Government Research Conference, Montreal, Canada.

Hurley, S. (1998). *Consciousness in Action*. Cambridge, MA: Harvard University Press.

Hutchby, I., & Wooffitt, R. (1997). *Conversation Analysis*. Oxford, MA: Blackwell.

Hutchins, E. (1995). *Cognition in the Wild*. Cambridge: MIT Press.

Hutchins, E. (1995). How a cockpit remembers its speeds. *Cognitive Science, 19*(3), 265–288.

Inokuchi, A., Washio, T., & Motoda, H. (2000). An apriori-based algorithm for mining frequent substructures from graph data. In *PKDD Conference*, (pp. 13–23).

International Human Genome Sequencing Consortium. (2004). Finishing the euchromatic sequence of the human genome. *Nature, 431*(7011), 931–945. doi:10.1038/nature03001

IQ2S. (2009). In *First Int'l Workshop on Information Quality and Quality of Service for Pervasive Computing.* Part of the *Seventh Annual IEEE Int'l Conf. on Pervasive Computing and Communications* (PerCom 2009), Galveston, TX, USA, March 9.

IQ2S. (2010). In *Second Int'l Workshop on Information Quality and Quality of Service for Pervasive Computing.* Part of the *Eighth Annual IEEE Int'l Conf. on Pervasive Computing and Communications* (PerCom 2010), Manheim, Germany, March 29.

Irwin, K., Yu, T., & Winsborough, W. H. (2006). On the modeling and analysis of obligations. In *Proc. of the 13th ACM Conf. on Computer and communications security,* (pp. 134–143).

Jacko, J., & Sears, A. (Eds.). (2008). *The Human-Computer Interaction Handbook: Fundamentals, Evolving Technologies and Emerging Applications* (2nd ed.). New York: Erlbaum Associates.

Jackson, D. (2002). Alloy: a lightweight object modelling notation. *ACM Transactions on Software Engineering and Methodology, 11*(2), 256–290. doi:10.1145/505145.505149

Jain, A., & Dubes, R. (1998). *Algorithms for Clustering Data.* Upper Saddle River, NJ: Prentice Hall.

Jammer-Kit. (n.d.). *Follower Jammer Kit.* Retrieved from http://www.polarizone.com/milzone.html

Janis, I. L. (1982). *Victims of Groupthink* (2nd ed.). Boston: Houghton Mifflin.

Jennings, N. R. (1993). Commitments and Conventions: The Foundation of Coordination in Multi-Agent Systems. *The Knowledge Engineering Review, 8*(3), 223–250. doi:10.1017/S0269888900000205

Jennings, N. R., Faratin, P., Lomuscio, A. R., Parsons, S., Sierra, C., & Wooldridge, M. (2001). Automated negotiation: prospects, methods and challenges. *Int. J. of Group Decision and Negotiation, 10*(2), 199–215. doi:10.1023/A:1008746126376

Jentsch, F. G., Sellin-Wolters, S., Bowers, C. A., & Salas, E. (1995). Crew coordinatin behaviors as predictors of problem detection and decision making times. In *Proceedings of the Human Factors and Ergonomics Society 39th Annual Meeting* (pp. 1350-1353). Santa Monica, CA: Human Factors and Ergonomics Society.

Jiang, X., & Bunke, H. (2008). Graph Matching. In *Case based Reasoning in Images and Signals.* Berlin: Springer.

Johnson, D., Hu, Y., & Maltz, D. (2007). *The Dynamic Source Routing Protocol (DSR) for Mobile Ad Hoc Networks for IPv4. RFC 4728.*

Jorgensen, C., & Binsted, K. (2005). *Web Browser Control Using EMG Based Sub Vocal Speech Recognition.* Paper presented at the 38th Annual Hawaii International Conference on System Sciences (HICSS'05), Big Island, Hawaii.

Juels, A., & Brainard, J. (1999). A Cryptographic Defense Against Connection Depletion Attacks. In *Network and Distributed Systems Security (NDSS).* Client Puzzle.

Jurafsky, D. (2006). Pragmatics and computational linguistics. In Horn, L. R., & Ward, G. (Eds.), *Handbook of Pragmatics.* Oxford, MA: Blackwell. doi:10.1002/9780470756959.ch26

Kakas, A. C., Michael, A., & Mourlas, C. (2000). ACLP: Abductive constraint logic programming. *The Journal of Logic Programming, 44*(1-3), 129–177. doi:10.1016/S0743-1066(99)00075-8

Kao, A., Poteet, S., & Xue, P. (2007). Conversation analysis of coalition communication in computational pragmatics. In *Proceedings of the First Annual Conference of the International Technology Alliance*, New York.

Kaplan, L. M., Damarla, T., & Pham, T. (2008). QoI for Passive Acoustic Gunfire Localization. *First Int'l Workshop on Quality of Information for Sensor Networks* (QoISN 2008). Atlanta, GA, Sept. 29.

Kaplan, L., & Cevher, V. (2007). Design considerations for a heterogeneous network of bearing-only sensors using sensor management. In IEEE Aerospace Conf, 2007, (pp. 1-14).

Karat, C., Karat, J., Brodie, C., & Feng, J. (2006). Evaluating Interfaces for Privacy Policy Rule Authoring. In *the Proceedings of the Conference on Human Factors in Computing Systems,* (pp. 83-92). New York: ACM Press.

Karat, J. (Ed.). (1991). *Taking Software Design Serously: Practical Techniques for Human-Computer Interaction Design.* New York: Academic Press.

Karat, J., Karat, C., Brodie, C., & Feng, J. (2005). Privacy in information technology: Designing to enable privacy policy management in organizations. *International Journal of Human-Computer Studies, 63*(1-2), 153–174. doi:10.1016/j.ijhcs.2005.04.011

Karkatzounis, K. (2004). *Performance evaluation of different jamming strategies over uncoded non-coherent fast frequency hopping mfsk communication systems.* Naval Postgraduate School Thesis.

Karlin, J., Forrest, S., & Rexford, J. (2006, November). Pretty Good BGP: Improving BGP by Cautiously Adopting Routes. In *The 14th IEEE International Conference on Network Protocols (ICNP).*

Karlof, C., & Wagner, D. (2002). Attacks and Countermeasures. In *Elsevier's Ad-Hoc Networks Journal, Special Issue on Sensor Network Applications and Protocols.* Secure Routing in Wireless Sensor Networks.

Karp, B. N., & Kung, H. T. (2000). GPSR: Greedy Perimeter Stateless Routing for Wireless Networks. In *Proceedings of the sixth annual ACM/IEEE International Conference on Mobile computing and networking (Mobicom).*

Kashima, H., Tsuda, K., & Inokuchi, A. (2003). *Marginalized Kernels between Labeled Graphs.* ICML.

Kauffman, S. (1995). *At Home in the Universe: The Search for Laws of Self-Organization and Complexity.* London: Penguin Books Ltd.

Keeney, R. L. (1994). Creativity in decision making with value-focused thinking. *Sloan Management Review, 35*(4), 33–41.

Kempton, W. (1986). Two theories of home heat control. *Cognitive Science, 10*, 75–90.

Kent, S., Lynn, C., & Seo, K. (2000, April). Secure Border Gateway Protocol (Secure-BGP). *IEEE Journal on Selected Areas in Communications, 18*(4), 582–592. doi:10.1109/49.839934

Kernighan, B. W., & Lin, S. (1970). An efficient heuristic procedure for partitioning graphs. *The Bell System Technical Journal, 49*, 291–307.

Kershner, R. (1939). The number of circles covering a set. *American Journal of Mathematics, 61*, 665–671. doi:10.2307/2371320

Khalili, A., Katz, J., & Arbaugh, W. A. (2003, January). Toward secure key distribution in truly ad-hoc networks. In *Symposium on Applications and the Internet Workshops*, (pp. 342–346).

Khisti, A., & Wornell, G. (2007). The MIMOME Channel. In *45th Allerton Conference on Communication, Control and Computing.*

Khisti, A., Tchamkerten, A., & Wornell, G. (2006). Secure Broadcasting with Multiuser Diversity. In *44th Allerton Conference on Communication, Control and Computing.*

Kirsh, D. (1995). The intelligent use of space. *Artificial Intelligence, 73*(1-2), 31–68. doi:10.1016/0004-3702(94)00017-U

Kirsh, D. (1996). Adapting the environment instead of oneself. *Adaptive Behavior, 4*(3/4), 415–452. doi:10.1177/105971239600400307

Kirsh, D. (2006). Distributed cognition: A methodological note. *Pragmatics & Cognition, 14*(2), 249–262. doi:10.1075/pc.14.2.06kir

Kirsh, D. (2009). Problem solving and situated cognition. In Ayded, M., & Robbins, P. (Eds.), *Cambridge Handbook of Situated Cognition*. Cambridge, UK: Cambridge University Press.

Kirsh, D., & Maglio, P. (1994). On distinguishing epistemic from pragmatic action. *Cognitive Science, 18*, 513–549.

Klein, A., Do, H. H., Hackenbroich, G., Karnstedt, M., & Lehner, W. (2007). Representing Data Quality for Streaming and Static Data. In *23rd Int'l Conf. on Data Engineering Workshops* (ICDE 2007), Istanbul, Turkey, April 15-20.

Klein, G. (1998). *Sources of Power: How People Make Decisions*. Cambridge: MIT Press.

Klein, G., & Hoffman, R. R. (2008). Macrocognition, mental models, and cognitive task analysis methodology. In Schraagen, J. M., Militello, L. G., Ormerod, T., & Lipshitz, R. (Eds.), *Naturalistic Decision Making and Macrocognition* (pp. 57–80). Aldershot, UK: Ashgate.

Klein, G., Ross, K. G., Moon, B. M., Klein, D. E., Hoffman, R. R., & Hollnagel, E. (2003). Macrocognition. *IEEE Intelligent Systems, May/June*, 81-85.

Kleinberg, J. (2000). The small-world phenomenon: an algorithm perspective. In *Proceedings of the thirty-second annual ACM Symposium on Theory of Computing*, (pp. 163–170). New York: ACM Press.

Kleinman, A. (1978). Concepts and a model for the comparison of medical systems as cultural systems. *Social Science & Medicine, 12B*, 85–93.

Kleinrock, L. (1975). Queueing Systems: *Vol. 1. Theory*. New York: John Wiley & Sons.

Klimoski, R., & Mohammed, S. (1994). Team mental model: Construct or metaphor. *Journal of Management, 20*, 403–437. doi:10.1016/0149-2063(94)90021-3

Knight, S.-A. (2007). *User perceptions of information quality in World Wide Web information retrieval behavior*. PhD Thesis, Edith Cowan University, Perth, Australia.

Koorapaty, H., Hassan, A. A., & Chennakeshu, S. (2000). Secure information transmission for mobile radio. *IEEE Communications Letters, 4*(2). doi:10.1109/4234.824754

Koskinen, H. (2004). On the Coverage of a Random Sensor Network in a Bounded Domain. In *Proceedings of 16th ITC Specialist Seminar*, (pp. 11-18).

Kowalski, R. A., & Sergot, M. J. (1986). A logic-based calculus of events. *New Generation Computing, 4*, 67–95. doi:10.1007/BF03037383

Krink, T., & Vollrath, F. (1997). Analysing Spider Web-building Behaviour with Rule-based Simulations and Genetic Algorithms. *Journal of Theoretical Biology, 185*(3), 321–331. doi:10.1006/jtbi.1996.0306

Krink, T., & Vollrath, F. (1998). Emergent properties in the behaviour of a virtual spider robot. *Proceedings of the Royal Society of London. Series B. Biological Sciences, 265*(1410), 2051–2055. doi:10.1098/rspb.1998.0539

Krink, T., & Vollrath, F. (1999). Virtual Spiders Guide Robotic Control Design. *IEEE Intelligent Systems, 14*(5), 77–84. doi:10.1109/5254.796093

Kudo, T., Maeda, E., & Matsumoto, Y. (2004). An Application of Boosting to Graph Classification. In NIPS Conf.

Kumar, R., Raghavan, P., Rajagopalan, S., Sivakumar, D., Tomkins, A., & Upfal, E. (2000). The Web as a Graph. In *ACM PODS Conference*.

Kumar, S., Lai, T., & Arora, A. (2007). Barrier Coverage with wireless sensors. *Wireless Networks, 13*, 817–834. doi:10.1007/s11276-006-9856-0

Kuramochi, M., & Karypis, G. (2001). Frequent subgraph discovery. In *ICDM Conference*, (pp. 313–320).

Kyasanur, P., & Vaidya, N. (2003). Detection and Handling {MAC} Layer Misbehavior in Wireless Networks. In *IEEE Symposium on Dependable Systems and Networks (DSN)*.

Lafferty, J., McCallum, A., & Pereira, F. (2001). Conditional random fields: Probabilistic models for segmenting and labeling sequence data. In *Proceedings of the Eighteenth International Conference on Machine Learning (ICML-2001),* Williamstown, MA.

Lakoff, G., & Johnson, M. (1980). *Metaphors we live by.* Chicago: University of Chicago Press.

Lal Mehta, M. (1991). *Random Matrices* (2nd ed.). San Diego, CA: Academic Press.

Landauer, T., Foltz, P. W., & Laham, D. (1998). Introduction to latent semantic analysis. *Discourse Processes, 25,* 259–284. doi:10.1080/01638539809545028

Larsson, A., Solodilova-Whiteley, I., Kao, A., Poteet, S., & Xue, P. (2008, September). A multidimensional approach to studying cultural differences and coping strategies in a multinational coalition environment. In *Proceedings of the 13th InternationalCommand and Control Research and Technology Symposium.,* Bellevue, WA.

Law, Y. W., Hoesel, L. V., Doumen, J., Hartel, P., & Havinga, P. (2005). Energy-efficient link-layer jamming attacks against wireless sensor network MAC protocols. In *SASN '05: Proceedings of the 3rd ACM workshop on Security of ad hoc and sensor networks,* (pp. 76–88), New York.

Lawrence, R. R. (1989). A tutorial on hidden markov models and selected applications in speech recognition. *Proceedings of the IEEE, 77,* 257–286. doi:10.1109/5.18626

Lazer, D., & Friedman, A. (2007). The Network Structure of Exploration and Exploitation. *Administrative Science Quarterly, 52*(4), 667–694. doi:10.2189/asqu.52.4.667

Lazer, D., Pentland, A., Adamic, L., Aral, S., Barabasi, A.-L., & Brewer, D. (2009). Social Science: Computational Social Science. *Science, 323*(5915), 721–723.

Lazos, L., & Poovendran, R. (2006). Stochastic coverage in heterogeneous sensor networks. *ACM Transactions on Sensor Networks, 2*(3), 325–358. doi:10.1145/1167935.1167937

Lee, M., Hsu, W., Yang, L., & Yang, X. (2002). XClust: Clustering XML Schemas for Effective Integration. In *ACM Conference on Information and Knowledge Management.*

Lee, S. H., Wong, S. H. Y., Chau, C. K., Varadarajan, S., Lee, K. W., Crowcroft, J., & Gerla, M. (2009). *Self-organizing Inter-Domain Routing for Heterogeneous MANETs.* UCLA Technical Report, Los Angeles, CA. Retrieved May 2009, from http://www.cs.ucla.edu/~shlee/papers/idrm.pdf

Leisch, F. (2004). FlexMix: A general framework for finite mixture models and latent class regression in R. *Journal of Statistical Software, 11*(8), 1–18.

Lenox T., Hahn, S., Lewis M., Payne T. and Sycara, K. (2000). Agent Based Aiding for Individual and Team Planning Tasks, *IEA 2000/HFES 2000 Congress.*

Lenox, T., Hahn, S., Lewis, M., Payne, T., & Sycara, K. (2000). Task Characteristics and Intelligent Aiding, Proceedings of the 2000 IEEE International Conference on Systems, Man, and Cybernetics, October 8-11, Nashville, TN pp. 1123-1127.

Leskovec, J., & Faloutsos, C. (2007). Scalable modeling of real graphs using Kronecker multiplication. *ICML Conference,* (pp. 497–504).

Leskovec, J., Kleinberg, J. & Faloutsos, C. (2007). Graph Evolution: Densification and Shrinking Diameters. *ACM Transactions on Knowledge Discovery from Data (ACM TKDD), 1*(1).

Leskovec, J., Krause, A., Guestrin, C., Faloutsos, C., VanBriesen, J., & Glance, N. S. (2007). Cost-effective outbreak detection in networks. In *KDD Conference,* (pp. 420–429).

Leskovec, J., Lang, K., Dasgupta, A., & Mahoney, M. (2008). Statistical Properties of Community Structure in Large Social and Information Networks. In *World Wide Web.* WWW.

Leskovec, J., McGlohon, M., Faloutsos, C., Glance, N., & Hurst, M. (2007). Cascading Behavior in Large Blog Graphs. In *SIAM International Conference on Data Mining (SDM).*

Levinson, S. C. (1983). *Pragmatics*. Cambridge, UK: Cambridge University Press.

Li, X., Chen, M., & Ratazzi, E. P. (2005). A Randomized Space-Time Transmission Scheme for Secret-Key Agreement. In *39st Annual Conference on Information Sciences and System (CISS)*.

Li, X., Hwu, J., & Ratazzi, E. P. (2006). Array Redundancy and Diversity for Wireless Transmissions with Low Probability of Interception. In *IEEE International Conference on Acoustics, Speech and Signal Processing (ICASSP)*.

Li, Z., Trappe, W., & Yates, R. (2007). Secret Communication via Multi-antenna Transmission. In *41st Annual Conference on Information Sciences and System (CISS)*.

Lian, W., Cheung, D. W., Mamoulis, N., & Yiu, S. (2004). An Efficient and Scalable Algorithm for Clustering XML Documents by Structure. *IEEE Transactions on Knowledge and Data Engineering, 16*(1).

Lin, G. & Noubir, G. (2005). On link layer denial of service in data wireless LANs. *Wireless Communications and Mobile Computing, 5*(3).

Liu, B., & Towsley, D. (2004). *A study of the coverage of large-scale sensor networks*. MASS.

Liu, B., Dousse, O., Wang, J., & Saipulla, A. (2008). *Strong barrier coverage of wireless sensor networks*. MobiHoc.

Liu, C. H., Leung, K. K., Bisdikian, C., & Branch, J. W. (2009). A New Approach to Architecture of Sensor Networks for Mission-Oriented Applications. *SPIE Defense & Security Symposium*, (DSS 2009), Orlando, FL, April 13-17.

Liu, Y., & Liang, W. (2005). *Approximate coverage in wireless sensor networks* (pp. 68–75). LCN.

Loftus, E. (1997). Creating false memories. *Scientific American, 277*(3), 70–75. doi:10.1038/scientificamerican0997-70

Lomuscio, A., Wooldridge, M., & Jennings, N. R. (2001). A Classification Scheme for Negotiation in Electronic Commerce. In Agent-Mediated Electronic Commerce: A European Perspective.

Lopez, V., Pasin, M., & Motta, E. (2005). *AquaLog: An Ontology-portable Question Answering System for the Semantic Web*. Paper presented at the 2nd European Semantic Web Conference (ESWC 2005), Heraklion, Greece.

Loscocco, P., & Smalley, S. (2001). Integrating flexible support for security policies into the linux operating system. In *Proceedings of the FREENIX Track: 2001 USENIX Annual Technical Conference*, Berkeley, CA, (pp. 29–42).

Lupu, E., & Sloman, M. (1999). Conflicts in policy-based distributed systems management. *IEEE Transactions on Software Engineering, 25*(6), 852–869. doi:10.1109/32.824414

Lupu, E., Dulay, N., Sloman, M., Sventek, J., Heeps, S., & Strowes, S. (2008). AMUSE: autonomic management of ubiquitous e-Health systems. *Concurrency and Computation, 20*(3), 277–295. doi:10.1002/cpe.1194

Ma, W., & Chuah, M. C. (2005). Comparisons of Interdomain Routing Schemes for Heterogeneous Ad Hoc Networks. In *Proceedings of the IEEE WOWMOM*.

Maglio, P., Matlock, T., Raphaely, D., Chernicky, B., & Kirsh, D. (1999). *Interactive skill in Scrabble*. Paper presented at the 21st Annual Conference of the Cognitive Science Society, Vancouver, British Columbia, Canada.

Mahajan, R., Wetherall, D., & Anderson, T. (2002). Understanding BGP misconfiguration. In *SIGCOMM '02: Proceedings of the 2002 conference on Applications, technologies, architectures, and protocols for computer communications*, (pp. 3–16), New York.

Markoff, J. (2009). The Cellphone, Navigating Our Lives. *New York Times*. Retrieved May 29, 2009, from http://www.nytimes.com/2009/02/17/science/17map.html

Martin, M., & Foltz, P. W. (2004). Automated team discourse annotation and performance prediction using LSA. In *Proceedings of the Human Language Technology and North American Association for Computational Linguistics Conference (HLT/NAACL)*, Boston.

Mason, S., Bashashati, A., Fatourechi, M., Navarro, K., & Birch, G. (2007). A comprehensive survey of brain interface technology designs. *Annals of Biomedical Engineering, 35*(2), 137–169. doi:10.1007/s10439-006-9170-0

Mason, W. A., Jones, A., & Goldstone, R. L. (2005). *Propagation of innovations in networked groups.* Paper presented at the 27th Annual Conference of the Cognitive Science Society, Stresa, Italy.

Maurer, U. M. (1993). Secret key agreement by public discussion from common information. *IEEE Transactions on Information Theory, 39*(3). doi:10.1109/18.256484

McCarthy, J. (1993). Notes on formalizing context. In *Proceedings of 1993 International Joint Conference on Artificial Intelligence.*

McHugh, A. P., Smith, J. L., & Sieck, W. R. (2008). Cultural variations in mental models of collaborative decision making. In Schraagen, J. M. C., Militello, L., Ormerod, T., & Lipshitz, R. (Eds.), *Naturalistic Decision Making and Macrocognition* (pp. 141–158). Aldershot, UK: Ashgate Publishing Limited.

McLachlan, G. J., & Peel, D. (2000). *Finite Mixture Models.* New York: Wiley. doi:10.1002/0471721182

Meddeb, A. (2010). Internet QoS: Pieces of the Puzzle. *IEEE Communications Magazine, 48*(1), 86–94. doi:10.1109/MCOM.2010.5394035

Megerian, S., Koushanfar, F., Qu, G., Veltri, G., & Potkonjak, M. (2002). Exposure in wireless sensor networks: Theory and practical solutions. *Wireless Networks, 8*(5), 443–454. doi:10.1023/A:1016586011473

Meguerdichian, S., Koushanfar, F., Potkonjak, M., & Srivastava, M. B. (2001a). *Coverage problems in wireless ad hoc sensor networks* (pp. 1380–1387). INFOCOM.

Meguerdichian, S., Koushanfar, F., Qu, G., & Potkonjak, M. (2001b). *Exposure in wireless ad hoc sensor networks* (pp. 139–150). MOBICOM.

Meguerian, S., Koushanfar, F., Potkonjak, M., & Srivastava, M. B. (2005). Worst and best-case coverage in sensor networks. *IEEE Transactions on Mobile Computing, 4*(1), 84–92. doi:10.1109/TMC.2005.15

Mehta, N. R., Medvidovic, N., & Phadke, S. (2000). Towards a taxonomy of software connectors. In *Proceedings of the 22nd ACM International Conference on Software engineering (ICSE)*, New York, (pp. 178–187).

Melnik, S., Garcia-Molina, H., & Rahm, E. (2002). Similarity Flooding: A Versatile Graph Matching Algorithm and its Application to Scheme Matching. In *ICDE Conference.*

Mézard, M., Parisi, G., & Zee, A. (1999). Spectra of euclidean random matrices. *Nuclear Physics B, 559*, 689–701.

Milner, R., Parrow, J., & Walker, D. (1992). A calculus of mobile processes, I. *Information and Computation, 100*(1), 1–40. doi:10.1016/0890-5401(92)90008-4

Minsky, M. (1986). *The Society of Mind.* New York: Simon & Schuster, Inc.

Misel, S. A. (1997). *Wow, AS7007!* Retrieved from http://www.merit.edu/mail.archives/nanog/ 1997-04/msg00340.html

Morris, C. (1938). Foundations of the Theory of Signs. In Carnap, R., & Morris, C. (Eds.), *International Encyclopedia of Unified Science* (pp. 77–138). Chicago: University of Chicago Press.

Mosier, K. L., & Chidester, T. R. (1991). Situation assessment and situation awareness in a team setting. *Designing for Everyone: Proceedings of the 11th Congress of the International Ergonomics Association*, Paris, 15-20 July, (pp. 798-800).

Mott, D., & Hendler, J. (2007). *Progress on the Collaborative Planning Model.* Paper presented at the 1st Annual Conference of the International Technology Alliance (ACITA), Maryland, USA.

MQ Telemetry Transport. Retrieved from http://mqtt.org/

Mueller, S. T., & Veinott, E. S. (2008). *Cultural mixture modeling: Identifying cultural consensus (and disagreement) using finite mixture modeling.* Paper presented at the Proceedings of the Cognitive Science Society, Washington, DC.

Mueller, S. T., Sieck, W. R., & Veinott, E. (2007). *Cultural Metrics: A Finite Mixture Models Approach* (Technical Report DAAD19-01-2-0009). Fairborn, OH: Klein Associates Division of ARA.

Myin, E., & O'Regan, J. K. (2009). Situated perception and sensation in vision and other modalities: a sensorimotor approach. In Ayded, M., & Robbins, P. (Eds.), *Cambridge Handbook of Situated Cognition*. Cambridge, UK: Cambridge University Press.

National Imagery Interpretability Rating Scale (NIIRS). (n.d.). Imagery Resolution Assessments and Reporting Standards (IRARS) Committee. Retrieved from http://www.fas.org/irp/imint/niirs.htm

Newell, A. (1980). Physical symbol systems. *Cognitive Science, 4*, 135–183.

Newell, A., & Simon, H. (1976). Computer science as empirical enquiry: Symbols and search. *Communications of the ACM, 19*, 113–126. doi:10.1145/360018.360022

Newman, M. (2003). The structure and function of complex networks. *SIAM Review, 45*, 167–256. doi:10.1137/S003614450342480

Newman, M. E. J. (2001). Clustering and preferential attachment in growing networks. *Physical Review E: Statistical, Nonlinear, and Soft Matter Physics, 64*(2), 025102.

Newman, M. E. J. (2002). The spread of epidemic disease on networks. *Physical Review E: Statistical, Nonlinear, and Soft Matter Physics, 66*, 016128.

Newman, M. E. J. (2006). Finding community structure in networks using the eigenvectors of matrices. *Physical Review E: Statistical, Nonlinear, and Soft Matter Physics, 74*(3), 036104.

Newman, M. E. J., Forrest, S., & Balthrop, J. (2002). Email networks and the spread of computer viruses. *Physical Review E: Statistical, Nonlinear, and Soft Matter Physics, 66*, 035101.

Newman, M. E., Watts, D. J., & Strogatz, S. H. (2002). Random graph models of social networks. *Proceedings of the National Academy of Sciences of the United States of America, 99*(Suppl 1), 2566–2572.

Nicelescu, D., & Nath, B. (2003). Ad Hoc Positioning (APS) using AOA. In *IEEE International Conference on Computer Communications (Infocom),* (pp: 1734-1743).

Nicolelis, M. (2001). Actions from thoughts. *Nature, 409*(6818), 403–407. doi:10.1038/35053191

Nisbett, R. E. (2003). *The geography of thought: How Asians and Westerners think differently.and why.* New York: The Free Press.

Nisbett, R. E., Peng, K., Choi, I., & Norenzayan, A. (2001). Culture and systems of thought: Holistic versus analytic cognition. *Psychological Review, 108*, 291–310. doi:10.1037/0033-295X.108.2.291

Noë, A. (2004). Action. In *Perception*. Cambridge, MA: MIT Press.

Noë, A. (2009). *Out of Our Heads: Why You Are Not Your Brain, and Other Lessons from the Biology of Consciousness*. New York: Hill & Wang.

Nofi, A. A. (2000). *Defining and Measuring Shared Situational Awareness* (No. [Final). Alexandria, VA: Center for Naval Analyses.]. *CRM, D0002895*, A1.

Nomikos, C., Rondogiannis, P., & Manolis Gergatsoulis, M. (2005). Temporal stratification tests for linear and branching-time deductive databases. *Theoretical Computer Science, 342*(2-3), 382–415. doi:10.1016/j.tcs.2005.05.014

Norenzayan, A., Smith, E. E., Kim, B. J., & Nisbett, R. E. (2002). Cultural preferences for formal versus intuitive reasoning. *Cognitive Science, 26*, 653–684.

Norman, D. (1988). *The design of everyday things.* New York: Double Day.

Norman, D. A. (1993). *Things That Makes Us Smart.* Reading, MA: Addison-Wesley.

Norman, D. A. (1998). *The Invisible Computer.* Cambridge, MA: MIT Press.

O'Regan, J. K., & Noë, A. (2001). A sensorimotor account of vision and visual consciousness. *The Behavioral and Brain Sciences, 24*(5), 939–973.

OASIS XACML TC. (2005). *extensible access control markup language (XACML) v2.0.*

Oliveira, R. V., Zhang, B., & Zhang, L. (2007). Observing the evolution of internet as topology. New York, NY, USA. In *Proceedings of the ACM SIGCOMM.*

Opyrchal, L., Prakash, A., & Agrawal, A. (2007). Supporting Privacy Policies in a Publish-Subscribe Substrate for Pervasive Environments. *Journal Of Networks, 2,* 17–26. doi:10.4304/jnw.2.1.17-26

Orasanu, J. M. (1990). Shared mental models and crew decision making. *Cognitive Science Laboratory Report No. 46.* Princeton, NJ: Princeton University.

Oren, N., Norman, T. J., & Preece, A. (2006). Loose lips sink ships: A heuristic for argumentation. *Proceedings of the Third International Workshop on Argumentation in Multi-Agent Systems,* OECD- Organization for Economic Co-operation and Development, (1980). *OECD Guidelines on the Protection of Privacy and Transborder Flow of Personal Data.* Paris, France, Accessed at www.oecd.org/home/.

Osland, J. S., & Bird, A. (2000). Beyond Sophisticated Stereotyping: Cultural Sensemaking in Context. *The Academy of Management Executive, 14*(1), 65–79.

Pach, J., & Agarwal, P. K. (1995). *Combinatorial Geometry* (3rd ed.). New York: Wiley-Interscience, New York.

Pach, J., & Tóth, G. (2007). Decomposition of multiple coverings into many parts. *SoCG, 2007,* 133–137.

Paillier, P. (1999). *Public-Key Cryptosystems Based on Composite Degree Residuosity Classes* (pp. 223–238). EUROCRYPT.

Papadias, D., Zhang, J., Mamoulis, N., & Tao, Y. (2003). Query Processing in Spatial Network Databases. In *VLDB Conference,* (pp. 802-813).

Papadimitratos, P., & Haas, Z. J. (2003, January). Secure link state routing for mobile ad hoc networks. In *IEEE Workshop on Security and Assurance in Ad Hoc Networks,* (pp. 379–383).

Paul, C. (2004). *Morphology and Computation.* Paper presented at the From Animals to Animats 8: Proceedings of the Eighth International Conference on Simulation of Adaptive Behavior, Los Angeles, CA.

Paul, C. (2006). Morphological Computation: A Basis for the Analysis of Morphology and Control Requirements. *Robotics and Autonomous Systems, 54,* 619–630. doi:10.1016/j.robot.2006.03.003

Pei, J., Han, J., Mortazavi-Asl, B., Pinto, H., Chen, Q., Dayal, U., & Hsu, M.-C. (2001). PrefixSpan: Mining sequential patterns efficiently by prefix-projected pattern growth. In *ICDE Conference,* (pp. 215-224).

Pei, J., Jiang, D., & Zhang, A. (2005). Mining Cross-Graph Quasi-Cliques in Gene Expression and Protein Interaction Data. In *ICDE Conference.*

Pei, J., Jiang, D., & Zhang, A. (2005). On Mining Cross-Graph Quasi-Cliques. In *ACM KDD Conference.*

Perez-Quinones, M. A. (1994). *Computational pragmatics in HCI: using dialog context in a multimodal interface.* Technical Report 20375, Naval Research Laboratory, Washington DC.

Perkins, C. E., & Bhagwat, P. (1994). Highly Dynamic Destination-Sequenced Distance Vector (DSDV) for Mobile Computers. In *Proceedings of the ACM SIGCOMM Conference on Communications Architectures, Protocols and Applications.*

Perkins, C. Royer, E., & Das, S. (2003). *Adhoc On-demand Distance Vector (AODV) Routing, RFC 3561.*

Pettit, C., Vecherin, S., & Wilson, D. K. (2009). *On the influence of problem definition on sensor placement optimization. Defense Transformation and Net-Centric Systems*. SPIE.

Pfeifer, R., & Bongard, J. (2007). *How the Body Shapes the Way We Think: A New View of Intelligence*. Cambridge, MA: MIT Press.

Pfurtscheller, G., Scherer, R., & Neuper, C. (2007). EEG-based brain-computer interface. In Parasuraman, R., & Rizzo, M. (Eds.), *Neuroergonomics: The Brain at Work*. New York: Oxford University Press.

Phillips, A. (2006). *Specifying and implementing secure mobile applications in the channel ambient system*. Ph.D. dissertation, Imperial College London.

Phillips, J. K., Klein, G., & Sieck, W. R. (2004). Expertise in judgment and decision making: A case for training intuitive decision skills. In Koehler, D. J., & Harvey, N. (Eds.), *Blackwell Handbook of Judgment & Decision Making* (pp. 297–315). Malden, MA: Blackwell. doi:10.1002/9780470752937.ch15

Pierce, L. G. (2002) Preparing and supporting adaptable leaders and teams for support and stability operations. In *Proceedings of the 11th ROK-US Defense Analysis Seminar(Manpower Policy, Session (4)*, Seoul, Korea, (pp. 97-129).

Pierce, L. G. (2002). Barriers to adaptability in a multinational team. In *Proceedings of the 45th Human Factors and Ergonomics Society Annual Meeting*, Baltimore, (pp. 225-229).

Pierce, L. G. (2007). October 2006 Defense Advanced Project Agency (DARPA) sponsored joint experiment between the Singapore (SN) Armed Forces and the U.S. Army Research Laboratory, Human Research and Engineering Directorate (ARL-HRED). *Unpublished white paper.*

Poisel, R. A. (2006). *Modern communications jamming principles and techniques*. Boston: Artech House Publishers.

Poltrock, S. E., Handel, M., Waggett, P., & Bowyer, H. (2008, September). A dynamic model of mission context. In *Proceedings of the 2nd Annual Conference of the International Technology Alliance*, London.

Poltrock, S.E., Handel, M., & Bowyer, H. (2009). *Collaborative behavior patterns and heuristic for recognizing these patterns*. Technical Report, International Technology Alliance, TA4.

Poteet, S., Patel, J., Giammanco, C., Whiteley, I., Xue, P., & Kao, A. (2008a). Words Are Mightier Than Swords… and Yet Miscommunication Costs Lives! In *Proceedings of the Second Annual Conference of the International Technology Alliance*, London.

Poteet, S., Xue, P., Patel, J., Kao, A., Giammanco, C., & Whiteley, I. (2008b). *Linguistic Sources of Coalition Miscommunication*. Paper presented at NATO HFM-142 Symposium, Copenhagen.

Putnam, H. (1975). The meaning of "meaning". In Gunderson, K. (Ed.), *Language, Mind and Knowledge* (pp. 131–193). Minneapolis, MN: University of Minnesota Press. doi:10.1017/CBO9780511625251

Pylyshyn, Z. (1984). *Computation and Cognition*. Cambridge, MA: MIT Press.

QoISN. (2008). *First Int'l Workshop on Quality of Information for Sensor Networks*. Part of the *Fifth IEEE Int'l Conf. on Mobile Ad-hoc and Sensor Systems* (MASS 2008), Atlanta, GA, Sept. 29.

Quinn, N. (2005). How to reconstruct schemas people share, from what they say. In Quinn, N. (Ed.), *Finding culture in talk*. New York: Palgrave MacMillan.

Radosavac, S., Baras, J. H., & Koutsopoulos, I. (2005). A framework for MAC misbehavior detection in wireless networks. In *ACM Workshop on Wireless Security (WiSe)*.

Raghavan, B., Panjwani, S., & Mityagin, A. (2007, April). Analysis of the SPV Secure Routing Protocol: Weaknesses and Lessons. In *ACM SIGCOMM. Computer Communication Review*, *37*(2).

Raghavan, S., & Garcia-Molina, H. (2003). Representing web graphs. In *ICDE Conference*, (pp. 405–416).

Ramasubramanian, V., Haas, Z. J., & Sirer, E. G. (2003). SHARP: A Hybrid Adaptive Routing Protocol for Mobile Ad Hoc Networks. In *Proceedings of the ACM MOBIHOC.*

Ranu, S., & Singh, A. K. (2009). GraphSig: A Scalable Approach to Mining Significant Subgraphs in Large Graph Databases. In *ICDE Conference.*

Rasmussen, L. J., Sieck, W. R., & Smart, P. (in press). What is a Good Plan? Cultural Variations in Expert Planners' Concepts of Plan Quality. *Journal of Cognitive Engineering and Decision Making.*

Rasmussen, L. J., Sieck, W. R., & Smart, P. R. (2008). US/UK Mental Models of Planning: The Relationship between Plan Detail and Plan Quality. In: *NATO RTO HFM-142 Symposium on Adaptability in Coalition Teamwork*, April 2008, Copenhagen, Denmark.

Rattigan, M., Maier, M., & Jensen, D. (2007). *Graph Clustering with Network Sructure Indices.* ICML.

Reeder, R., Bauer, L., Cranor, L., Reiter, M., Bacon, K., How, K., & Strong, H. (2008). "Expandable Grids for Visualizing and Authoring Computer Security Policies," In Proceedings of the Conference on Human Factors in Computing Systems – CHI 2008, ACM Press.

Reeder, R., Karat, C., Karat, J., & Brodie, C. (2007). Usability Challenges in Security and Privacy Policy-Authoring Interfaces. In. *Proceedings of INTERACT, 2007*, 141–155.

Réka, A., & Albert-László, B. (2000). Dynamics of complex systems: Scaling laws for the period of boolean networks. *Physical Review Letters, 84*(24), 5660. doi:10.1103/PhysRevLett.84.5660

Rekhter, Y. & Li, T. (1995). *A Border Gateway Protocol 4 (BGP-4). RFC 1771.*

Rekhter, Y., Li, T. & Hares, S. (2006, January). *A Border Gateway Protocol 4 (BGP-4).* RFC 4271, Draft Standard.

Reynolds, T. J., & Gutman, J. (1988). Laddering theory, method, analysis, and interpretation. Journal of Advertising Research, Feb/March, 11-31.

Rhodes, B., & Maes, P. (2000). Just-in-time information retrieval agents. *IBM Systems Journal, 39*(3-4), 685–704.

RIPE NCC. (2008, February). *YouTube Hijacking: A RIPE NCC RIS case study.* Retrieved from http://www.ripe.net/news/study-youtube-hijacking.html

Risch, J. S., Rex, D. B., Dowson, S. T., Walters, T. B., May, R. A., & Moon, B. D. (1999). The STARLIGHT information visualization system. In Card, S., Mackinlay, J., & Shneiderman, B. (Eds.), *Lectures on Conversation.* Oxford, UK: Basil Blackwell.

Robbins, P., & Ayded, M. (2009). *The Cambridge Handbook of Situated Cognition.* New York: Cambridge University Press.

Roberts, D., Lock, G., & Verma, D. C. (2007). Holistan: A Futuristic Scenario for International Coalition Operations. In *4th Conference on Knowledge Systems for Coalition Operations (KSCO).*

Romney, A. K., Batchelder, W. H., & Weller, S. C. (1987). Recent applications of cultural consensus theory. *The American Behavioral Scientist, 31*(2), 163–177. doi:10.1177/000276487031002003

Romney, A. K., Weller, S. C., & Batchelder, W. H. (1986). Culture as consensus: a theory of culture and informant accuracy. *American Anthropologist, 88*, 313–338. doi:10.1525/aa.1986.88.2.02a00020

Ross, N., & Medin, D. (2005). Ethnography and experiments: Cultural models and expertise effects elicited with experimental research techniques. *Field Methods, 17*, 131–149. doi:10.1177/1525822X04271681

Rowlands, M. (2006). *Body Language: Representation in Action.* Cambridge, MA: MIT Press.

Sacks, H. (1992). *Lectures on Conversation.* Oxford, UK: Basil Blackwell.

Samet, H., Sankaranarayanan, J., & Alborzi, H. (2008), Scalable Network Distance Browsing in Spatial Databases. In *ACM SIGMOD Conference.*

Sarma, A. D., Gollapudi, S., & Panigrahy, R. (2008). Estimating Page Rank on graph streams. In *ACM PODS Conference*.

Schaeffer-Filho, A. Lupu, E., Dulay, N., Keoh, S. L. Twidle, K., Sloman, M., Heeps, S., Strowes, S. & Sventek, J. (2007). Towards supporting interactions between self-managed cells. In *Proceedings of the 1st IEEE International Conference on Self-Adaptive and Self-Organizing Systems (SASO)*, Boston, (pp. 224–233).

Schaeffer-Filho, A. Lupu, E., Sloman, M., Keoh, S.-L. Lobo, J., & Calo, S. (2008, July). A role-based infrastructure for the management of dynamic communities. In *Proceedings of the 2nd International Conference on Autonomous Infrastructure, Management and Security (AIMS)*, Bremen, Germany (LNCS pp. 1–14). Berlin: Springer.

Schaeffer-Filho, A., Lupu, E., & Sloman, M. (2009). Realising management and composition of self-managed cells in pervasive healthcare. In *Proceedings of the 3rd International Conference on Pervasive Computing Technologies for Healthcare*, London.

Schaeffer-Filho, A., Lupu, E., Sloman, M., & Eisenbach, S. (2009, July). Verification of policy-based self-managed cell interactions using alloy. In *Proceedings of the 10th IEEE International Symposium on Policies for Distributed Systems and Networks*, London.

Schmid, S., Eggert, L., Brunner, M., & Quittek, J. (2004). TurfNet: An Architecture for Dynamically Composable Networks. In *Proceedings of the 1st IFIP International Workshop on Autonomic Communication (WAC)*.

Schmitt, R. (2005). Systematic metaphor analysis as a method of qualitative research. *Qualitative Report, 10*(2), 358–394.

Schraefel, M. C., Smith, D. A., Owens, A., Russell, A., Harris, C., & Wilson, M. L. (2005). *The evolving mSpace platform: leveraging the Semantic Web on the trail of the memex.* Paper presented at the Hypertext 2005 Conference, Salzburg, Austria.

Schwartz, S. H. (1994). Cultural dimensions of values: Towards an understanding of national differences. In Kim, U., Triandis, H. C., Kagitcibasi, C., Choi, S. C., & Yoon, G. (Eds.), *Individualism and collectivism: Theory, method, and applications* (pp. 85–119). Newbury Park, CA: Sage.

Searle, J. R. (1969). *Speech Acts. An Essay in the Philosophy of Language*. Cambridge, UK: Cambridge University Press.

Sebastiani, F. (2002). Machine learning in automated text categorization. [CSUR]. *ACM Computing Surveys, 34*(1), 1–47. doi:10.1145/505282.505283

Serafin, R., & Di Eugenio, B. (2004, July). FLSA: Extending latent semantic analysis with features for dialog act classification. In *ACL04, 42nd Annual Meeting of the Association for Computational Linguistics*, Barcelona.

Shadbolt, N., Hall, W., & Berners-Lee, T. (2006). The Semantic Web revisited. *IEEE Intelligent Systems, 21*(3), 96–101. doi:10.1109/MIS.2006.62

Shaw, M., DeLine, R., Klein, D. V., Ross, T. L., Young, D. M., & Zelesnik, G. (1995). Abstractions for software architecture and tools to support them. *IEEE Transactions on Software Engineering, 21*(4), 314–335. doi:10.1109/32.385970

Sieck, W. R., & Mueller, S. T. (2009). Cultural variations in collaborative decision making: Driven by beliefs or social norms? In *Proceedings of the International Workshop on Intercultural Collaboration* (pp. 111-118), Palo Alto, CA.

Sieck, W. R., Grome, A. P., Smith, J., & Rababy, D. A. (in press). Expert cultural sensemaking in the management of Middle Eastern crowds. In Mosier, K. L., & Fischer, U. M. (Eds.), *Informed by Knowledge: Expert Performance in Complex Situations*. San Francisco: Taylor and Francis.

Sieck, W. R., McHugh, A. P., & Smith, J. L. (2006). Use of cognitive field research methods to investigate cultural groups: The case of individual decision making in Middle Eastern crowds. In R. Sun & N. Miyake (Eds.), *Proceedings of the 28th Annual Conference of the Cognitive Science Society*, (pp. 2164-2168).

Sieck, W. R., Smith, J., & Rasmussen, L. J. (2008). *Expertise in making sense of cultural surprises.* Paper presented at the Interservice/Industry Training, Simulation, and Education Conference (I/ITSEC), Orlando, FL.

Sieck, W. R., Smith, J., Grome, A. P., Veinott, E. S., & Mueller, S. T. (2009). *Violent and Peaceful Crowd Reactions in the Middle East: Cultural Experiences and Expectations.* Paper presented at the International Academy for Intercultural Research, Honolulu, HI.

Simon, R., & Zurko, M. E. (1997). Separation of duty in role-based environments. In CSFW, (pp. 183–194). Washington, DC: IEEE Computer Society.

Single Integrated Air Picture *(SIAP) Attributes, ver. 2.0.* (2003). SIAP System Engineering Task Force (SE TE), Technical Report 2003-029, Aug. 2003.

Smart, P. R., Engelbrecht, P. C., Braines, D., Hendler, J. A., & Shadbolt, N. R. (2008). The Extended Mind and Network-Enabled Cognition (No. ITA/P12/NEC). Southampton, UK: School of Electronics and Computer Science, University of Southampton.

Smart, P. R., Engelbrecht, P., Braines, D., Strub, M., & Hendler, J. (2009). *Cognitive Extension and the Web.* Paper presented at the Web Science Conference: Society On-Line, Athens, Greece.

Smart, P. R., Huynh, T. D., Mott, D., Sycara, K., Braines, D., Strub, M., et al. (2009). *Towards an Understanding of Shared Understanding in Military Coalition Contexts.* Paper presented at the 3rd Annual Conference of the International Technology Alliance (ACITA'09), Maryland, USA.

Smart, P. R., O'Hara, K., Engelbrecht, P., Giammanco, C., & Braines, D. (in press). *Networks, Memory and the Extended Mind.* Paper presented at the 1st ITA Workshop on Network-Enabled Cognition, Maryland, USA.

Smart, P. R., Russell, A., Liang, S. F., Shadbolt, N. R., Booth, C., Briscombe, N., et al. (2009). *Using Semantic Technologies to Improve Information Exploitation in Military and Civilian Application Contexts.* Paper presented at the Knowledge Systems for Coalition Operations Conference (KSCO'09), Southampton, UK.

Smart, P. R., Sycara, K., & Huynh, T. D. (in press). *Exploring the Dynamics of Belief Propagation and Collective Problem Solving in Simulated Agent Networks.* Paper presented at the 1st ITA Workshop on Network-Enabled Cognition, Maryland, USA.

Smith, R. G. (1988). The contract net protocol: high-level communication and control in a distributed problem solver. In Distributed Artificial Intelligence, (pp. 357–366).

Sperber, D. (1985). Anthropology and psychology: Towards an epidemiology of representations. *Man, 20,* 73–89. doi:10.2307/2802222

Sperber, D. (1996). *Explaining culture: A naturalistic approach.* Malden, MA: Blackwell.

Sporns, O. (2002). Network analysis, complexity, and brain function. *Complexity, 8*(1), 56–60. doi:10.1002/cplx.10047

Sporns, O., Chialvo, D., Kaiser, M., & Hilgetag, C. (2004). Organization, development and function of complex brain networks. *Trends in Cognitive Sciences, 8*(9), 418–425. doi:10.1016/j.tics.2004.07.008

Srivatsa, M., Ko, B.-J., Beygelzimer, A., & Madduri, V. (2008). Topology Discovery and Link State Detection using Routing Events. In *IEEE Symposium on Reliable Distributed Systems (SRDS).*

Stajano, F., & Anderson, R. (1999). The Resurrecting Duckling: Security Issues for Ad-Hoc Wireless Networks. In *Workshop on Security Protocols,* (LNCS 1796, pp. 172-194). Berlin: Springer-Verlag.

Stasser, G., & Titus, W. (1985). Pooling of unshared information in group decision making: biased information sampling during group discussion. *Journal of Personality and Social Psychology, 48,* 1467–1478. doi:10.1037/0022-3514.48.6.1467

Sterelny, K. (2004). Externalism, Epistemic Artefacts and the Extended Mind. In R. Schantz (Ed.), The Externalist Challenge: New Studies on Cognition and Intentionality (pp. 239-254). Berlin, Germany: de Gruyter.

Stix, G. (2008). Jacking into the brain - is the brain the ultimate computer interface? *Scientific American, 299*(5), 56–61. doi:10.1038/scientificamerican1108-56

Stolcke, A., Reies, K., Cocarro, N., Shriberg, E., Bates, R., & Jurafsky, D. (2000). Dialogue Act Modeling for Automatic Tagging and Recognition of Conversational Speech. *Computational Linguistics, 26*(3), 339–373. doi:10.1162/089120100561737

Strasser, M., Popper, C., Capkun, S., & Cagalj, M. (2008). Jamming resistant key establishment using uncoordinated frequency hopping. In *IEEE Symposium on Security and Privacy.*

Strauss, C., & Quinn, N. (1997). *A Cognitive Theory of Cultural Meaning.* Cambridge, UK: Cambridge University Press.

Sukthankar, G., Sycara, K., & Giampapa, J. A. (2008). "A Model of Human Teamwork for Agent-Assisted Search Operations", In *Proceedings of the NATO Human Factors & Medicine Panel Symposium on Adaptability in Coalition Teamwork*, Copenhagen, Denmark.

Sycara, K., & Lewis, M. (2004). Integrating Agents into Human Teams. In Salas, E. (Ed.), *Team Cognition* (pp. 203–233). Erlbaum Publishers.

Tablan, V., Damljanovic, D., & Bontcheva, K. (2008). *A Natural Language Query Interface to Structured Information.* Paper presented at the 5th European Semantic Web Conference (ESWC'08), Tenerife, Spain.

Tao, T., Vu, V., & Krishnapur, M. (2008). *Random matrices: Universality of esds and the circular law.*

Tarsitano, M., & Jackson, R. (1997). Araneophagic jumping spiders discriminate between detour routes that do and do not lead to prey. *Animal Behaviour, 53*(2), 257–266. doi:10.1006/anbe.1996.0372

Thompson, A., Harvey, I., & Husbands, P. (1996). Unconstrained Evolution and Hard Consequences. In Sanchez, E., & Tomassini, M. (Eds.), *Toward Evolvable Hardware: The Evolutionary Engineering Approach* (pp. 136–165). Berlin, Germany: Springer-Verlag.

Thornley, D. J., Young, R. I., & Richardson, J. (2009, April 13-17). Toward Mission-specific Service Utility Estimation Using Analytic Stochastic Process Models. In *SPIE Defense & Security Symposium*, (DSS 2009), Orlando, FL.

Thul, A. (1910). On the densest packing of congruent circles in the plane. (in Norwegian), *Skr. Vidensk-Selsk. Christiania, 1*, 3–9.

Tian, Z., & Giannakis, G. (2007). Compressed sensing for wideband cognitive radios. In *IEEE Conference on Acoustics, Speech and Signal Processing (ICASSP).*

Tollefsen, D. P. (2006). From extended mind to collective mind. *Cognitive Systems Research, 7*(2-3), 140–150. doi:10.1016/j.cogsys.2006.01.001

Tribble, E. (2005). Distributing Cognition in the Globe. *Shakespeare Quarterly, 56*(2), 135–155.

Tsay, A. A., Lovejoy, W. S., & Karger, D. R. (1999). Random Sampling in Cut, Flow, and Network Design Problems. *Mathematics of Operations Research, 24*(2), 383–413.

Twidle, K., Lupu, E., Dulay, N., & Sloman, M. (2008 June). Ponder2 - A Policy Environment for Autonomous Pervasive Systems. In Policy, (pp. 245 – 246).

Unified Modeling Language (UML). (n.d.). Open Management Group (OMG). Retrieved from http://www.uml.org/

United States Army Research Laboratory. (2008). *Cognition and Neuroergonomics Collaborative Technology Alliance: Draft Program Announcement (No. W911NF-08-R-0014).* USA: United States Army Research Laboratory.

Van Boven, L., & Thompson, L. (2003). A look into the mind of the negotiator: Mental models in negotiation. *Group Processes and Interpersonal Relations, 6*, 387–404. doi:10.1177/13684302030064005

Van Essen, D., Anderson, C., & Olshausen, B. (1994). Dynamic routing strategies in sensory, motor, and cognitive processing. In Koch, C., & Davis, J. L. (Eds.), *Large-Scale Neuronal Theories of the Brain* (pp. 271–299). Cambridge, MA: MIT Press.

Vasconcelos, W. W., Kollingbaum, M. J., & Norman, T. J. (2007). Resolving Conflict and Inconsistency in Norm-Regulated Virtual Organizations. Procs. of the International Joint Conference on Autonomous Agents and Multiagent Systems (AAMAS'07). Honolulu, Hawai'I, USA.

Vecherin, S., Wilson, D. K., & Pettit, C. (2009). *Optimal sensor placement with terrain-based constraints and signal propagation effects. Unattended ground, sea, and air sensor technologies and applications.* SPIE.

Vendruscolo, M., Najmanovich, R., & Domany, E. (1999). Protein folding in contact map space. *Physical Review Letters, 82*(3), 656–659.

Verma, D. (Ed.). (2009). International Technology Alliance in Network and Information Sciences Biennial Program Plan 2009, May 12th 2009 - May 11th 2011.

Verma, D., Brown, T., Bar-Noy, A., Toce, A., Kaplan, L., & Nixon, M. (2008, December). The Base Zone Protection Problem. ASC.

Verma, D., Brown, T., Kaplan, L., & Nixon, M. (2008, September).The Green Zone Protection Problem. *ACITA 2008.*

Verma, D., Ko, B. J., Zerfos, P., Lee, K.-w., He, T., Duggan, M., Stewart, K., Swami, A., & Sofra, N. (2009). Understanding the Quality of Monitoring for Network Management. *British Computer Society: The Computer Journal.* doi: 10.1093/comjnl/bxp058

Verma, D., Wu, C., Brown, T., Bar-Noy, A., Shamoun, S., & Nixon, M. (2008). Location dependent heuristics for sensor coverage planning. *ACITA, 2008*, 346–347.

Wan, T., Kranakis, E., & Oorschot, P. C. (2005 February). Pretty Secure BGP (psBGP). In *The 12th Annual Network and Distributed System Security Symposium.*

Wang, G., Cao, G., & La Porta, T. F. (2006). Movement-assisted sensor deployment. *IEEE Transactions on Mobile Computing, 5*(6), 640–652. doi:10.1109/TMC.2006.80

Wang, H., He, H., Yang, J., Xu-Yu, J., & Yu, P. (2006). Dual Labeling: Answering Graph Reachability Queries in Constant Time. In *ICDE Conference.*

Wang, Q., Yu, T., Li, N., Lobo, J., Bertino, E., Irwin, K., & Byun, J.-W. (2008). On the Correctness Criteria of Fine-Grained Access Control in Relational Databases. In VLDB'07, Vienna, Austria.

Wang, R. Y., & Strong, D. M. (1996). Beyond accuracy: What data quality means to data consumers. *Journal of Management Information Systems, 12*(4), 5–33.

Wang, X., & Reiter, M. K. (2003). Defending Against Denial-of-Service Attacks with Puzzle Auctions. In *IEEE Symposium on Security and Privacy.*

Wang, Y., Chakrabarti, D., Wang, C., & Faloutsos, C. (2003). *Epidemic Spreading in Real Networks: An Eigenvalue Viewpoint* (pp. 25–34). SRDS.

Wang, Y.-C., & Tseng, Y.-C. (2008, September 9-19). Distributed Deployment Schemes for Mobile Wireless Sensor Networks to Ensure Multilevel Coverage. In IEEE Transactions on Parallel and Distributed. Systems.

Watts, D. J. (2003). *Six Degrees: The Science of a Connected Age*. London: William Heinemann.

Watts, D. J., & Strogatz, S. H. (1998). Collective dynamics of "small word" networks. *Nature, 393*, 440–442. doi:10.1038/30918

Wheeler, M. (2005). *Reconstructing the Cognitive World.* Cambridge, MA: MIT Press.

Wheeler, M., & Clark, A. (1999). Genic representation: reconciling content and causal complexity. *The British Journal for the Philosophy of Science, 50*(1), 103–135. doi:10.1093/bjps/50.1.103

White, R. (2003, September). Securing BGP Through Secure Origin BGP. *Internet Protocol Journal, 6*(3).

Wilcox, S., & Jackson, R. (2002). Jumping Spider Tricksters: Deceit, Predation, and Cognition. In Bekoff, M., Allen, C., & Burghardt, G. M. (Eds.), *The Cognitive Animal.* Cambridge, MA: MIT Press.

Wilson, R. A. (1994). Wide computationalism. *Mind, 103*(411), 351–372. doi:10.1093/mind/103.411.351

Wilson, R. A. (2000). The mind beyond itself. In Sperber, D. (Ed.), *Misrepresentations: A Multidisciplinary*

Perspective (pp. 31–52). New York: Oxford University Press.

Wilson, R. A. (2004). *Boundaries of the Mind: The Individual in the Fragile Sciences: Cognition*. New York: Cambridge University Press. doi:10.1017/CBO9780511606847

Wilson, R. A., & Clark, A. (2009). Situated Cognition: Letting Nature Take its Course. In Ayded, M., & Robbins, P. (Eds.), *Cambridge Handbook of Situated Cognition*. Cambridge, UK: Cambridge University Press.

Witt, P. N., Reed, C. F., & Peakall, D. B. (1968). *A Spider's Web: Problems in Regulatory Biology*. Heidelberg, Germany: Springer.

Wu, C., & Verma, D. (2007). *A sensor placement algorithm for redundant covering based on riesz energy minimization*. ISCAS.

Wu, C.-H., Lee, K.-C., & Chung, Y.-C. (2007). A delaunay triangulation based method for wireless sensor network deployment. *Computer Communications*, *30*(14-15), 2744–2752. doi:10.1016/j.comcom.2007.05.017

Wu, F., Huberman, B., Adamic, L., & Tyler, J. (2004). Information Flow in Social Groups. *Physica A. Statistical and Theoretical Physics*, *337*(1-2), 327–335.

Wun, A., & Jacobsen, H. (2007). *A Policy Management Framework for Content-based Publish/Subscribe Middleware* (pp. 368–388). IFIP Middleware.

Xu, W., Trappe, W., Zhang, Y., & Wood, T. (2005). The feasibility of launching and detecting jamming attacks in wireless networks. In *ACM Symposium on Mobile Ad-Hoc Networks (MobiHoc)*.

Yan, X., & Han, J. (2002). gSpan: Graph-Based Substructure Pattern Mining. In *ICDM Conference*.

Yan, X., & Han, J. (2003). CloseGraph: Mining Closed Frequent Graph Patterns. In *ACM KDD Conference*.

Yan, X., Cheng, H., Han, J., & Yu, P. S. (2008). Mining Significant Graph Patterns by Scalable Leap Search. In *SIGMOD Conference*.

Yan, X., Yu, P. S., & Han, J. (2004). Graph Indexing: A Frequent Structure-based Approach. In *SIGMOD Conference*.

Yan, X., Yu, P. S., & Han, J. (2005). Substructure Similarity Search in Graph Databases. In *SIGMOD Conference*, (pp. 766-777).

Yates, J. F., Ji, L.-J., Oka, T., Lee, J.-W., Shinotsuka, H., & Sieck, W. R. (in press). Indecisiveness and thoroughness: Cultural variations in customs, values, and expectations. *Journal of Cross-Cultural Psychology*.

Yates, J. F., Lee, J.-W., Sieck, W. R., Choi, I., & Price, P. C. (2002). Probability judgment across cultures. In Gilovich, T., Griffin, D., & Kahneman, D. (Eds.), *Heuristics and Biases: The Psychology of Intuitive Judgment* (pp. 271–291). Cambridge, UK: Cambridge University Press.

Younis, M., & Akkaya, K. (2008, June). Strategies and Techniques for Node Placement in Wireless Sensor Networks: A Survey. *Ad Hoc Networks*, *6*(4), 621–655. doi:10.1016/j.adhoc.2007.05.003

Zafer, M., Ko, B.-J., & Ho, I. W. (2008, May). Cooperative Transmit-Power Estimation under Wireless Fading. In *ACM Symposium on Mobile Ad-Hoc Networks (Mobihoc)*, Hong Kong.

Zahedi, S., Srivastava, M. B., & Bisdikian, C. (2008, Nov. 17-19). A Computational Framework for Quality of Information Analysis for Detection-oriented Sensor Networks. San Diego, CA: MILCOM.

Zahedi, S., Szczodrak, M., Ji, P., Mylaraswamy, D., Srivastava, M. B., & Young, R. (2008, Nov. 17-19). Tiered Architecture for On-Line Detection, Isolation and Repair of Faults in Wireless Sensor Networks. *MILCOM 2008*, San Diego, CA.

Zaki, M. J., & Aggarwal, C. C. (2003). XRules: An Effective Structural Classifier for XML Data, *KDD Conference*.

Zapata, M. G. (2006, September). *Secure ad hoc on-demand distance vector (SAODV) routing*. Internet Draft. Retrieved from http://personals.ac.upc.edu/guerrero/saodv.html

Zapata, M. G., & Asokan, N. (2002, September). Securing Ad hoc Routing Protocols. In *ACM Workshop on Wireless Security (WiSe)*, (pp. 1–10).

Zhang, H., & Hou, J. (2005) Maintaining Sensing coverage and connectivity in large sensor networks. In *Ad Hoc and Sensor Networks*, (pp.89-124).

Zhang, H., & Hou, J. (2006). *Is deterministic deployment worse than random deployment for wireless sensor networks?* INFOCOM.

Zhou, C., Towsley, D., & Gong, W. (2004). *Email Virus Propagation Modeling and Analysis*. UMass Technical Report TR-CSE-03-04.

Zhou, Z., Das, S., & Gupta, H. (2004). Connected {K}-coverage problem in sensor networks. In ICCCN 2004, (pp. 373--378).

Zhu, F., Mutka, M. W., & Ni, L. M. (2005). Service discovery in pervasive computing environments. *IEEE Pervasive Computing / IEEE Computer Society [and] IEEE Communications Society*, *4*(4), 81–90. doi:10.1109/MPRV.2005.87

Zou, Y., & Chakrabarty, K. (2003). Sensor deployment and target localization based on virtual forces. In INFOCOM 200.

Zou, Y., & Chakrabarty, K. (2003). *Uncertainty-aware sensor deployment algorithms for surveillance applications* (pp. 2972–2976). GLOBECOM.

Zou, Y., & Chakrabarty, K. (2004). Sensor deployment and target localization in distributed sensor networks. *ACM Trans. Embedded Comput. Systems*, *3*(1), 61–91. doi:10.1145/972627.972631

Zou, Y., & Chakrabarty, K. (2004). Uncertainty-aware and coverage-oriented deployment for sensor networks. *Journal of Parallel and Distributed Computing*, *64*(7), 788–798. doi:10.1016/j.jpdc.2004.03.019

About the Contributors

Dinesh Verma is a researcher and senior manager in the Networking technology area at IBM T J Watson Research Center, Hawthorne, New York. He received his doctorate in Computer Networking from University of California Berkeley in 1992, the bachelors' in Computer Science from Indian Institute of Technology, Kanpur, India in 1987, and a Masters in Management of Technology from Polytechnic University, Brooklyn, NY in 1998. He holds over twenty-eight US patents related to computer networks, and has authored over fifty papers and six books in the area. He is the program manager for the US/UK International Technology Alliance in Network Sciences. He is a fellow of the IEEE, and has served in various program committees and technical committees. His research interests include topics in wireless networks, network management, distributed computing, and autonomic systems.

* * *

Charu Aggarwal is a Research Staff member at the IBM T. J. Watson Research Center in Yorktown Heights, New York. He completed his B.S. from IIT Kanpur in 1993 and his Ph.D. from MIT in 1996. He has since worked in the field of performance analysis, databases, and data mining. He has published over 120 papers in refereed conferences and journals, and has been granted over 45 patents. Because of the commercial value of the above-mentioned patents, he has received several invention achievement awards and has thrice been designated a Master Inventor at IBM. He is a recipient of an IBM Corporate Award (2003) for his work on bio-terrorist threat detection in data streams, a recipient of the IBM Outstanding Innovation Award (2008) for his scientific contributions to privacy technology, and a recipient of an IBM Research Division Award (2008) for his scientific contributions to data stream research. He has served on the program committees of most major database/data mining conferences, and served as program vice-chairs of the SIAM Conference on Data Mining, 2007, the IEEE ICDM Conference, 2007, the WWW Conference 2009, and the IEEE ICDM Conference, 2009. He served as an associate editor of the IEEE Transactions on Knowledge and Data Engineering Journal from 2004 to 2008. He is an action editor of the Data Mining and Knowledge Discovery Journal , an associate editor of the ACM SIGKDD Explorations, and an associate editor of the Knowledge and Information Systems Journal. He is a senior member of the IEEE, and a life-member of the ACM.

Dakshi Agrawal is a research staff member at IBM T. J. Watson Research Center in the Policy & Networking Department. He has been a core team member in developing the Policy Management Toolkit for IBM Autonomic Computing, and received an IBM Research Division Award and Invention Achievement Award for this contribution in the project. Dr. Agrawal received a Ph.D. in 1999 from the

University of Illinois–Urbana-Champaign (UIUC), Urbana, IL in electrical engineering. He worked as a Visiting Assistant Professor at UIUC during 1999-2000 before joining T. J. Watson Research Center as a Research Staff Member. Dr. Agrawal has more than 40 publications in international conferences and journals in the area of digital communication theory, distributed computing systems, and digital security and privacy. He has been granted or has applied for more than 10 patents with the US Patent Office.

Arosha Bandara was appointed Lecturer in Computing at The Open University in October 2006. Previously he was a postdoctoral researcher at Imperial College London, where he completed his M. Eng (1998) and Ph.D (2005). His research interests are in formal and semi formal techniques for the analysis and refinement of policies as well as software engineering aspects of policy-driven systems. In addition to serving on the program committees of international conferences such as NOMS, IM and AIMS, Dr. Bandara has chaired the program committee of IEEE Symposium on Policies for Distributed Systems and Networks (POLICY 2009) and also served on its program committee (2004-2008). He is a co-investigator on the UK EPSRC funded PRiMMA project and the Policy-based Security Management project funded by the International Technology Alliance (ITA).

Mandis Beigi is a Senior Software Engineer at IBM Research. Currently she is also pursuing her doctorate degree at Columbia University conducting research in sensor data analysis and learning algorithms for anomaly detection. Previously, she has worked on Policy based Management for Autonomic Computing and various policy analysis algorithms such as policy transformation. Some of her policy related works is on policy enablement of QoS servers, IPSec and Intrusion Detection and various systems such as Storage File Systems. She has also worked on MPLS, quality of service, network virtualization, and network monitoring and management.

Chatschik Bisdikian is a Research Staff Member at the IBM T. J. Watson Research Center, Hawthorne, NY, USA. He holds a Ph.D. degree in Electrical Engineering from the University of Connecticut, Storrs, CT, USA. He has been with IBM Research since 1989 and has worked in numerous projects covering a variety of research topics in communications, networking, pervasive computing, IPTV services, computer system management, sensor networks, and so on. In 2004, he was elected IEEE Fellow for his contributions to the development, modeling, and analysis of communication protocols and wireless personal area networks. He has authored over 100 peer-reviewed papers in the aforementioned areas, has 13 patents issued, and co-authored the book *Bluetooth Revealed* (Prentice Hall). He has served as the Editor-in-Chief of *IEEE Network Magazine*, where he currently serves as the Senior Technical Editor. He also serves in the editorial board of the *Pervasive and Mobile Computing.*

Dave Braines is an Emerging Technologies Specialist and Technical Staff Member working for the IBM UK Emerging Technology Services team. He has a background in leading-edge application and infrastructure development, generally focused on Web-based technologies and architectures. He has interests in the technological, social and cultural aspects of emerging Web usage in groups and communities. Currently, Dave Braines is a project champion for the International Technology Alliance (ITA) research programme, where he is co-leading a research team carrying out fundamental research on the topic of "Network Enabled Shared Understanding". Dave Braines is also a visiting research fellow at the School of Electronics and Computer Science at the University of Southampton.

Peter Brass is Associate Professor at the City College of New York, of CUNY; his current research subjects are algorithms, geometry, and their applications in sensor networks and robotics. Past work included graph theory, discrete and computational geometry, shape matching, and quantum logics. He received his PhD in mathematics 1992 from the Technical University of Braunschweig, Germany, spent a postdoc period at the University of Greifswald, and was Heisenberg Research Fellow at the Free University of Berlin. He wrote two books, `Advanced Data Structures' (Cambridge University Press 2008) and `Research Problems in Discrete Geometry' (with J. Pach, W. Moser, Springer 2005) and about 80 research papers.

Carolyn Brodie is a former research staff member at IBM Research. She received her Ph.D. degree in computer science from the University of Illinois at Urbana-Champaign, where she developed a methodology for the design of military planning tools. Her additional research interests include personalization of Web sites and the use of collaboration tools to enhance information flow in organizations. At IBM her research focused on the design and development of usable policy management technologies.

Ted Brown is the Executive Officer of the Ph.D, Program in Computer Science and the Executive Director of the CUNY Institute for Software Design and Development, both positions he has held for ten years. Prior to this he was chair of the Computer Science Department at Queens College of CUNY. He does research in algorithm design and development and has over 30 papers. His recent interest has been in proper placement of sensors. He received his Ph.D. in Operations Research from NYU in 1971.

Seraphin Calo is a Research Staff Member at IBM Research and currently manages the Policy Lifecycle Technologies group within that organization. He received the M.S., M.A., and Ph.D. degrees in electrical engineering from Princeton University, Princeton, New Jersey. He has worked, published, and managed research projects in a number of technical areas, including: queueing theory, data communications networks, multi-access protocols, expert systems, and complex systems management. He has been very active in international conferences, particularly in the systems management and policy areas. Most recently, he served as the General Chair of IM 2009 (The Eleventh IFIP/IEEE International Symposium on Integrated Network Management). Dr. Calo has authored more than sixty technical papers and has several United States patents (five issued and three pending). He has received four IBM Research Division awards, and five IBM Invention Achievement awards. His current research interests include: distributed applications, services management, and policy based computing.

Chin K Chau (Sid Chau) is a post-doctoral researcher at University of Cambridge. He comes from Hong Kong. His research lies in the areas of wired/wireless networking and communications, specifically in the topics of Networking and Stochastic Models, Networking and Policies, and applying game theory to networking.

Robert Craven is a Research Associate in the Department of Computing at Imperial College London. He studied philosophy at Cambridge and Oxford, then took an M.Sc. on the mathematical foundations of computer science and wrote a Ph.D. on temporal reasoning in artificial intelligence, both at Imperial College. His current research interests include policy languages, policy refinement, deontic logic, and anything related to norm-governed distributed systems.

Jon Crowcoft is the Marconi Professor of Communications Systems in the Computer Lab, at the University of Cambridge. He was a fellow of Wolfson College Until the end of September 2001, and a professor in the Department of Computer Science University College London. He graduated in Physics from Trinity College, Cambridge University in 1979, and got an MSc in Computing in 1981, and PhD in 1993 both from UCL. He is a fellow of the ACM, the British Computer Society, the IE[ET], The Royal Academy of Engineering and the IEEE. He has published profusely in the field of computer networks and has received several awards including the 2009 ACM SIGCOMM Award.

Paula Engelbrecht is a part-time PhD student at the School of Psychology, University of Southampton. Her PhD research is concerned with the cognitive process of comparison and its effects on perception and decision making. She is also interested in the extended mind hypothesis; in particular, how it relates to human memory and decision making. Paula Engelbrecht is currently employed as a research scientist at Ordnance Survey, Britain's national mapping agency. To date her work has involved knowledge elicitation for ontology development; conducting user tests of Rabbit, a controlled natural language for ontology development; and research on the perception of vague geographical boundaries.

Cheryl Giammanco, Ph.D. is the Human Dimensions Network Science Team Leader in the Cognitive Sciences Branch of the U.S. Army Research Laboratory (ARL) Human Research and Engineering Directorate and the Program Manager, Tactical Human Integration of Networked Knowledge Army Technology Objective (THINK ATO). Dr. Giammanco is transitioning to serve as the International Technology Alliance (ITA) U.S. Government Technical Area 4 Lead (Distributed Network Enabled Cognition). Her research interests include social/cognitive networks, group processes, and human cognitive performance for collaboration and decision making. Dr. Giammanco's research focuses on warfighter cognitive performance for collaborative planning and Battle Command decision making in dynamic network-enabled coalition operations. Dr. Giammanco is also the Army panel member of The Technical Cooperation Program, Command, Control, Communication, and Information Systems Technical Panel 4: Dynamic Planning and Scheduling.

Matthew P. Johnson is a PhD student in Computer Science at the City University of New York Graduate Center. His research interests include algorithms with applications to sensor networks.

Anne Kao is a Technical Fellow in Boeing Research and Technology. She has been leading various Text Mining and Text Analysis projects and tasks since 1991. Dr. Kao is an internationally recognized expert in Text Mining and Social Network Analysis. She is the co-editor of a book on Natural Language Processing and Text Mining published by Springer in 2007, as well as a program committee member or reviewer for various major ACM and IEEE conferences on text mining, data mining and visual analytics. Currently, she is leading a coalition communication task under the International Technology Alliance on the study of cultural differences in coalition command, control and communication, jointly funded by US Army Research Lab and UK Ministry of Defence. Dr. Kao has a BA, MA and Ph.D. in Philosophy, specializing in philosophy of language, as well as an MS in Computer Science from San Diego State University.

Lance M. Kaplan received the B.S. degree with distinction from Duke University in 1989 and the M.S. and Ph.D. degrees from the University of Southern California in 1991 and 1994, respectively, all in Electrical Engineering. Dr. Kaplan held a National Science Foundation Graduate Fellowship and a USC Dean's Merit Fellowship from 1990-1993. He has worked at the Georgia Tech Research Institute (1987-90), Hughes Aircraft Company (1994-96), and on faculty in the Department of Engineering at Clark Atlanta University (1996-04). Currently, he is a team leader in the Networked Sensing and Fusion branch of the U.S. Army Research Laboratory. Dr. Kaplan also serves the IEEE Aerospace and Electronic Systems (AES) Society as Associate Editor-In-Chief for the IEEE Transactions on AES, Tutorials Editor for the IEEE AES Magazine, and a member of the Board of Governors. His current research interests include signal and image processing, information/data fusion, and resource management.

Clare-Marie Karat is recently retired from IBM Reaearch. Until 2009, she was a research staff member in the policy technologies group. She received her B.A. degree in psychology from Stanford University and her Ph. D. degree in social psychology from the University of Colorado. She joined IBM after leading the applied research area of the Texas Office of the Inspector General. Her research interests at IBM included policy, privacy, security, usability, and personalization. She led the SPARCLE policy team and research on which the IBM Secure Perspective policy management product is based. She is a recipient of the ACM SIGCHI Lifetime Service Award.

John Karat is a research staff member in the policy technologies group. He received his B.S. degree in mechanical engineering from Lehigh University and his Ph.D. degree in cognitive psychology from the University of Colorado. He joined IBM in 1982 after working for Johns Hopkins University as a data analyst and the United Nations in Bangladesh as a population studies consultant. His research interests at IBM include policy, privacy, security, usability, and personalization. He is a recipient of the ACM SIGCHI Lifetime Service Award and named an ACM Distinguished Scientist.

Kang-Won Lee is a research staff member and Manager of Wireless Networking Group at IBM Watson Research Center. He is also an industrial technical area leader for Network Theory of the ITA (International Technology Alliance for Network and Information Science) program, a long term research program jointly funded by US ARL and UK MOD. Dr. Lee's main research interest includes wireless and dynamic networks, distributed systems, and policy-based autonomic systems. He published more than 50 technical articles in premier venues. Dr. Lee received a Ph.D. degree in Computer Science from University of Illinois at Urbana-Champaign in 2000, and received a B.S. and a M.S. from Seoul National University in 1992 and 1994, respectively. Dr. Lee has been a recipient for the C.W. Gear Award (1999), IBM Research Division Award (2003), and IBM Outstanding Technical Achievement Award (2007). In 2004 – 2005, he served as secretary of IEEE TCCC (Technical Committee on Computer Communications). In 2004, he was President of KOCSEA (Korean Computer Scientists and Engineers Association in America). He now serves on the advisory board of KOCSEA. Dr. Lee is a senior member of ACM and IEEE.

Jorge Lobo joined IBM T. J. Watson Research Center in 2004. Previous to IBM he was principal architect at Teltier Technologies, a start-up company in the wireless telecommunication space acquired by Dynamicsoft and now part of Cisco System. Before Teltier he was tenured associate professor of CS at the University of Illinois at Chicago and member of the Network Computing Research Department at Bell Labs. In Teltier he developed a policy server for the availability management of Presence Servers. The servers were successfully tested inside two GSM networks in Europe. He also designed and co-developed PDL, one of the first generic policy languages for network management. A policy server based on PDL was deployed for the management and monitoring of Lucent first generation of softswitch networks. Jorge Lobo has more than 50 publications in international journals and conferences in the areas of Networks, Databases and AI. He is co-author of two books, and is co-founder and member of the steering committee for the IEEE International Workshop on Policies for Distributed Systems and Networks. He has a PhD in CS from University of Maryland at College Park, and a MS and a BE from Simon Bolivar University, Venezuela.

Emil Lupu is Reader in Adaptive Computing Systems in the Department of Computing at Imperial College London where he leads several research projects in the areas of pervasive computing, trust and security and policy-based network and systems management. He has numerous publications in these areas, serves on the editorial boards of the Journal of Network and Systems Management and the International Journal of Network Management, and on the program committee of several conferences.

Jiefei Ma is a PhD student in the Department of Computing, Imperial College London, where he obtained his Master of Engineering degree in Computing (Artificial Intelligence) with First Class honours. He is currently participating in the International Technology Alliance project sponsored by the U.S. Army Research Laboratory and the U.K. Ministry of Defence, and is mainly working on distributed policy analysis. His current research/study mainly focuses on automated reasoning and multi-agent systems. He is particularly interested in applying novel A.I. techniques in real life applications such as pervasive computing.

Andrew McDonald is a Consultant Engineer at Roke Manor Research, a Siemens company based in Romsey, Hampshire, England. He has worked at Roke for the last 9 years in the fields of networking and security. His work has been principally focussed in the field of IP networking in the mobile and wireless domains, for commercial, defence and national security applications. Before joining Roke, Andrew studied at the University of Cambridge and with the Information Security Group at Royal Holloway, University of London.

Tim Norman is Professor and the Head of Computing Science (Research) in the Department of Computing Science, University of Aberdeen. He is leader of the Aberdeen Intelligent Software Agents Research Theme. Prof Norman's principal research field is multi-agent systems. He has interests in computational models of norms (or policies) for specifying and reasoning about the ideal behaviour of complex distributed systems. He is interested in how systems may be monitored for compliance with norms, and how these techniques can be applied to support for human decision making (e.g. in collaborative planning teams). A further focus of research interest is in computational models of argumentation, and, in particular, formal models of delegation, and social and dialogical commitment. Prof Norman is Area Chair and member of the Senior Programme Committee of the International Joint Conference on Autonomous Agents and Multi-Agent Systems and member of the Programme Committee of the International Conference on Computational Models of Argument.

Stephen R. Poteet is a computational linguist with over nineteen years of experience in both artificial intelligence and statistical approaches to text mining. He designed and implemented the pragmatics/semantics component of a natural language understanding system for processing battlefield intelligence.

In addition, he has designed and implemented systems for conversational analysis and opinion extraction, aviation maintenance and prognostics, safety, business intelligence, and political intelligence using various text mining technologies including text classification and clustering, trend analysis, latent semantic analysis, and information extraction. He holds three US patents and has six patents pending and he co-edited a book on "Natural Language Processing and Text Mining" with Anne Kao which was published by Springer in 2007. He is currently performing on two government contracts, one regarding culture and networking influences on coalition communication and one related to social network analysis.

Louise J. Rasmussen, Ph.D. is a senior scientist in the Culture & Cognition Group at Applied Research Associates. Her research focuses on cultural cognition, particularly on how culturally determined understandings of a task domain can influence behaviors, performance, and expectations in work situations. Dr. Rasmussen currently leads several research projects in intercultural interaction and collaborative decision making. Dr. Rasmussen received a Ph.D. in Psychology with an emphasis on discourse analysis and human factors from Wright State University in 2007

Alessandra Russo is a Senior Lecturer in the Department of Computing, Imperial College London where she obtained her PhD in mathematical logic. Her research interests revolve around the areas of Artificial Intelligence, Software Engineering and Policy-based Management Systems. Her current work mainly focuses on the development of formal approaches and reasoning techniques for the elaboration, analysis, refinement and revisions of specifications, including requirements and design specifications for complex systems and specifications of security policies. She has published over 70 papers at international conferences and journals in the areas of Artificial Intelligence, Software Engineering and Management of Distributed Systems, two of which have been awarded prizes for best application papers. She has been principal investigator of the project regarding "Reasoning Techniques for Analysis and Refinement of Policies for Quality of Service Management", and co-investigator on various projects regarding analysis, refinement and integration of policies for distributed systems, for QoS management, and for privacy rights management in the context of mobile applications. She has served on various program committees of international conferences, and is currently Editor-in-Chief of the IEE Software Journal.

Alberto Schaeffer-Filho received his BSc and MSc degrees in Computer Science at the Federal University of Rio Grande do Sul, Brazil, in 2002 and 2005 respectively. He is currently a PhD student at the Department of Computing, Imperial College London. His research interests include self-management of ubiquitous systems, policy-based management, self-adaptive and autonomous systems. He has investigated software engineering principles, patterns and model-checking techniques for composing and federating autonomous policy-based systems. Alberto is a student member of the IEEE.

Katia Sycara is a Research Professor in the School of Computer Science at Carnegie Mellon University and holds the Sixth Century Chair (part time) in Computing Science at the University of Aberdeen in the U.K. She is also the Director of the Laboratory for Agents Technology and Semantic Web Technologies at CMU. She holds a B.S. in Applied Mathematics from Brown University, M.S. in Electrical Engineering from the University of Wisconsin and Ph.D. in Computer Science from Georgia Institute of Technology. She holds an Honorary Doctorate from the University of the Aegean (2004). Prof. Sycara is a Fellow of the Institute of Electrical and Electronic Engineers (IEEE), Fellow of the American Association for Artificial Intelligence (AAAI) and the recipient of the 2002 ACM/SIGART

Agents Research Award. She is also the recipient of the Outstanding Alumnus Award from the University of Wisconsin in 2005. She is a member of the Scientific Advisory Board of France Telecom, and a member of the Scientific Advisory Board of the Greek National Center

Simon Shamoun is currently student in the Ph.D. Program in Computer Science at the CUNY Graduate Center. His currently line of research is in algorithmic problems in sensor and ad hoc networks. Prior research areas include high level vision and graph algorithms. He received his B.A. in computer science from New York University in 1996 and his M.S. in computer science from Columbia University in 2000. Besides working in the private sector, he was also an adjunct lecturer in computer science at Queens College, CUNY, for six years, primarily teaching computer architecture and discrete structures.

Winston R. Sieck, Ph.D. is a principal scientist at Applied Research Associates, where he leads the Culture & Cognition Group. He conducts research on culture and decision making, and develops applications for cultural analysis, cultural understanding and influence, and multicultural collaboration. His work has addressed such specific topics as ideological extremism, cultural network analysis, crowd reactions in the Middle East, cultural variations in collaborative planning and decision making, cultural sense-making expertise, situated cultural training, influence effectiveness across cultures, cultural variations in reasoning thoroughness, confidence in beliefs and decisions, explanation in cognition. Dr. Sieck received a Ph.D. in Psychology with emphasis on cognition, culture, and decision making from the University of Michigan in 2000, and a masters in statistics from the same university in 1995.

Morris Sloman is Deputy Head of the Department of Computing, Imperial College London. His research interests include autonomic management of pervasive and distributed systems, adaptive security management, privacy and security for pervasive systems. He has many journal and conference publications and is a member of the editorial boards of the Journal of Network and Systems Management and IEEE Transactions on Network and Services Management. He is on the steering committees for the conferences on Policies for Distributed Systems and Networks, Integrated Management (IM), Network Operations and Management (NOMS).

Paul Smart is a Senior Research Fellow at the School of Electronics and Computer Science, University of Southampton. His research interests include the development and exploitation of Web-based technologies, particularly for the purposes of enhanced situation awareness, information exchange and collaborative problem-solving applications. He is also interested in the relationship between mind, technology and culture, and is currently leading a research project to investigate the relationship between network structures and socio-cognitive processes. Dr Smart works as part of the International Technology Alliance (ITA), a consortium of government, academic and industrial organizations, which undertakes fundamental research in the network and information sciences. He is also a member of the Web Science Research Institute (WSRI), which is a joint endeavour between the Computer Science and Artificial Intelligence Laboratory (CSAIL) at the Massachusetts Institute of Technology (MIT) and the School of Electronics and Computer Science at the University of Southampton.

Mani Srivastava is on the faculty at UCLA where he is a Professor and Vice Chair in the Electrical Engineering Department, and is also affiliated with the Computer Science Department and the Center for Embedded Networked Sensing. Prior to joining UCLA, he received is PhD from UC Berkeley in

1992, and worked at Bell Labs Research for four years. He currently serves at the Editor-in-Chief of the IEEE Transactions on Mobile Computing, and is a Fellow of the IEEE. His research interests are in embedded wireless systems, power-aware systems, wireless networks, and pervasive sensing, more details on which can be found on the web site of UCLA's Networked and Embedded Systems Laboratory (http://nesl.ee.ucla.edu).

Mudhakar Srivatsa is a Research Staff Member at IBM T.J. Watson Research Center. In May 2007, he graduated with a PhD in Computer Science from Georgia Tech, where he worked with Prof. Ling Liu on secure distributed systems. Earlier in 2002, hegraduated with a bachelors in Computer Science and Engineering from IIT Madras. His research interests broadly lie in the field of secure and scalable distributed systems. His work focuses on developing integrated systems and security solutions that address application specific security requirements while preserving its performance and scalability. In the past, he had developed algorithmic techniques and scalable security tools in the form of composable and customizable plug-ins for existing distributed systems such as: VoIP Networks and Publish/Subscribe Networks.

Michael Strub is a US Army Research Lab Exchange Scientist assigned to the MoD dstl laboratory at Porton Down, UK. His research interests include cognition, especially as it relates to command information processing and decision making in a networked environment. He is also interested in the military applications associated with investigating the fundamental relationship between network structures and human cognitive processing. Dr Strub has been a government technical area leader for the International Technology Alliance (ITA), a consortium of government, academic and industrial organizations, which undertakes fundamental research in the network and information sciences. He is a fellow of the American Psychological Association and the Association for Psychological Science, and a member of the Human Factors and Ergonomics Society.

David J. Thornley read Engineering and Computer Science at the University of Oxford, and for a PhD at Imperial College London. His work generally comprises the use of timed stochastic models for system performance analysis, and machine learning or nonlinear model fitting to incorporate system behaviours for which ab initio modeling is intractable itself, or contributes to a state space explosion. His publications concern signal processing, biochemical modeling, DNA sequencing, machine learning, queuing and network performance theory, and more recently tactical information system modeling for predicting quality of information and utility characteristics.

Starsky H.Y. Wong is a research staff member in the Wireless Networking Group at IBM T. J. Watson Research Center. He received a Ph.D. degree in Computer Science from the University of California, Los Angeles in 2007, and received a B.S. and a M.S. from the Chinese University of Hong Kong 2000 and 2002, respectively. He has been a recipient for the Croucher Foundation Scholarships in 2002. His research interests include wireless and mobile systems, wireless and internet security, distributed algorithms and large-scale network measurement and network analysis.

David Wood is a Senior Software Engineer at IBM Research, where he is building technologies in support of policy-based systems. He has been with IBM since 1992 working in areas including parallel computing, data visualization, visual and spoken user interfaces, and location-based services and architectures. He has authored multiple patents and technical papers. He obtained a BA in Physics from UC San Diego in 1985 and a Master's degree in Computer Science from New York University in 1989.

Ping Xue is a research scientist in Boeing Research and Technology, specializing in computational linguistics and natural language processing. He received a Ph.D. in linguistics from The University of Victoria, Canada in 1991. Before he joined Boeing, he worked as a research fellow at The Center for the Study of Language and Information, Stanford University, and then an Assistant Professor at Simon Fraser University, Canada. In the past 20 years, Dr. Xue has worked on grammar formalism, linguistic theory, and a number of computational and language processing research projects including machine translation, automatic grammar checking, information extraction, information integration, metadata and terminology management. He is currently working on two government sponsored research contracts, both of which are related to the computational analysis of socio-cultural aspects of human language and communication.

Robert I. Young gained a PhD in Computer Generated Holography from Northumbria University in 1993. He has nearly twenty years research experience in remote sensing using multiple sensor and ISTAR collector systems and supporting advanced sensor development through the realization of novel signal processing algorithms and information fusion techniques. His research interests include a variety of sensor and data fusion topics especially understanding the fundamental bounds on performance for the detection, location, recognition, tracking, status/activity and intent of targets using co-modal and cross-modal sensing. More recently, he has worked on matching data products to information requirements and reasoning under conditions of epistemic uncertainty.

Eric Yu-En Lu is a research associate in Computer Laboratory at Cambridge University. His primary research interests include high-dimensional data analysis and mining, stochastic modeling, and very large stream databases. With collaborators, he has also treaded upon topics in structural biology and quantum physics. He is a fellow of Wolfson College and Cambridge Overseas trust. Prior to his PhD in Cambridge, he obtained his BA and Masters from National Taiwan University, Taiwan.

Petros Zerfos is a Research Staff Member at the IBM Research. He received his Ph.D. (2005) and M.Sc. (2002) degrees in Computer Science from the University of California, Los Angeles, and M.Eng. (1999) in Electical & Computer Engineering from the National Technical University of Athens, Greece. He has worked as a senior research scientist at the Strategic Research Group of Deutsche Telekom Laborories in Berlin, Germany, as a research intern at the ICSI Center for Internet Research at UC Berkeley, CA, USA, and as a network engineer at the Network Operations Center of the National Technical University of Athens, Greece. His research interests lie in the areas of network and service management, high performance software architectures for predictive analytics and event management, as well as next generation mobile infrastructure.

Index

D

E

J

L

M

N

O

P

Q

R

S

T

U

V

W

Z